Unravelling Single Cell Genomics
Micro and Nanotools

RSC Nanoscience & Nanotechnology

Series Editors:
Professor Paul O'Brien, *University of Manchester, UK*
Professor Sir Harry Kroto FRS, *University of Sussex, UK*
Professor Harold Craighead, *Cornell University, USA*

Titles in the Series:
1: Nanotubes and Nanowires
2: Fullerenes: Principles and Applications
3: Nanocharacterisation
4: Atom Resolved Surface Reactions: Nanocatalysis
5: Biomimetic Nanoceramics in Clinical Use: From Materials to Applications
6: Nanofluidics: Nanoscience and Nanotechnology
7: Bionanodesign: Following Nature's Touch
8: Nano-Society: Pushing the Boundaries of Technology
9: Polymer-based Nanostructures: Medical Applications
10: Molecular Interactions in Nanometer Layers, Pores and Particles: New Findings
    at the Yoctovolume Level
11: Nanocasting: A Versatile Strategy for Creating Nanostructured Porous Materials
12: Titanate and Titania Nanotubes: Synthesis, Properties and Applications
13: Raman Spectroscopy, Fullerenes and Nanotechnology
14: Nanotechnologies in Food
15: Unravelling Single Cell Genomics: Micro and Nanotools

How to obtain future titles on publication:
A standing order plan is available for this series. A standing order will bring delivery of
each new volume immediately on publication.

For further information please contact:
Book Sales Department, Royal Society of Chemistry, Thomas Graham House,
Science Park, Milton Road, Cambridge, CB4 0WF, UK
Telephone: +44 (0)1223 420066, Fax: +44 (0)1223 420247, Email: books@rsc.org
Visit our website at http://www.rsc.org/Shop/Books/

Unravelling Single Cell Genomics
Micro and Nanotools

Edited by

Nathalie Bontoux
Agilent Technologies, Massy, France

Luce Dauphinot and Marie-Claude Potier
CRICM; CNRS UMR 7225, INERM UMR 975, UPMC, Hôpital de la Pitié-Salpétrière, Paris, France

RSCPublishing

RSC Nanoscience & Nanotechnology No. 15

ISBN: 978-1-84755-911-1
ISSN: 1757-7136

A catalogue record for this book is available from the British Library

Published by The Royal Society of Chemistry,
Thomas Graham House, Science Park, Milton Road,
Cambridge CB4 0WF, UK

Registered Charity Number 207890

For further information see our web site at www.rsc.org

Preface

The sequencing of the first genome in 1977 by Sanger and colleagues opened the way to a whole new field of biology: genomics, *i.e.* the study of genes and their function. This new field has boomed since then with the development of high-throughput techniques, such as DNA microarrays, which aim to increase information while driving costs and time per analysis down. In less than 20 years, tremendous progress has been made: whereas the first microarrays, spotted on nylon membranes, allowed profiling of a few tens of genes at the same time, current microarrays allow the detection of a whole transcriptome with more than 2 million probes per array. In the same way, the cost of sequencing the human genome dropped in less than 10 years from over hundreds of millions to a few thousands of dollars. In terms of resolution, current techniques often require a few nanograms of starting material; that is, a few thousands cells.

The cell constitutes the basic unit of all known living organisms. In this context and given the high cellular heterogeneity of some tissues such as the brain, it is quite striking that despite all the technological advances of these last decades, high-throughput data at the single cell level has been reported only in a handful of genomic studies. Indeed, single cell assays remain extremely difficult to carry out and prior amplification of the genetic content is required. Because standard laboratory techniques are not adapted to single cell manipulation, single cell transcriptome and genome analysis currently imply a more than a million-fold dilution of the cell content in microliter volumes. This dilution may impact the assay's sensitivity as it increases the risk of non-specific reactions and/or contamination and put the sensitivity and reliability of single cell analysis at risk.

To understand genomics at the basic and fundamental level of the cell, micro- and nano-tools could be appropriate. Microfluidics lab-on-a-chip seems particularly promising as they allow reactions to be performed at the scale of the

RSC Nanoscience & Nanotechnology No. 15
Unravelling Single Cell Genomics: Micro and Nanotools
Edited by Nathalie Bontoux, Luce Dauphinot and Marie-Claude Potier
© Royal Society of Chemistry 2010
Published by the Royal Society of Chemistry, www.rsc.org

cell, thus at higher concentrations, and the integration of multiple reactions on a single microchip could reduce the risks of contamination and the consumption of reagents.

This book was developed to help scientists understand the latest developments in microfluidics for genomics. After an introduction to molecular and cell biology (Chapters 1 to 3), the need for single cell analysis (Chapters 4 to 6) and the latest developments in this field (Chapters 7 to 10) are reviewed. In the subsequent chapters, microfluidic devices are introduced (Chapter 11) and their application to genomics and proteomics is discussed (Chapters 12 to 15). The development of a microfluidic device for single cell transcriptome analysis is detailed to illustrate key steps, pitfalls, and advantages of such lab-on-a-chips (Chapter 16). The last chapters cover droplet microfluidics (Chapter 17) and discuss detection techniques (Chapter 18) as these hold great promise for easier cell manipulation and increased sensitivity.

In an effort to provide a comprehensive overview, these chapters have been contributed to by experienced scientists with various backgrounds in biology, physics, and chemistry, all working in reference laboratories in Europe. We would like to acknowledge their contribution and warmly thank them.

Contents

RSC Nanoscience & Nanotechnology No. 15
Unravelling Single Cell Genomics: Micro and Nanotools
Edited by Nathalie Bontoux, Luce Dauphinot and Marie-Claude Potier
© Royal Society of Chemistry 2010
Published by the Royal Society of Chemistry, www.rsc.org

Chapter 8 Looking at the DNA of a Single Cell 73
Bernhard Polzer and Christoph A. Klein

Chapter 9 Gene Analysis of Single Cells 81
Bruno Cauli and Bertrand Lambolez

Chapter 13 DNA Analysis in Microfluidic Devices and their Application to Single Cell Analysis 185

Yann Marcy and Angélique Le Bras

Chapter 14 Gene Expression Analysis on Microchips 196

Max Chabert

CHAPTER 1
An Introduction to Molecular Biology

LUCE DAUPHINOT

CRICM, CNRS UMR7225, INSERM UHRS975, UPHC, Hôpital de la Pitié Salpétrière, Paris, France

Abstract

The cell constitutes the basic structure of all living organisms (*cellula* in latin means small chamber). The typical diameter of a cell is 10–100 micrometers (μm), its volume around 10 picoliters (pl) and its mass around 1 nanogram (ng).

Cells can be divided in two main groups. Prokaryotic cells, such as bacteria, lack nucleus and are unicellular organism, characterized by a relatively simple organization with only one compartment containing a circular DNA molecule. Eukaryotic cells are characterized by a nucleus and a cytoplasm containing many sub-cellular compartments. The nucleus is surrounded by a nuclear envelope with nuclear pores that allow the transport of macromolecules between the nucleus and the cytoplasm. The DNA molecule is localized inside the nucleus and organized in chromosomes. Some eukaryote organisms are unicellular such as yeasts, but the most part are pluricellular, with the most complex organism being human, with more than 10 000 billion cells.

1.1 DNA Structure and Gene Expression

Eukaryotic cells contain a nucleus while prokaryotic cells do not (Figures 1.1 and 1.2). The genetic information of the cell is stored as a double-helix DNA molecule inside the nucleus (the model proposed by Watson and Crick in 1953).

RSC Nanoscience & Nanotechnology No. 15
Unravelling Single Cell Genomics: Micro and Nanotools
Edited by Nathalie Bontoux, Luce Dauphinot and Marie-Claude Potier
© Royal Society of Chemistry 2010
Published by the Royal Society of Chemistry, www.rsc.org

(A)

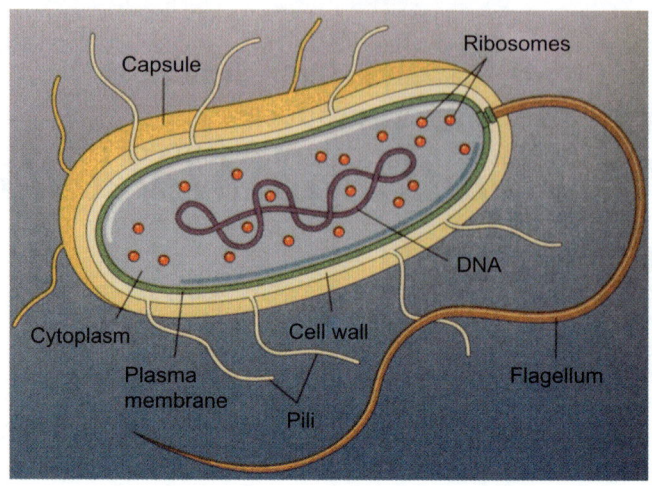

(B)

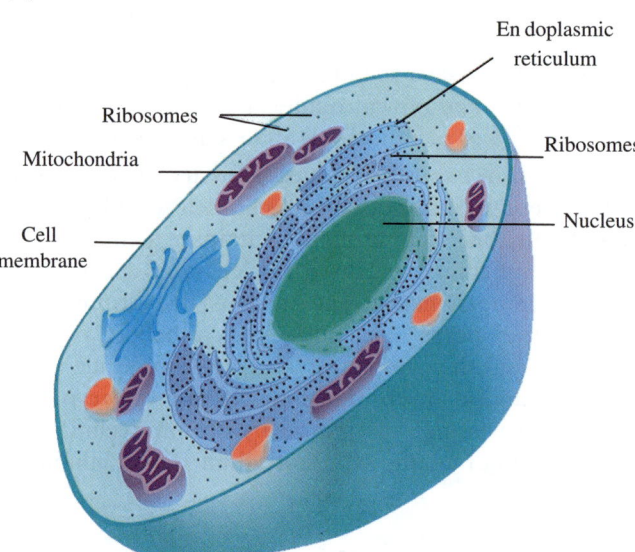

Figure 1.1 Schematics of prokaryotic (A) and eukaryotic cells (B).

The DNA molecule is made up of four different units called nucleotides, consisting of a sugar (deoxyribose) with a phosphate group linked to one of the four following bases (Figure 1.3A): adenosine (A), cytosine (C), thymine (T) or guanine (G). The double-helix DNA molecule is constituted of two complementary strands linked by hydrogen bonds between A–T and C–G

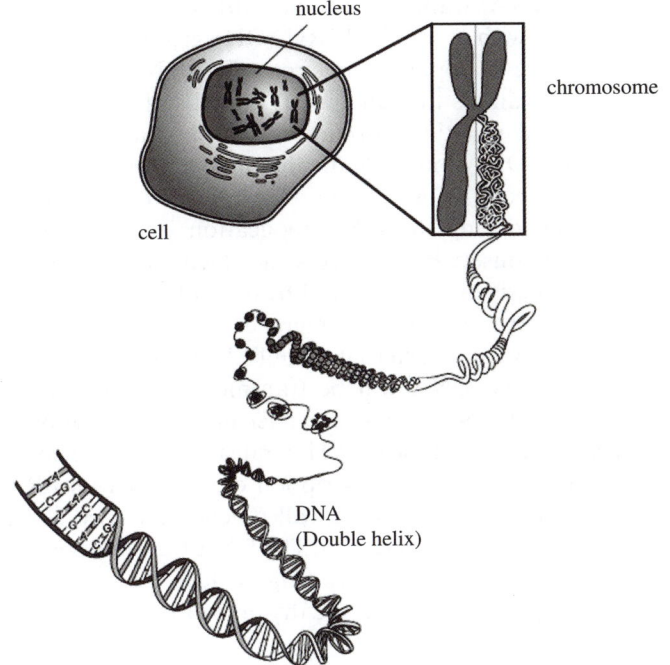

Figure 1.2 Organization of the DNA into chromosomes inside the nucleus of eukaryotic cells.

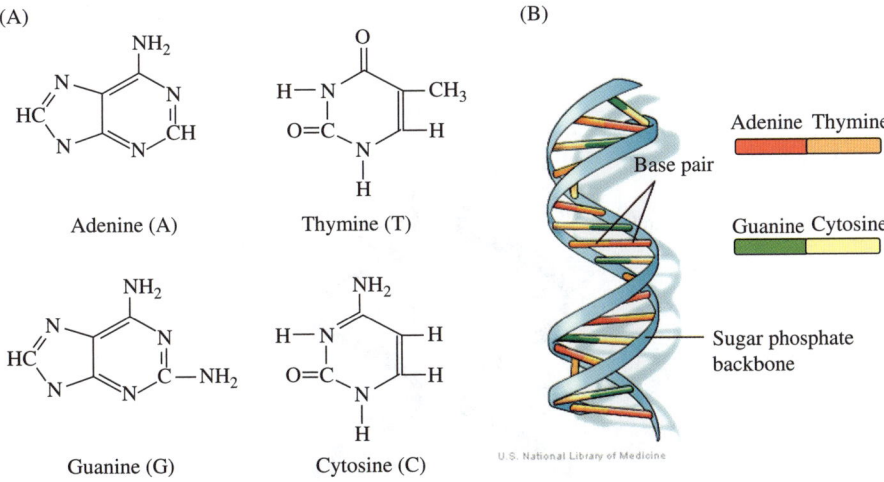

Figure 1.3 Structure of the DNA molecule. (A) The double helix of DNA is composed of four nucleotides consisting of a sugar (deoxyribose) with a phosphate group linked to one of the four bases: adenine (A), cytosine (C), thymine (T) or guanine (G). (B) The two complementary strands are linked by base pairs of A–T and C–G.

(Figure 1.3B). Each DNA strand has a 5′ end with a free phosphate group at its extremity and 3′ end with a free hydroxyl (OH) group. In a double helix of DNA, both strands are in opposite directions. The DNA molecule constitutes the genome of the cell and is the same in all the cells of a given organism; it is used as the storage of genetic information. The human genome consists of 3×10^9 base pairs of a DNA molecule divided in 23 chromosomes pairs. When the cell divides, this genetic information is transmitted to the daughter cells: the genome is duplicated during the DNA replication step. The double helix is replicated by a semi-conservative process in which each strand is used as a template for the synthesis of a new strand by a DNA polymerase (Figure 1.4).

During its lifetime, the cell divides, interacts and communicates with other cells, and responds to various external stimuli. During all these processes, the information stored in the DNA will be transmitted by two other molecules: RNA and proteins. The RNA is only a carrier of the information whereas the proteins are the key effectors of the cell. The transmission of genetic information starts inside the nucleus with the transcription of the DNA into RNA. During this step, specific regions of the genome, called genes, are used as templates for the RNA transcript synthesis (Figure 1.5). The RNA transcripts are quite similar to the DNA with only a few differences (Figure 1.6): they are single-stranded, contain a different sugar (ribose), and the thymine (T) is replaced by uracil (U).

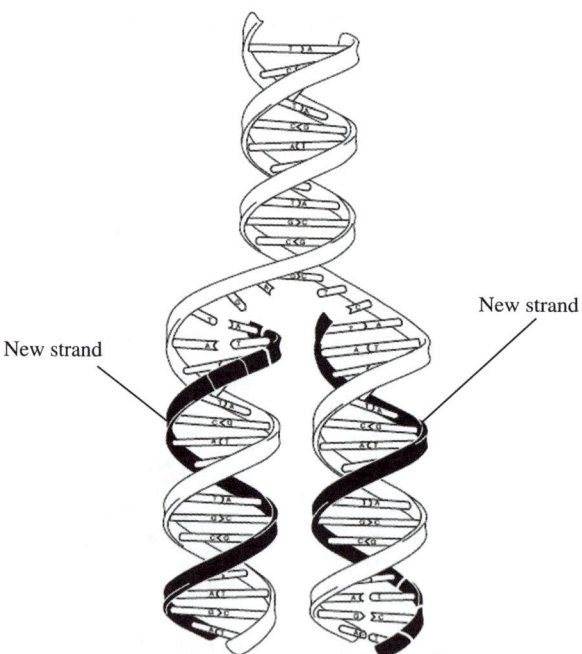

New strand

New strand

Figure 1.4 DNA replication. The DNA molecule is replicated in a semi-conservative way in which each parental strand is used as a template for the synthesis of a new strand.

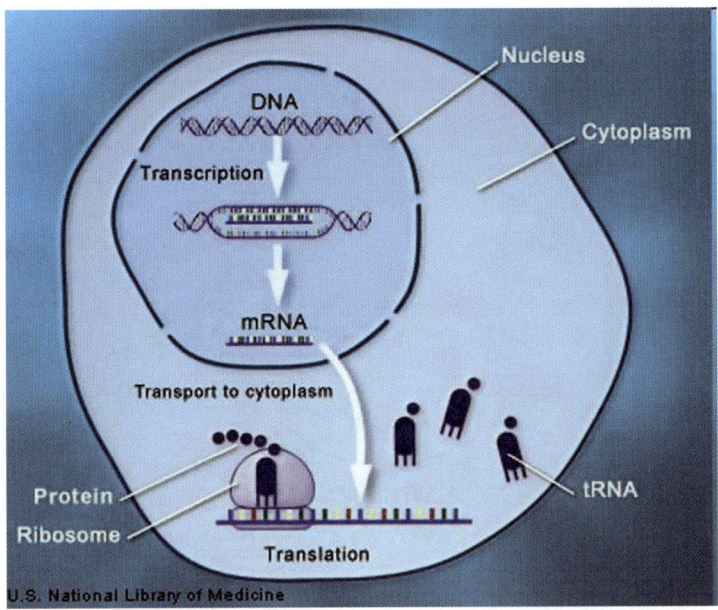

Figure 1.5 Schematics of the different steps of gene expression from DNA to protein. The gene is first transcribed into mRNA inside the nucleus. Then, mRNA migrates to the cytoplasm where it is used to guide protein synthesis during translation in partnership with transfer RNA (tRNA) and ribosomal RNA (rRNA) in the ribosome sub-units.

These RNA molecules, called messenger RNA (mRNA), are the intermediates for protein synthesis. Their sizes range from 100 base pairs (bp) to more than 6000 bp depending on the size of the gene. After synthesis, mRNA molecules migrate to the cytoplasm through the nuclear pores, where they guide protein synthesis during the translation step (Figure 1.5). During translation, the sequence of the mRNA is read out and converted to amino acids. Despite their crucial role, mRNAs are in a minority in the cell and represent only 1–5% of the total RNA. Transfer RNA (tRNA) and ribosomal RNA (rRNA), which are implicated in the translation process (Figure 1.5), represent more than 90% of the total RNA of a cell. In humans, the RNA content of a cell is around 6–10 picograms (pg) which means that there is less than 1 pg of mRNA.

The simplest cells have fewer than 500 genes whereas the most complex ones have more than 25 000 different genes. A gene is said to be "expressed" when its DNA sequence is transcribed into mRNA. All the genes that are expressed at a given time constitute the transcriptome of a cell, *i.e.* all the mRNAs present in the cell. At a time point, there are only around 5000 genes expressed in a cell with one to thousands of copies of each gene, depending on its level of expression. The transcriptome is thus a kind of instantaneous picture of the cell. In the cell, gene expression is highly regulated in order to adjust the mRNA and protein levels on demand.

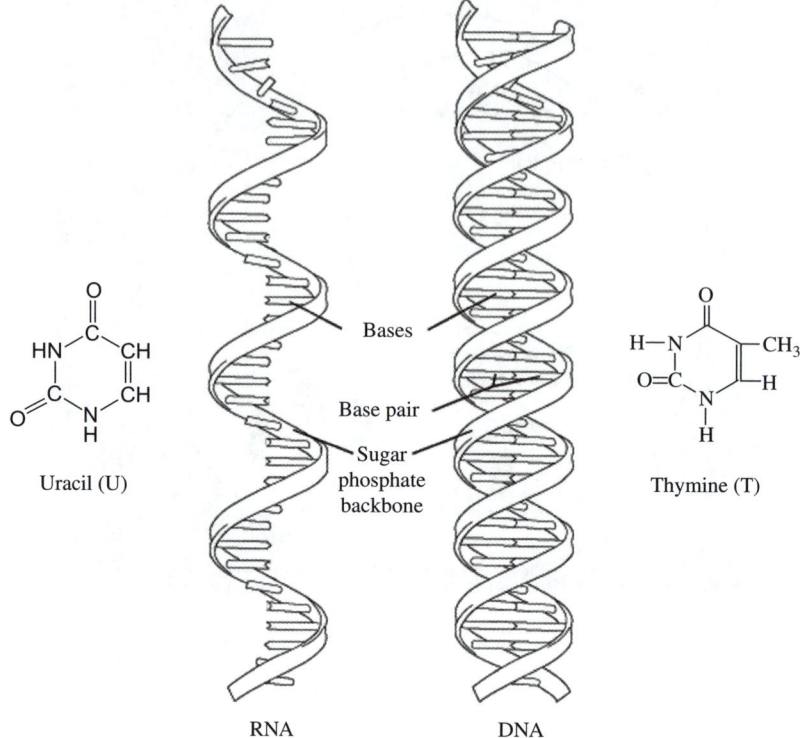

Figure 1.6 Comparison of the structure of DNA and RNA molecules. RNA and DNA molecules are almost the same except that RNA is single-stranded, with a ribose instead of deoxyribose and uracil replacing thymine.

1.2   Molecular Biology Tools for Nucleic Acid Studies

1.2.1   DNA Engineering

Until 1970, studying DNA proved to be very difficult, mostly because of its size. With the emergence of DNA cloning technologies in 1972 followed by Southern blotting[1] and DNA sequencing,[2,3] this difficulty was rapidly bypassed.

Southern blotting is based on the ability of two complementary sequences of DNA to hybridize together with high specificity. This technique allows the detection of a specific single-stranded sequence of DNA among a mixture of millions of different ones, thus providing a powerful tool to identify and characterize genes. DNA fragments are first cut into smaller fragments by restriction enzymes, then separated according to their size by electrophoresis and transferred on a nylon membrane. Once the DNA is bound to the membrane, it is placed in solution with a specific radiolabeled probe that will hybridize to its complementary strand. After washing, the specific hybridization is revealed by autoradiography (Figure 1.7). A similar technique has been

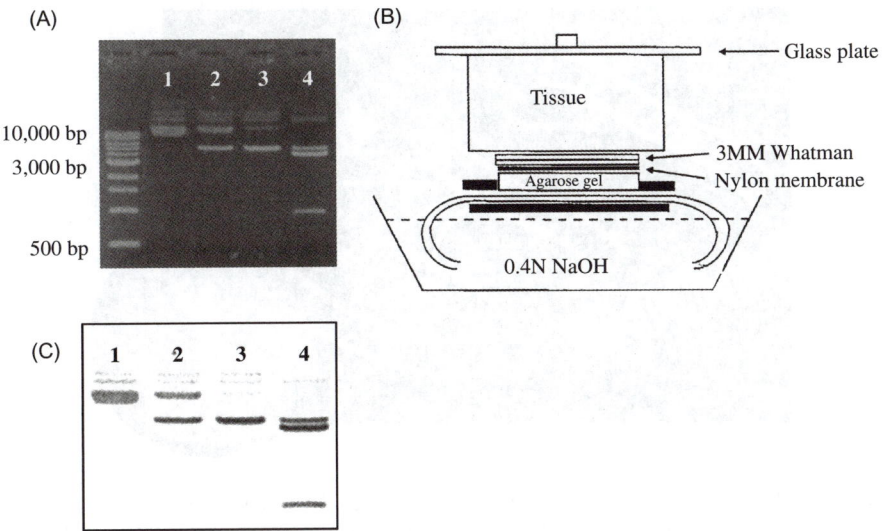

Figure 1.7 Southern blotting. DNA fragments are first cut into smaller fragments by restriction enzymes and separated on an agarose gel according to their size by electrophoresis (A). DNA fragments are then transferred on a nylon membrane by capillarity (B). Once the DNA is bound to the membrane, it is hybridized with a specific radiolabeled probe that will allow the detection of its complementary DNA sequence. The hybridization is revealed by autoradiography (C).

developed for RNA (northern blotting) and protein (western blotting) analysis. RNA molecules or protein extracts are separated by electrophoresis and then revealed by, respectively, a specific DNA probe or antibody. These approaches have been improved to characterize genes or mRNAs directly inside the cell: it is called *in situ* hybridization. Fluorescence *in situ* hybridization (FISH) consists of the hybridization of a fluorescent probe (corresponding to a specific DNA fragment) directly on the chromosomes of a cell, allowing localization of a gene on a chromosome (Figure 1.8). This technique proved to be a very powerful tool in clinical research, especially in cancer, to look for and characterize chromosomal abnormalities such as deletion, duplication or translocation. The same approach can be used to quantify mRNA in cells on fixed tissues.

1.2.2 Polymerase Chain Reaction

In 1985, fundamental progress was made in DNA analysis with the invention of the polymerase chain reaction (PCR)[4] for which Mullis was awarded the Nobel Prize for chemistry in 1993. PCR allows the exponential amplification of a given DNA sequence using specific primers matching on both DNA strands (Figure 1.9). The reaction starts with an initial denaturation step of 1–2 min at 95 °C to obtain single-stranded DNA molecules, followed by a variable number

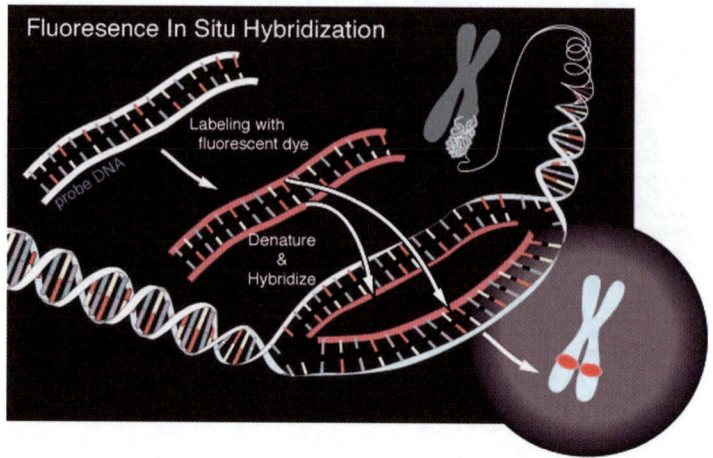

Figure 1.8 The fluorescence *in situ* hybridization (FISH) technique. FISH is used to localize a gene directly onto the chromosomes. A specific DNA probe is synthesized, labeled with a fluorescent dye and then hybridized on interphase or metaphase cells fixed on a glass slide.

(*N*) cycles of amplification. Each cycle of amplification consists of three steps: (1) denaturation to obtain single-stranded molecules, (2) primers annealing to complementary sequences, and (3) elongation of new DNA strands starting from the primers by a DNA polymerase. Once the elongation is done, the amplification cycle starts again at the denaturation step and each DNA strand will be used as a new template. This approach can also be used to study RNA but RNA must be converted into complementary DNA (cDNA) by reverse transcription (RT) prior to PCR amplification (Figure 1.10).

PCR is an extremely sensitive technique and can generate millions of copies of a DNA fragment starting from very small amounts. This provided the opportunity to easily produce DNA probes for cloning, sequencing and genetic engineering, and this method is also widely used for genetic diagnosis and forensic analysis. However, the main limitation of the technique was the impossibility of quantification since it produces nearly the same amount of DNA molecules independently of the initial quantity (Figure 1.11). This was solved in 1992 with the development of real-time quantitative PCR (qPCR).[5] The protocol is the same as for PCR but DNA quantification is achieved during the reaction by monitoring a fluorescence signal that is proportional to the amount of DNA synthesized. Moreover, the initial amount of DNA in the sample can be deduced from the number of amplification cycles required to obtain a given level of fluorescence.[6]

There are two common methods for qPCR quantification (Figure 1.12). First, a fluorescent dye such as SybrGreen, which binds to the double helix of DNA, can be used. In this case, the emission of fluorescence increases with the number of DNA molecules produced (Figure 1.12A). However, the dye

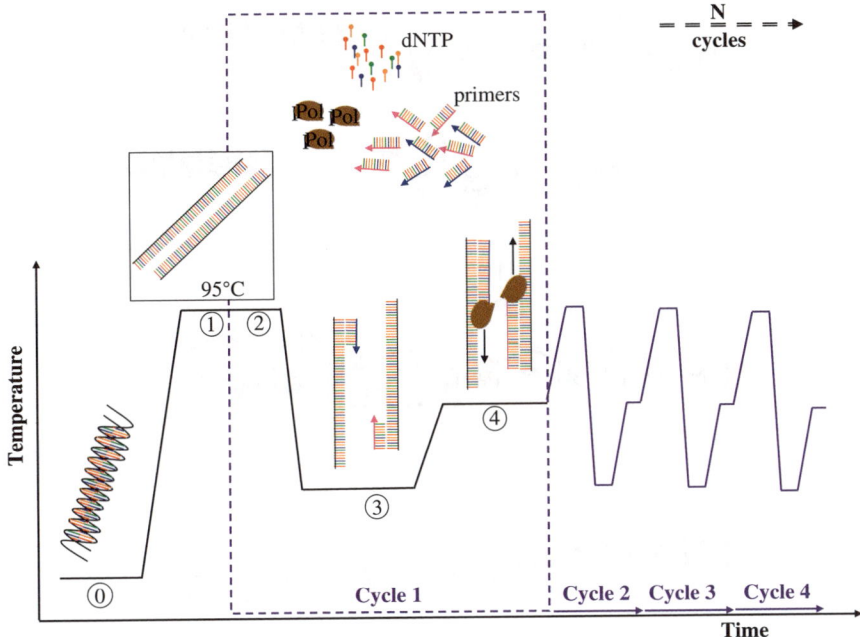

Figure 1.9 Schematics of the polymerase chain reaction (PCR) cycles. Starting from double-stranded DNA, the first step (0) consists of an initial denaturation step for 1–2 min at 95 °C to obtain single-stranded DNA molecules (1). This step is then followed by N cycles of amplification. Each cycle of amplification is composed of three steps: (2) denaturation to obtain single-stranded molecules, (3) forward (pink) and reverse (blue) primers annealing to their complementary sequences, and (4) synthesis of new DNA strands starting from the primers by DNA polymerase. Once the elongation is done, a new cycle starts in step (2) to generate single-stranded DNA molecules that will be used as a new template. The number of DNA molecules is then multiplied by 2 at each cycle of amplification.

is not specific for the amplified sequence and can also bind to non-specific products. As such, it is essential to use primers that give rise to a unique DNA amplification product. Second, a double-labeled probe specific for the amplified DNA fragment, called a Taqman probe,[7] can be used. The probe, located between two primers, is linked at the 5′ end to a fluorescent dye and at the 3′ end to a quencher that inhibits its emission of fluorescence. During the annealing step, the probe and the primers will hybridize to their complementary sequence. During the elongation step, the probe is degraded by 5′–3′ exonuclease activity of the DNA polymerase, which sets the dye free from the quencher (Figure 1.12B) and induces the emission of fluorescence. Today, qPCR is widely used in diagnosis, single-nucleotide polymorphism (SNP), genetic analysis, and pathogen detection. The most important improvement relies on gene expression analysis using RT-qPCR, currently the most accurate and the most sensitive method to detect and quantify mRNA.[8]

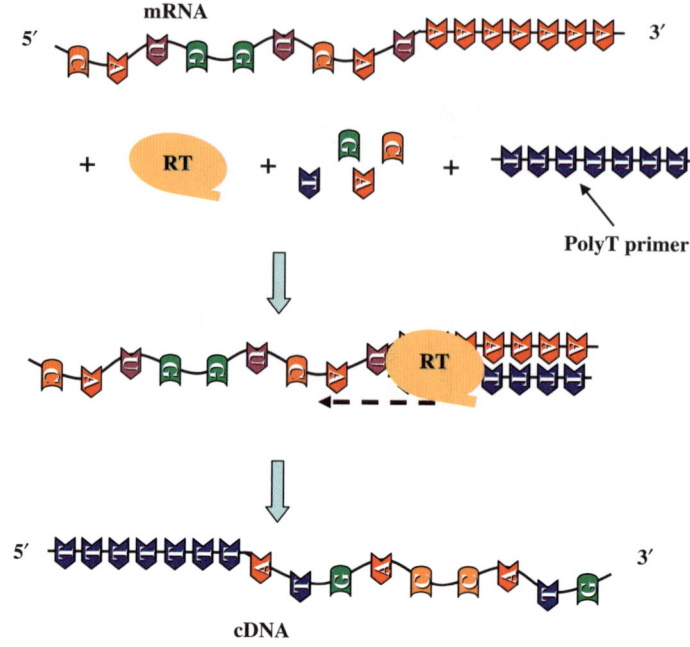

Figure 1.10 Reverse transcription. Messenger RNA (mRNA) is converted into complementary DNA (cDNA) by the reverse transcriptase (RT), starting from a polyT primer that hybridizes with the polyA tail of the mRNA molecules. cDNA is a single-strand DNA molecule with the complementary sequence of the mRNA that can thus be used as a new template for PCR amplification.

1.2.3 DNA Microarrays

In 1995, another technological breakthrough was achieved in gene expression analysis with the development of DNA microarrays by Pat Brown and his colleagues at Stanford University.[9] This new method was based on an old concept of molecular biology (the specific hybridization of two complementary strands of DNA) but opened the way to high-throughput gene profiling. The first microarrays were made of a hundred features and were used to monitor the expression of few genes. Since the achievement of the genome sequence in 2001, DNA microarrays allowing the analysis of the whole transcriptome (more than 25 000 genes for human or mouse) are now available.[10]

DNA microarrays consist of a solid matrix (nylon membrane, glass slide or silicone wafers) on which are covalently linked DNA oligonucleotides or PCR products, called probes. Each probe is specific for one DNA of interest: the design is made with high stringency *in silico* to prevent any cross-hybridization,[11] then millions copies of these probes are spotted on the chip. Usually, the probes are designed, synthesized and then printed onto the chip but some companies have developed specific technologies that allow for *in situ* synthesis

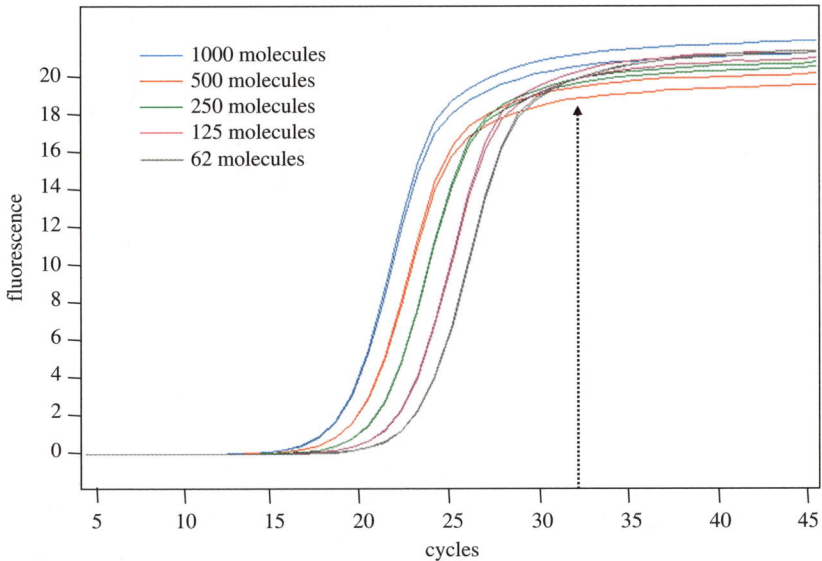

Figure 1.11 Sigmoidal PCR amplification curves. Because the quantity of DNA is multiplied by 2 at each cycle of amplification, the expected PCR curves are exponential. In practice, looking at the number of DNA molecules synthesized across the time, we observe that the curves are sigmoidal with a plateau phase that is independent of the initial quantity of DNA. Indeed, starting from 62 molecules of DNA or 1000 molecules will give rise to the same amount after 30–35 cycles of amplification, which corresponds to the number of cycles usually needed to generate enough DNA to work with. This phenomenon is due to depletion of one of the reagents.

of probes (www.affymetrix.com, www.nimblegen.com, www.agilent.com). When the main goal is to analyse gene expression profiling, mRNA isolated from the different samples that have to be compared are used as targets. They will recognize and hybridize specifically to the complementary sequence of the probes.

There are two kinds of microarray experiments: one-color and dual-color. In the first case, mRNA samples are all labeled with the same dye and each hybridized on a single microarray. The probe–target hybridization signals are then analyzed and compared altogether. In contrast, in dual-color experiments, RNA is extracted from two tissues or cell types (*e.g.* cancer cells versus normal cells), then independently labeled with two different dyes (cyanine 5 and cyanine 3) and loaded together on the microarrays (Figure 1.13). As the targets hybridize to their complementary probe in a quantitative way, measuring the fluorescence of each dye allows the relative level of expression of each gene in the samples to be determined.

DNA microarrays constitute an extremely powerful tool to study differential gene expression, although many problems remain, such as the standardization

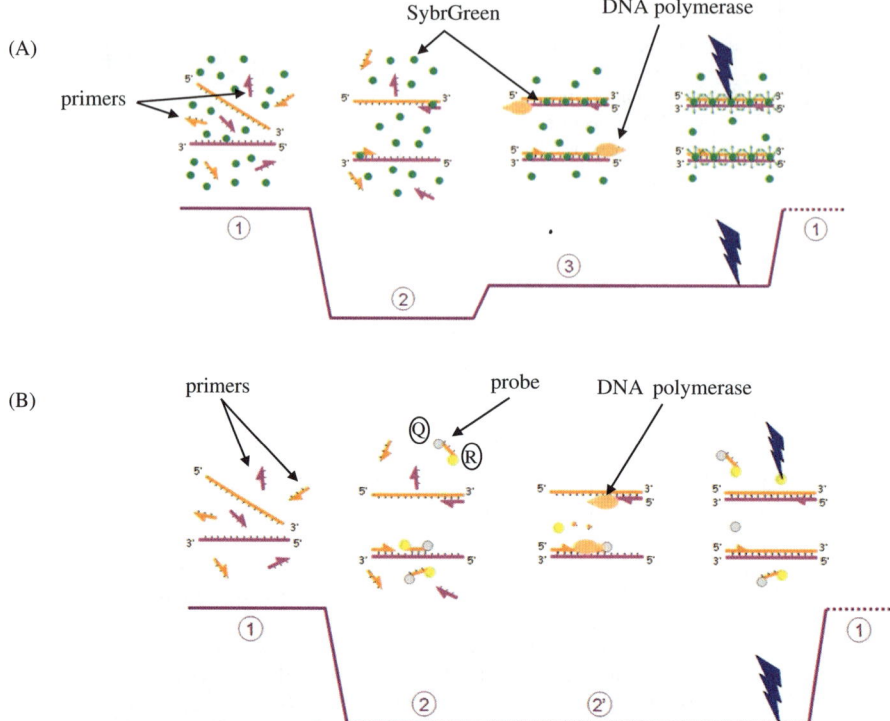

Figure 1.12   Real-time quantitative PCR (qPCR). (A) Using a fluorescent dye such as SybrGreen, which binds to the double helix of DNA, the emission of fluorescence increases with the number of DNA molecules produced. However, the dye is not specific to the amplified sequence and can also bind to non-specific products, so it is essential to use primers that give rise to a unique DNA amplification product. (B) Use of a double-labeled probe specific to the amplified DNA fragment, called a Taqman probe. The probe, located between two primers, is linked at the 5′ end to a fluorescent dye and at the 3′ end to a quencher that inhibits its emission of fluorescence. During the annealing step, the probe and the primers will hybridize to their complementary sequence. During the elongation step, the probe will be degraded by the 5′–3′ exonuclease activity of the DNA polymerase, which will set the dye free from the quencher and induce the emission of fluorescence.

of experiments and the interpretation of results. Data analysis represents a real challenge for bioinformatics.[12,13] First, the collection of raw data through image processing is required; then a normalisation step to remove the systematic bias is necessary; and, finally, statistical tests to determine the significant differences of expression between two or more samples have to be carried out. For each of these steps, many tools and software are available.

Today, the use of microarrays is greatly expanding. DNA microarrays are widely used for gene expression profiling, pathogen detection, resequencing

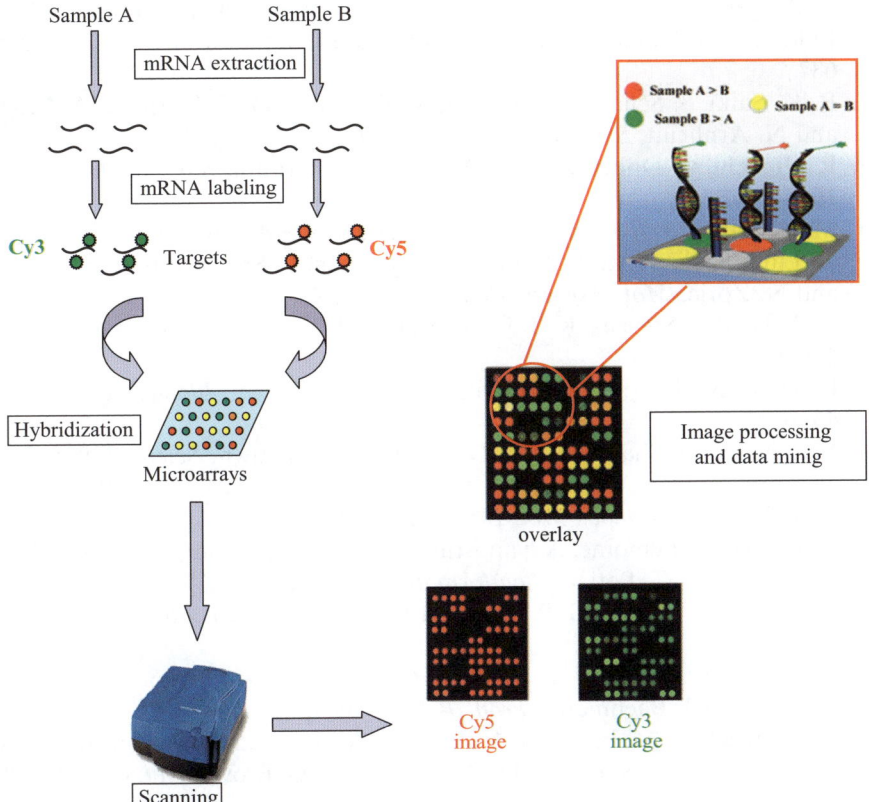

Figure 1.13 Schematics of a dual-color microarray experiment. mRNA is extracted from two tissues or cell lines that have to be compared and independently labeled with two different dyes (cyanine 5 and cyanine 3). The fluorescent targets are then loaded together on the microarrays and will hybridize to their complementary probe on the slide. By scanning the slide with both wavelengths (532 nm and 635 nm), a fluorescent image is obtained for each fluorochrome reflecting the level of expression of each gene in each sample. After images processing and data mining, the overlay allows for comparison of the samples and identification of differences in gene expression.

or genotyping, as well as for other applications such as protein expression or chromosomal copy-number changes analysis using protein arrays and comparative genomic hybridization (CGH) arrays respectively.[14,15]

References

1. E. M. Southern, *J. Mol. Biol.*, 1975, **98**, 503.
2. A. M. Maxam and W. Gilbert, *Proc. Natl. Acad. Sci. U.S.A.*, 1977, **74**, 560.

3. F. Sanger, G. M. Air, B. G. Barrell, N. L. Brown, A. R. Coulson, C. A. Fiddes, C. A. Hutchison, P. M. Slocombe and M. Smith, *Nature*, 1977, **265**, 687.

4. R. K. Saiki, S. Scharf, F. Faloona, K. B. Mullis, G. T. Horn, H. A. Erlich and N. Arnheim, *Science*, 1985, **230**, 1350.

5. R. Higuchi, G. Dollinger, P. S. Walsh and R. Griffith, *Biotechnology*, 1992, **10**, 413.

6. M. Kubista, J. M. Andrade, M. Bengtsson, A. Forootan, J. Jonak, K. Lind, R. Sindelka, R. Sjoback, B. Sjogreen, L. Strombom, A. Stahlberg and N. Zoric, *Mol. Aspects Med.*, 2006, **27**, 95.

7. C. A. Heid, J. Stevens, K. J. Livak and P. M. Williams, *Genome Res.*, 1996, **6**, 986.

8. H. D. VanGuilder, K. E. Vrana and W. M. Freeman, *Biotechniques*, 2008, **44**, 619.

9. M. Schena, D. Shalon, R. W. Davis and P. O. Brown, *Science*, 1995, **270**, 467.

10. U. Bilitewski, *Methods Mol. Biol.*, 2009, **509**, 1.

11. G. Golfier, S. Lemoine, A. van Miltenberg, A. Bendjoudi, J. Rossier, S. Le Crom and M. C. Potier, *Bioinformatics*, 2009, **25**, 128.

12. D. B. Allison, X. Cui, G. P. Page and M. Sabripour, *Nat. Rev. Genet.*, 2006, **7**, 55.

13. C. Steinhoff and M. Vingron, *Brief Bioinform.*, 2006, **7**, 166.

14. T. Joos and J. Bachmann, *Front Biosci.*, 2009, **14**, 4376.

15. J. R. Pollack, C. M. Perou, A. A. Alizadeh, M. B. Eisen, A. Pergamenschikov, C. F. Williams, S. S. Jeffrey, D. Botstein and P. O. Brown, *Nat. Genet.*, 1999, **23**, 41.

CHAPTER 2

The Central Dogma in Molecular Biology

LAILI MAHMOUDIAN

CRICM, UPMC – INSERM URMS975 – CNRS UMR7225, Hôpital de la Pitié Salpétrière, Paris, France

Abstract

The central dogma tells that the genetic information coded by DNA molecules, first transferred to RNA molecules as intermediated molecules, is then transferred to protein molecules. In other words it says that the genetic information is saved as sequences of nucleic acids but the function has to be expressed in the form of proteins. There are three steps in the conversion of the genetic information to the proteins: replication, transcription and translation. Figure 2.1 shows the principals of central dogma.

2.1 Replication

Replication is a process that copies the double-stranded DNA of a cell to two identical copies (Figure 2.1). DNA replication is a semi-conservative process, which means two original strands of the DNA molecules separate and each act as a template for a new complementary strand. Two identical DNA molecules will be produced from a single, double-stranded DNA molecule. Basically, in a cell, DNA replication starts at specific locations, called origins. During DNA replication first, an enzyme called helicase unwinds the DNA strand. This enzyme breaks the hydrogen bonds between the two strands of the DNA molecule. The resulting structure is two branching single-stranded DNA

RSC Nanoscience & Nanotechnology No. 15
Unravelling Single Cell Genomics: Micro and Nanotools
Edited by Nathalie Bontoux, Luce Dauphinot and Marie-Claude Potier
© Royal Society of Chemistry 2010
Published by the Royal Society of Chemistry, www.rsc.org

Figure 2.1   The central dogma of molecular biology. The DNA molecules in a cell are self-replicating molecules. The RNA molecules in the cell are made on a DNA template and all protein molecules are determined by RNA molecules. The important message of this figure is that the arrows are unidirectional; that is, RNA sequences are never copied on protein templates; likewise, RNA never acts as a template for DNA.

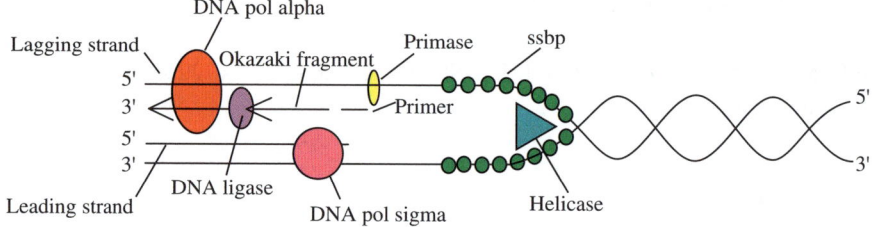

Figure 2.2   A eukaryotic replication fork. The leading strand is synthesized in the direction in which the fork moves, and the lagging strand is synthesized in the reverse direction in the form of Okazaki fragments. Okazaki fragments are then joined together by the ligase enzyme. The single-stranded binding proteins (ssbps) hold the single-stranded DNA and prevent rejoining of the DNA fragments before finishing the replication.

molecules, which are called a replication fork. DNA polymerase is the enzyme that generates the two complementary strands by adding the nucleotides (units of DNA) from the 5′ end to the 3′ end of the DNA strand.

Figure 2.2 shows a simple scheme of a replication fork. Each replication fork consists of two new DNA strands, which are called leading and lagging strands. The leading strand is the newly replicated DNA strand that is synthesized in the 5′ to 3′ direction. DNA polymerase α is able to synthesize the leading strand continuously in the same direction in which the replication fork is moving through the DNA strand.

On the other hand, the lagging strand is the DNA strand that is in the opposite direction of the replication fork from the 3′ to 5′ direction. Since DNA polymerase cannot synthesize in this direction, replication of the lagging strand takes place in short segments called the Okazaki fragments. An enzyme called primase makes short-length RNA primers which are the short strands of RNA that serves as the starting point of DNA replication. DNA polymerase δ can use the free 3′-OH of the RNA primer to synthesize DNA in the 5′ to 3′ direction. The RNA fragments then are removed from the lagging strands and new deoxyribonucleotides are added to fill the gaps. An enzyme called DNA ligase then joins the desoxyribonuleotides together and completes the synthesis of the lagging strand.

2.2 Transcription

Transcription is synthesis of RNA on a DNA template. In this process the DNA nucleotide sequence information is transcribed into an RNA sequence. The transcribed RNA molecule is called messenger RNA (mRNA) because it carries the genetic information from a gene (DNA) out of the nucleus into the cytoplasm of the cell.

The mechanism of transcription includes three steps: initiation, elongation, and termination. Figure 2.3 shows the mechanism of transcription. In the initiation step first, RNA polymerase recognizes specific sequences in the DNA molecule, called a promoter, and bind to it. The promoter identifies the start of the gene. Then, RNA polymerase unwinds the DNA strand, so a transcription bubble generates and the complementary bases assemble as in DNA except in RNA uracil (U) is used instead of thymine (T) in DNA. In the elongation step, the mRNA transcript grows while RNA polymerase moves along the DNA strand and finally a terminator (a special sequence in the DNA) indicates when

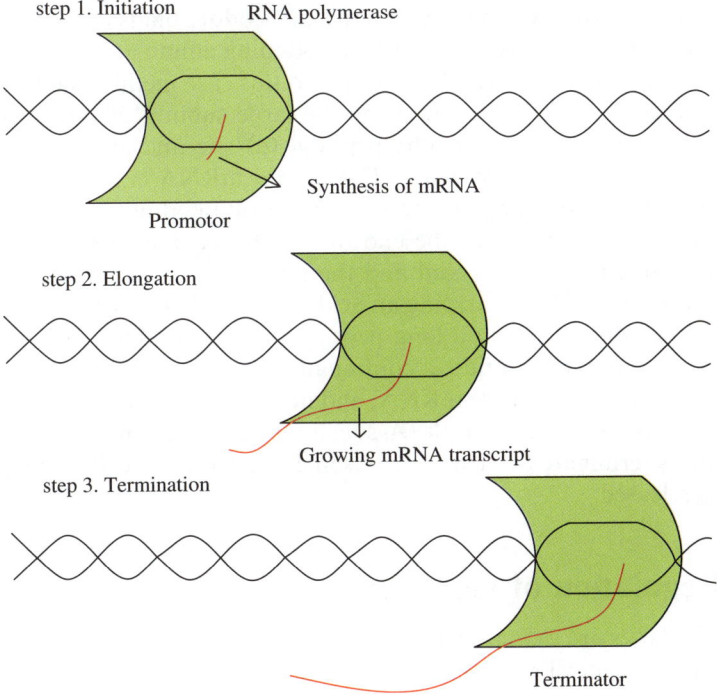

Figure 2.3 Mechanisms of transcription in eukaryotes.

transcription will stop. This step is called termination. The mRNA produced is called mRNA transcript. After transcription the DNA strands rejoin.

2.3   Translation

Translation is the production of proteins (sequences of amino acids) using mature mRNA produced in the transcription. This process takes place in the cytoplasm where the ribosomes are located. In eukaryotes ribosomes are made of small (40S) and large subunits (60S) and surround mRNA. In the living cells the genetic information can be translated based on the genetic code rule. In this rule, triplets of nucleic acid sequences can be translated to single amino acids. Each triplet is called a codon. Usually, each codon specifies a single amino acid of the 20 amino acids used in the synthesis of proteins.

The translation process requires several macromolecules, including mRNA, the small and large subunits of a ribosome, tRNA (the RNA molecule which carries the anti-codon that is complementary to mRNA codon), and the release factor. The process includes three stages: initiation, elongation, and termination. Figure 2.4 shows the mechanism of translation. Initiation starts when a complex of eukaryotic initiation factors (EIF4) bind to the 5′ cap of mRNA. There are four EIFs in the eukaryotes (EIFs G, A, B, and E) which are collectively known as EIF4F. They hold mRNA for the small subunit of the ribosome to be attached to it. Then, the 40S subunit moves to the translation initiation site and tRNA, which carries the anti-codon, binds to the first mRNA codon, which typically is AUG. The corresponding amino acid for the codon is attached to the end of the tRNA molecule, in this case methionine (methionine corresponds to the AUG codon). Then the large subunit of the chromosome binds to the complex to create the peptidyl (p) site and aminoacyl (A) site. The first tRNA occupies the p-site. The second mRNA enters the A-site and carries the complementary sequences to the second codon. The methionine is then transferred and binds to the end of the tRNA that exists in the A-site. Then, the first mRNA is released and the ribosome moves along the mRNA strand and the next tRNA carrying an acid aminoacyl-tRNA enters. This stage is called elongation in which a long peptide chain is produced. As elongation continues, the growing peptide is continually transferred to the A-site tRNA, the ribosome moves along the mRNA, and new aminoacyl-tRNA enters. When a stop codon is encountered in the A-site, a release factor enters the A-site and translation is terminated. Then the ribosome dissociates and the newly formed protein is released.

2.4   Regulation of Gene Expression

Many eukaryotes are estimated to have between 20 000 and 25 000 genes. Some are expressed in all cells at all times but others are only expressed when a cell enters a special differentiation step or when the conditions inside or outside of the cell change. So, there should be precise control stages for gene expression in living cells. Gene expression can be controlled at several stages, resulting in

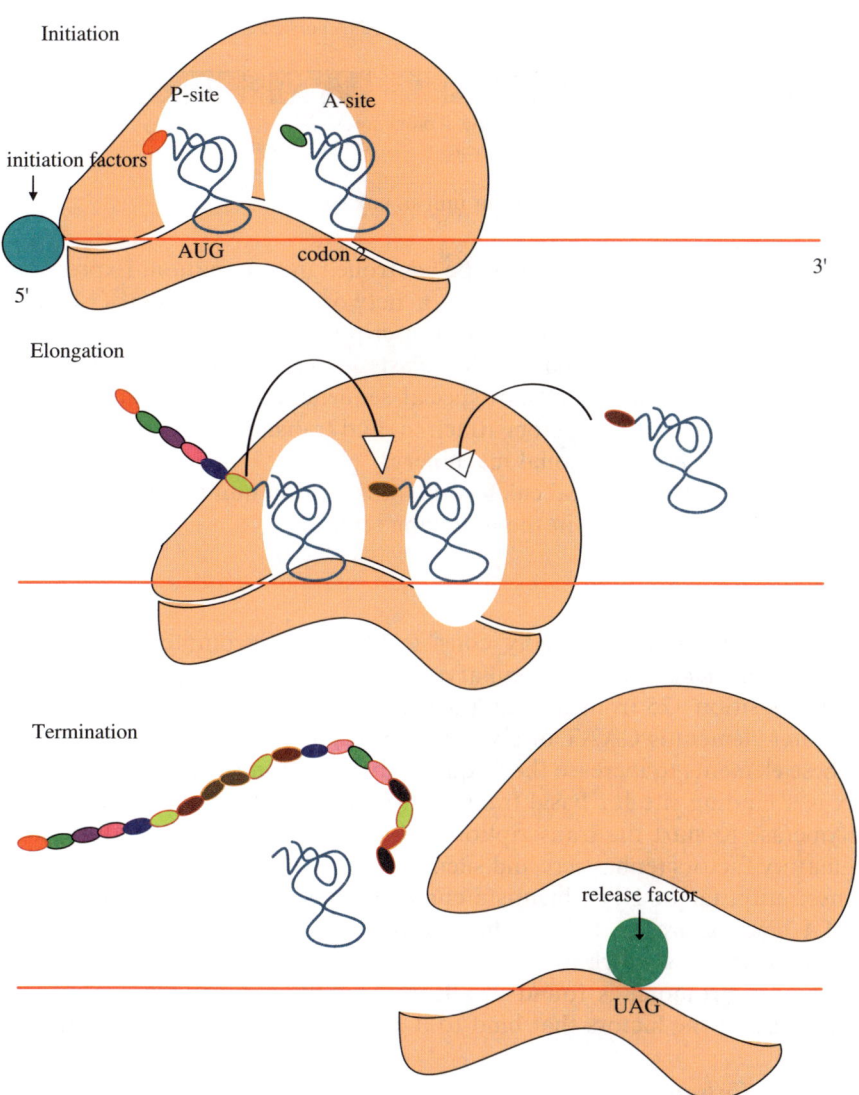

Figure 2.4 Mechanisms of translation in eukaryotes. The small subunit of the ribosome is 40S and the large subunit is 60S.

transcriptional, post-transcriptional, translational, and post-translational control. These regulatory mechanisms will be explained in more detail in the following sections.

2.4.1 Transcriptional Control

Transcription is often controlled at the initiation step. In this way a cell saves energy by not producing unnecessary transcripts. Transcription is not normally

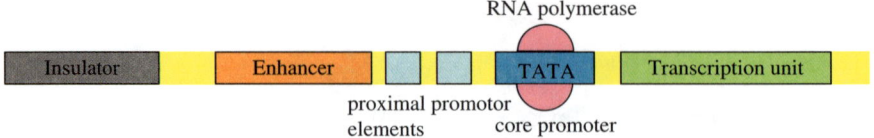

Figure 2.5   The general transcription unit in eukaryotes.

controlled in elongation but it may be controlled in termination. Expression of protein-coding genes is mediated by a network of thousands of sequence-specific DNA-binding proteins called transcription factors (TFs). Transcription factors interpret the information that exists in the promoter of the gene and other regulatory elements (REs) (special sequences, usually upstream of the gene, which transcription factors attach to) and transmit the proper response to the RNA pol II transcriptional machinery.

Figure 2.5 shows a typical eukaryotic transcription unit consisting of several parts which are explained in the following sections.

2.4.1.1   Promoters

A eukaryotic promoter usually consists of a core promoter and proximal promoter elements. A typical eukaryotic core promoter consists of a TATA box (at position –25 from starting point of the gene). An example of proximal promoter element is CAAT or a GC box. Special regulatory proteins can bind to these elements to increase the frequency of initiation transcription. A TATA box is a binding site for basal transcription factors (TFs) which recruit RNA polymerase to start the transcription. CAAT or a GC box is binding site for regulatory factors (enhancers and silencers). A TATA box can be identified by transcription factor IID which is a complex of 10 different proteins including TATA-binding protein (TBP) which recognizes and binds to the TATA box and other proteins which bind to TBP and each other but not to the DNA.

The core promoter is found in all protein-coding genes but the upstream promoter and the factors that bind to it are different from gene to gene.

2.4.1.2   Enhancers

Enhancers are regions of DNA that are thousands of base pairs away from the gene they control. Special proteins called enhancer-binding proteins bind to these regions of DNA to increase the rate of transcription of the gene.

Enhancers can be located upstream, downstream or even within the gene they control.

2.4.1.3   Silencers

Silencers are regulatory sequences that are capable of binding transcription regulation factors termed repressors. Upon binding of a repressor, transcription initiation is prevented and thus mRNA synthesis will be decreased or suppressed.

2.4.1.4   Insulators

An insulator is a regulatory DNA sequence that works as an enhancer-blocking element, thereby preventing an enhancer from binding to a promoter of some other genes in the same region. This strategy prevents a gene from being influenced by the activation (or repression) of its neighbor.

2.4.1.5   Example of Transcriptional Regulations

In humans and mice, the gene for insulin-like growth factor 2 (*Igf2*), inherited from the father, is active, while the allele inherited from the mother is not active. This phenomenon is called imprinting. The mechanism of inactivation in the maternal allele is that in both the mother's and father's promoter of *Igf2* there is an insulator, but in the case of father, the insulator has been methylated. So, the regulator protein is not able to bind to the insulator and the enhancer is free to turn on the *Igf2* promoter and the gene can be expressed.

2.4.2   Post-transcriptional Modifications

In eukaryotic cells the newly produced mRNA transcript must be further modified before being used. The more stable mature mRNA molecules can survive better for translation. There are three main post-transcriptional modifications in eukaryotes: capping, polyadenylation, and splicing.

- *Capping*. Modification of the 5′ end of mRNA in eukaryotes is called capping. The cap consists of a methylated GTP linked to the mRNA. Capped mRNA can efficiently be translated to the corresponding protein.
- During *polyadenylation* a polyA tail (150–200 adenines) is added to the 3′ end.
- *Splicing*. The newly transcribed mRNA molecule contains regions that do not hold a genetic message. These regions are called introns and must be removed. The process for removing the introns is called mRNA splicing. The remaining portions of mRNA are called exons. Then, they spliced together to make a mature mRNA transcript. Figure 2.6 shows the mechanism of splicing and the position where transcription and splicing take place.

2.4.3   Translational Control

Translation can be controlled in three steps: initiation, elongation, and termination. Initiation is the most highly regulated step since it involves a large number of initiation factors and accessory proteins. Figure 2.7 shows a typical human protein coding mRNA (including several regions) and some regulatory proteins.

As shown in Figure 2.7, there are regions at the beginning of mRNA which do not code for the proteins. These are called untranslated regions (UTRs). Regulatory proteins (activators or repressors) can bind to the UTRs to enhance or restrict binding of ribosome and though control the translation. Translation

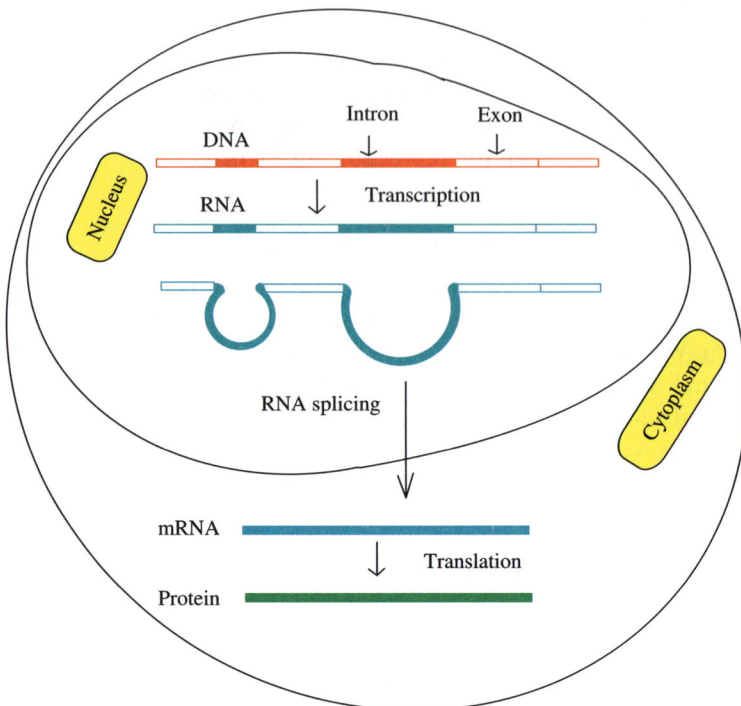

Figure 2.6 Transcription and mRNA splicing in a eukaryotic cell. These steps take place in the nucleus but the mature mRNA is then transported to the cytoplasm of the cell. Then the protein synthesis machinery can synthesize the protein from the mature mRNA in the cytoplasm.

in eukaryotes can be controlled by several methods such as phosphorylation of initiation factor eIF-2α. Phosphorylation of eIF4E-binding-protein or by the stimulation of the mRNA-binding proteins. For example, phosphorylation of initiation factor e-IE2α can inhibit the translation of hemoglobin or phosphorylation of eIF4E-binding-protein stimulates the translation by increasing the cap-binding efficiency. Insulin (a hormone that stimulates growth in humans) can stimulate the release of eIF4E from eIF4E-binding-proteins and in this way increase the translation rate in the cells.[1] Other regulatory molecules are small single-stranded RNA molecules (miRNAs) that bind by base pairing to 3'UTRs of mRNA and inhibit translation.[2] On the other hand, mRNA molecules might be degraded in the cytoplasm, which affects the amount of protein that can be expressed.

2.4.4 Post-translational Control

Usually, newly synthesized proteins are rarely active. They need to be modified to be functional. There are four main types of post-translational modifications, as follows.

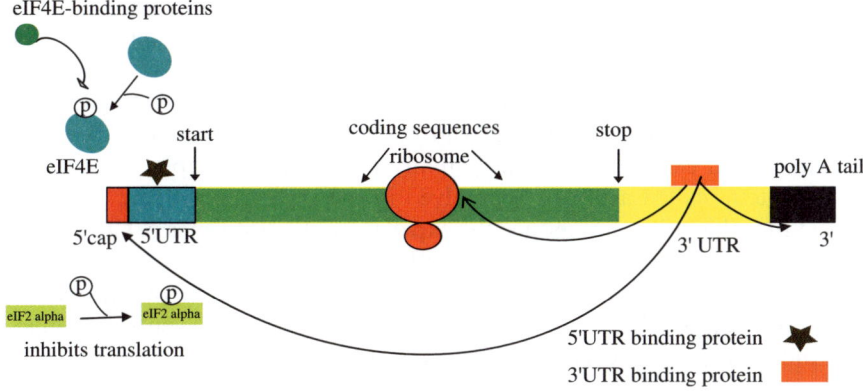

Figure 2.7 Protein coding unit in eukaryotes. Some examples of translation regulatory mechanisms have shown. For example, eIF4E can induce the translation in the phosphorylated form. On the other hand eIF4E-binding protein can inactivate eIF4E protein. Or eIF2α can inhibit the translation in the phosphorylated form. There are also untranslated regions (UTRs) close to the coding sequence. Special proteins can bind to these regions to enhance or suppress the translation by affecting the 5′ cap or polyA tail or the ribosome.

- *Protein folding.* The newly synthesized polypeptide needs to fold to keep its correct tertiary structure.
- *Proteolytic cleavage.* Some proteins should be cut, in part using enzymes called proteases. Protease cuts some segments in the protein molecule to make an active molecule.
- *Chemical modification.* Some amino acids in the newly synthesized polypeptide might be modified by methylation, phosphorylation, and glycosylation.
- *Intenin splicing.* Inteins are intervening sequences in some polypeptides which should be removed and the exteins are the remaining parts that ligate to make the active protein.

2.5   Limitations of the Central Dogma

Although the central dogma can generally describe events in the conversion of information in a cell to functional molecules, there are exceptions that cannot partly or completely be matched with it. For example, in tobacco mosaic virus (TMV) the genetic material has been found to be RNA. Thus, it is possible to circumvent DNA and the first step of the central dogma. As explained above, different mature mRNA can be generated using the same mRNA transcript (alternative splicing). Therefore the sequences in protein are encoded in DNA discontinuously. Thus the sequences in protein are not solely from DNA as explained in the central dogma. Several types of mRNA editing were found in eukaryotes after the first demonstration of central dogma. For instance in the human parasite *Trypanosome bruceri*, there are 551 uracils all over the mRNA transcript that codes for NADH dehydrogenase subunit 7, and 88 are deleted.[3] So, there is a huge difference between what is in the sequence of DNA and what is the final sequence of the protein. Another example concerns the disease that infects neuronal cells in sheep and humans and is called scrapie. In this virus, only a protein, called a prion, is involved in the replication of the virus. It is believed that this prion can induce the expression of silent genes of the host and cause some post-translational changes in the proteins of the normal cells in the host.[4] These findings have shown that in nature, transfer of protein–protein information is possible. Although there are exceptions to the central dogma, it has a wide role in the development of molecular biology.

2.6 Single Cells and their Complexity

Only less than 2% of the human genome in a cell is protein-coding sequences. In each particular cell, a unique combination of regulatory elements and transcription factors (that bind to regulatory elements) signifies that a special gene can be expressed in a particular cell. For example, in each DNA fragment, there are different exons and few close promoters and terminators, although a single cell can express differently by using different exons, promoters, and terminators depending on the conditions outside the cell or on the regulatory elements that control gene expression. In other words, one gene can make multiple proteins. Therefore by using all regulatory steps from transcription to translation, a cell population in a tissue will be heterogeneous. This issue will be important when, for example, a single cell starts to divide in an uncontrolled manner and cause the generation of tumors in the body. This fact highlights the importance of developing methods for the analysis of gene expression at a single cell level. Microfluidic devices recently proved to be very convenient and useful tools for the analysis of the gene expression at the single cell level.[5,6] Microfluidic devices provide a very small growth environment, comparable to the dimensions of the environment that surrounds a cell, in order to analyze cells under conditions that are similar to those that occur naturally. Also, these tools provide a controlled area for monitoring the small changes in a single cell.[7–9]

References

1. A. Pause, G. J. Belsham, A. C. Gingras, O. Donzé, T. A. Lin, J. C. Lawrence and N. Sonenberg, *Nature*, 1994, **371**, 747–748.
2. R. S. Pillai, S. N. Bhattacharyya and W. Filipowicz, *Trends Cell Biol.*, 2007, **17**, 118.
3. D. J. Koslowski, *Cell*, 1990, **62**, 901–911.
4. S. B. Prusiner, *Science*, 1982, **216**, 136–144.
5. N. Bontoux, L. Dauphinot, T. Vitalis, V. Studer, Y. Chen, J. Rossier and M.-C. Potier, *Lab Chip*, 2008, **8**, 443–450.

6. J. F. Zhong, Y. Chen, J. S. Marcus, A. Scherer, S. R. Quake, C. R. Taylor and L. P. Weiner, *Lab Chip*, 2008, **8**, 68–74.
7. R. J. Taylora, D. Falconnetb, A. Niemisto, S. A. Ramseya, S. Prinza, I. Shmulevicha, T. Galitskia and C. L. Hansen, *Proc. Natl. Acad. Sci. U. S. A.*, 2009, **106**, 3758–3763.
8. J. Olofsson, H. Bridle, A. Jesorka, I. Isaksson, S. Weber and O. Orwar, *Anal. Chem.*, 2009, **81**, 1810–1818.
9. N. M. Toriello, E. S. Douglas, N. Thaitrong, S. C. Hsiao, M. B. Francis, C. R. Bertozzi and A. Mathies, *Proc. Natl. Acad. Sci. U. S. A.*, 2008, **105**, 20173–20178.

CHAPTER 3

From Unicellular to Multicellular Organisms: Tells from Evolution and from Development

TANIA VITALIS

CNRS-UMR 7637, Laboratoire de Neurobiologie, ESPCI, 10 rue Vauquelin, 75005, Paris, France

Abstract

Both evolution from unicellular to multicellular organisms and development of complex organisms from one cell to the final body plan require specific cellular features. Unicellular organisms or fertilized eggs are virtually pluripotent. As body plans complexify and pluricellularity emerges cells will become specified and more and more differentiated and they will have to realise specific tasks. Therefore, organisms will have to develop means of cell–cell communication, cells will have to maintain cohesion and adhesion between each other, organisms will have to find ways to reproduce and to repair or regenerate specific cells or tissues. We will review points that characterise this passage from unicellular to pluricellular organisms using specific examples taken from evolution and development.

3.1 Cells from Evolution

Multicellular organisms are organisms consisting of more than one cell, and have differentiated cells that perform specialized tasks. Within the plant and

RSC Nanoscience & Nanotechnology No. 15
Unravelling Single Cell Genomics: Micro and Nanotools
Edited by Nathalie Bontoux, Luce Dauphinot and Marie-Claude Potier
© Royal Society of Chemistry 2010
Published by the Royal Society of Chemistry, www.rsc.org

animal kingdoms several levels of organization can be observed. For instance, sponges consist of multiple specialized cellular types and pluripotent cells cooperating together for a common goal.[1] These cell types include digestive cells (the choanocytes), secreting cells (the sclerocytes), tubular pore cells (the porocytes), and epidermal cells (the pinacocytes) (Figure 3.1). Although these cells cooperate and form a multicellular structure, the sponge, they are not interconnected and do not form tissues. More complex organisms such as diploblasts (jellyfish, coral, and hydra) display a higher level of organization. In these organisms, cells are interconnected and form specific tissues.[2,3] For instance, in the jellyfish an epidermis and a nerve net perform protective and sensory functions.

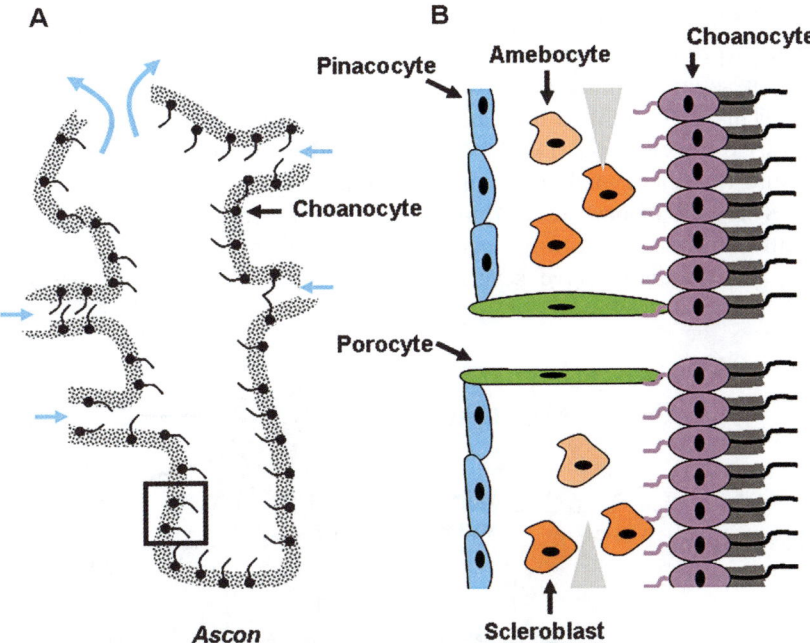

Figure 3.1 Body plan organization of the sponges. (A) Water (blue) enters numerous small pores (ostia) that are surrounded by donut-shaped cells (porocytes) that open and close to control water flow. For the sponge depicted here, water flows directly into an open chamber (the spongocoel) and leaves the spongocoel by the osculi. The interior of the spongocoel is lined with flagellated cells (choanocytes) which possess a flagellum that creates currents to force the water and nutritive particles to enter the sponge. (B) High magnification of the boxed area shown in (A). Amebocytes carry the nutrients to the other cells, compensating for the lack of a circulatory system. In addition, amebocytes can also undergo developmental changes and turn into any cell type that may be required. They also produce the sponge's skeleton which is constituted of spicules and the fibers of flexible protein.

In the most complex organisms, differentiated cells are not only organized in tissues but also in organs. The flatworm nervous system, which is studied extensively, is constituted by grouping ganglion cells[3] together (Figure 3.2). In the most complex organisms, such as mammals, groups of organs, or organ systems, act together to perform complex and related functions. In this way, the mammalian brain is able to integrate sensory signals from the periphery (sensory system), integrate the signals in the cerebral cortex, and respond to it by stimulating the appropriate neuronal circuits by, for instance, inducing specific movements[4] (Figure 3.3).

Several evolutionary hypotheses have been proposed to explain how such a level of organization occurred. In the **symbiotic theory** the first multicellular organism is thought to have occurred from symbiosis, a cooperation of different species of unicellular organisms with different functions. From this theory, these different cells would have become so dependent on each other that they would have been unable to survive independently, leading to their genome being incorporated into one, multicellular organism. Each organism would become a separate lineage of differentiated cells within a newly generated species. This is thought to have occurred with mitochondria in animals endosymbiosis. However, the DNAs of mitochondria are distinct from genomic DNA and this theory is highly controversial.

The **syncytial theory** states that a single organism could have developed internal membrane partitions around each of its nuclei. In support of this theory, ciliates could have several nuclei, and in the fruit fly, *Drosophila*

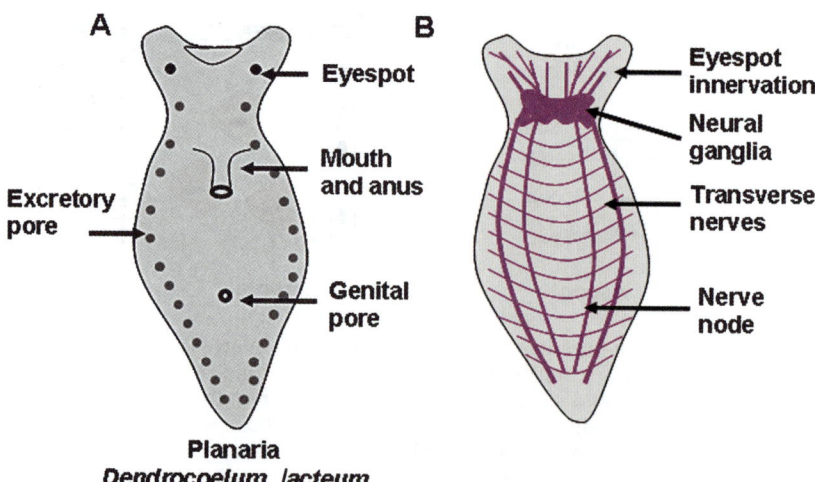

Planaria
Dendrocoelum lacteum

Figure 3.2   The Planaria nervous system. (A) Planaria is a flat worm (a Platy-helminthe) that possesses a bilateral symmetry and is composed of three fundamental cell layers (it is triphoblastic). (B) The nervous system of the planaria is composed of a cerebral ganglia from which longitudinal nerves radiate and communicate *via* transverse nerves. In addition, Eyespot innervations are also part of the system.

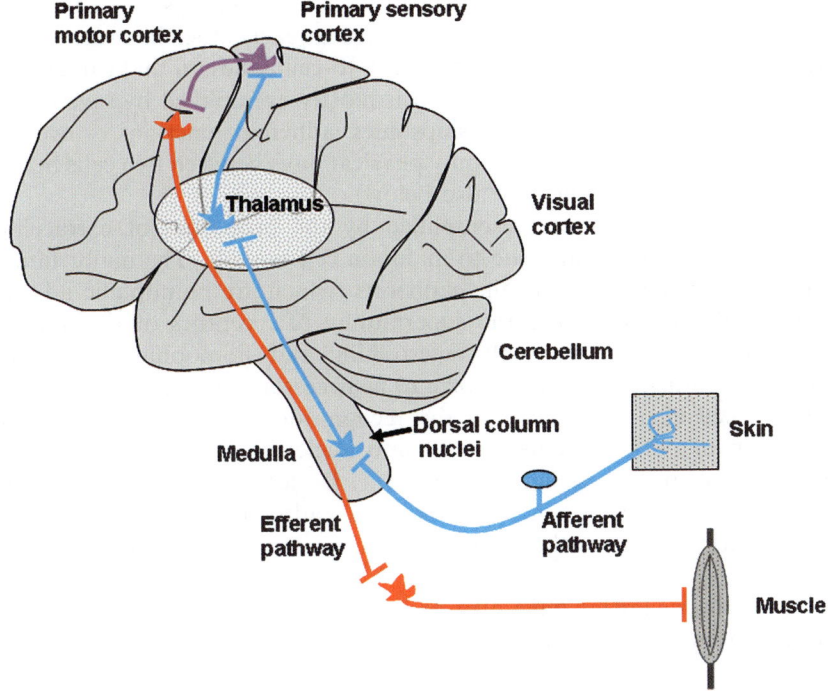

Figure 3.3 The mammalian brain as an example of an organ system. The somato-
sensory sytem and the motor system cooperate to carry out most behavioral tasks. Sensory inputs ascend through the spinal cord to a synaptic
relay in the dorsal column nuclei in the brain stem, then to a synaptic relay
in the thalamus, and eventually reach the primary sensory cortex. The
direct motor pathway descends from the primary motor cortex through
the brain stem to the motor neurons of the spinal cord, and from this to
the muscle. (Adapted from Kandel *et al.*[4] Redrawn with permission from
Appleton and Lange.)

melanogaster, the first 13 cell divisions of the embryo occur within a syncytium
prior to cellularization. At this stage the embryonic syncytium is made up of
80 000 cells.[5] It should be noted that this is an unusual development and that
most of the time within a syncytium the different nuclei appear to support
different functions. Typically, the macronucleus serves the organism's needs
while the micronucleus serves reproductive functions.

The last explanation seems to be the **colonial theory**. In this theory multiple
organisms of the same species led to a multicellular organism. Cells appeared
not to separate following incomplete cytokynesis. Most researchers are in favor
of this last theory, which remains to be completely demonstrated. Indeed,
Dictiostelium (an amoeba) and *Volvox* cells group together during food
shortages, forming a colony that moves to a new location.

The passage from unicellularity to multicellularity has induced the necessity
to solve several problems. In order to reproduce, muticellular organisms have

developed a specific cell lineage, germ cells (*i.e.* sperm and eggs cells). They have also had to develop specific means to adhere to each other in order to form tissues and organs. Cells have developed cell-to-cell adhesion that can be either physical or chemical. Physical adhesion is mostly exemplified by several junctions known as tight junctions, desmosomes, adherent junctions or even gap junctions (Figure 3.4) that are not only physical links between two cells but also a means of communication[6] (see also below).

Chemical adhesion is best exemplified by the properties of extracellular matrix molecules that contribute to maintain cell bounds. The major families of cell adhesion molecules are the cadherins (which are homophilic adhesion molecules), the selectins, and the superfamiliy of receptors of the immuno-globin that participate in both heterophilic and homophilic interactions (Figure 3.5). Cadherins are well known to be necessary to hold epithelial sheets together. They also participate in the recognition of cell types and of the organisation of cells into clusters and into nuclei in the mammalian brain.[7] Deficiency of a specific cadherin subtype (*i.e.* E-cadherin) leads to the detach-ment of cells from each other. In addition, cadherins are required for the subsequent formation of other types of junctions such as desmosomes and gap junctions.[6]

Pluricellular organisms have also had to develop several ways of commu-nicating between cells in order to grow, coordinate developmental processes, and maintain homeostasis. This specific issue will be discussed in detail in the

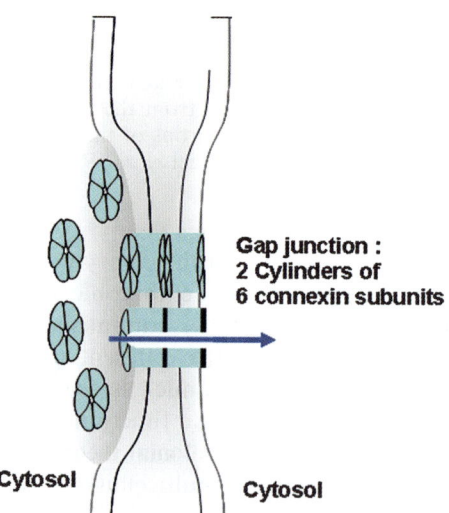

Figure 3.4 Schematic drawing of gap junctions. Each opposite cell contributes half a channel, called a connexin. Each connexin is formed from six protein subunits, the connexins, and is about 1.5–2 nm in diameter. Each connexin spans the membrane and contacts a matched connexin across the gap between the cells. At these sites, the cells are only 3.5 nm apart.

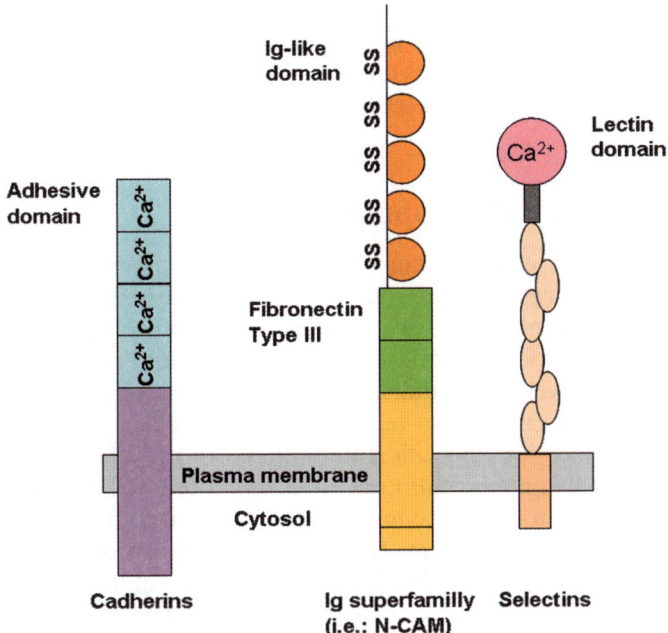

Figure 3.5 The major families of cell adhesion molecules. Three major families of adhesion molecules could be identified; the cadherins, the selectins, and members of the immunoglobin (Ig) superfamily. They are transmembrane receptors, typically constituted by an intracellular domain that interacts with the cytoskeleton, a transmembrane domain, and an extracellular domain that interacts either with other cell adhesion molecules of the same kind (homophilic binding) or with other cell adhesion molecules (heterophilic binding).

following chapter. Finally, they had to find methods of regenerating (repairing and re-growing) specific tissues. This last issue is well resolved in the less complex organisms, which most of the time possess cells that are either able to dedifferentiate or give rise to most cell types. This is the case for animals possessing bilaminar structures, such as hydrae and sponges, and of the less developed metazoans possessing the characteristic three laminar structures also found in mammals. Typically, the section in two of sponges (bilaminar structured) and flat worms (metazoan) gave rise to two animals possessing all the characteristics of the sectioned animals. These animals are actually intensively studied for their regenerating properties in the hope of applying the scientific discoveries to more specified animals and, ultimately, to humans. Indeed, mammals have lower capacities of regeneration. The skin and the liver are indeed able to intensively regenerate but some organs, such as the brain, contain only a few newly generated cells and their number is, with some exception, insufficient to overcome neuronal cell death. In addition, nervous system regeneration has very poor performance. Each year nervous (central and peripheral) system injuries affect more than 90 000 people. Neural tissue engineering is becoming a growing field of research. Behind this is the need to understand the genetic programs that would be necessary to induce in order to increase the potential of neural tissue regeneration *in vivo*.

3.2 Cells from Development

One of the features characterizing development of mammals is the passage from a unicellular organism, the egg, to a multicellular organism that becomes more and more complex. During development, cells proliferate, become specified, and finally differentiate into tissues and organs. During this developmental process, as well as in adults, the distance from one cell to the other has dramatically increased. Cell-to-cell communication has become a need to instruct cell specification and differentiation and to maintain homeostasis.

Some cell-to-cell communication requires direct cell–cell contact. Some cells can form gap junctions that connect their cytoplasm to the cytoplasm of adjacent cells. Gap junctions are formed by the apposition by each cell of a cylinder constituted by six dumbbell-shaped connexin subunits[8] (Figure 3.4). In cardiac muscle, gap junctions allow the propagation of an action potential from the cardiac pacemaker region of the heart to spread and coordinately cause contraction of the heart. Gap-junction channels also allow the passage of low molecular weight mitogens, morphogens, and secondary messengers such as cAMP or IP3, which are important regulators of gene expression.[9] Evidence suggests that these molecules coordinate transcriptional activity of a cluster of coupled cells. Another contact-dependent signaling known to play key role during development is the Notch–Delta signaling (Figure 3.6). This signaling requires two adjacent cells to make physical contact. One cell will produce more of a cell surface protein that activates the Notch receptor on the adjacent cell. This, in turn, activates a feedback loop that will reduce Notch expression in the cell that will differentiate. By contrast, Notch will increase in the cell that will continue to divide. This communication is used in various species (*e.g.* in a wide range of cellular interactions in *Drosophila*) and during mammalian development to maintain a stem cell or progenitor niche.

One of the first communications between cells results in the specification of specific cell types. In order to do so, restricted sets of cells have the function to secrete morphogenes. Being secreted in restricted locations these morphogenes form gradients of morphogenes. In this respect, one cell according to its position is going to detect a concentration of a cocktail of specific morphogenes. The "reactions" of such cells will be proportional to the concentration of the morphogene they will be in contact with. In mammal embryos, dorsalizing factors antagonize ventralizing factors determining the ventro-dorsal axis of the body. In addition, anteriorizing and posteriorizing signals define the anterior–posterior body axis. This will have the consequence of inducing the expression of specific genes – mainly transcription factors – that will induce the specific activation of characteristic gene sets within a cell, leading the cells to become

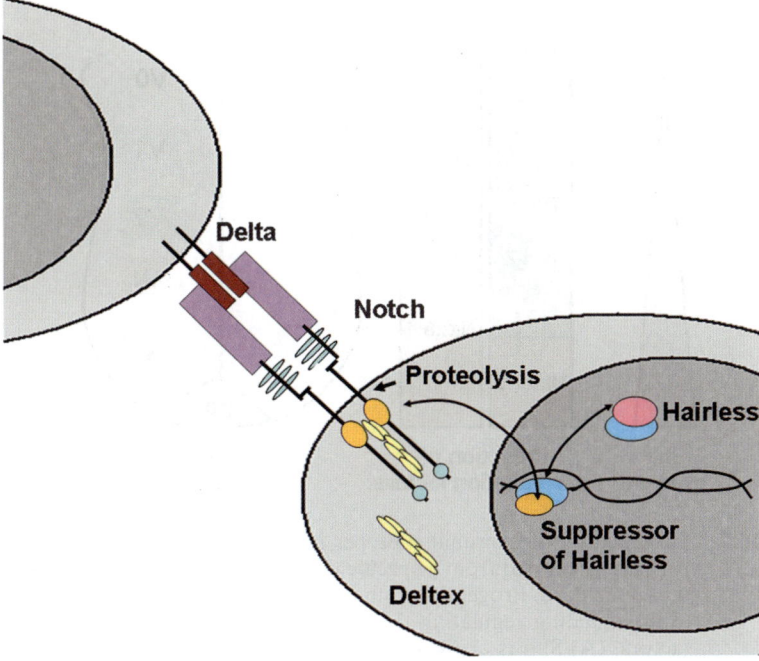

Figure 3.6   Notch–Delta signaling. The Notch protein sits like a trigger spanning the cell membrane, with part of it inside and part outside. Ligand protein (*i.e.* Delta) binding to the extracellular domain induces proteolytic cleavage and release of the intracellular domain, which enters the nucleus to modify gene expression. Hairless gene expressions could be modified that way.

specified. The specification of motor neurons in the developing spinal cord occurs similarly; it has been studied extensively and is illustrated in Figure 3.7. Motor neurons are generated from a specific domain of ventral neural progenitors of the spinal cord. Progenitor domains are defined by graded sonic hedgehog signaling regulating the combinatorial expression of transcription factors. Sonic hedgehog (Shh) is a morphogene that is secreted by the floor plate and the notochord. Progressively more dorsal progenitor domains are exposed to a decreasing concentration of Shh protein and the concentration of Shh determines the neuronal subtype generated. Shh either represses or induces the expression of transcription factors according to its concentration. Progenitor gene expression domains are refined and maintained by negative cross-regulatory interactions between those proteins that share the same boundary. Therefore, a gradient of a morphogene is able to give rise to sharp progenitor domains.[10–12]

In mature organisms, different types of communication could be identified. Cell-to-cell communication could be related to either fast electrical communication or slower chemical communication. In the central nervous system,

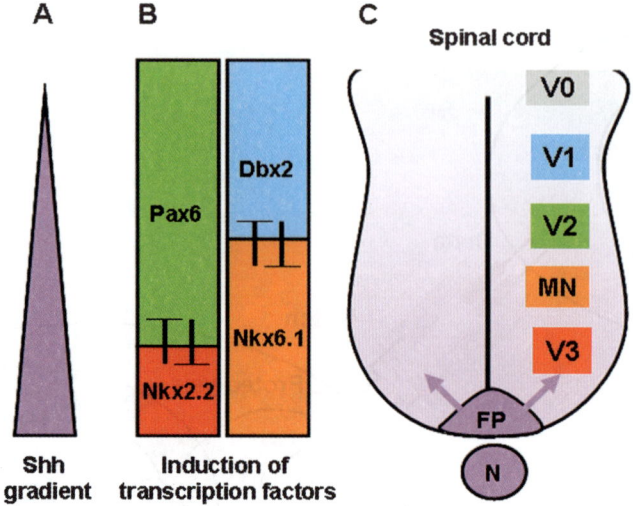

Figure 3.7   Specification of neuronal subtypes in the rodent spinal cord. Motor neurons are generated from a specific domain of ventral neural progenitors of the spinal cord. Progenitor domains are defined by graded sonic hedgehog (Shh) signaling regulating the combinatorial expression of transcription factors. (A) Shh is expressed by the notochord and floor plate. Five distinct ventral neuronal subtypes arise at distinct positions along the dorso-ventral axis of the neural tube. Progressively more dorsal progenitor domains are exposed to a decreasing concentration of Shh protein and the concentration of Shh determines the neuronal subtypes generated. The concentration gradient of Shh regulates the ventral expression domains of a series of transcription factors in ventral progenitor cells. Shh either represses or induces expression at different concentration thresholds. (B) Progenitor gene expression domains are refined and maintained by negative cross-regulatory interactions between those proteins that share the same boundary. (C) The domains determine the neuronal subtype that arises from each domain.

neurons communicate through synapses that are either "electric" or "chemical" in nature. In a chemical synapse, the axon terminal of the presynaptic cell contains vesicles filled with a specific neurotransmitter, such as serotonin or dopamine (Figure 3.8). In most cases, the dendrite, the cell body of the post-synaptic cell or a tissue (*i.e.* muscle in the case of the neuro-muscular junction), contains receptors that will detect the exocytosis of the neurotransmitter that will be released following the modifications of the electric properties of the presynaptic membrane.

Long-distance communication between cells could also be achieved through the release of hormones. In vertebrates, hormones will flow in the blood before reaching their targets. Hormones are also key players during development. One of the best examples to demonstrate their efficiency in directing developmental processes is the metamorphosis of *Xenopus*. During early metamorphosis the level of thyrotropin increases inducing high thyroid hormone (T3 and T4)

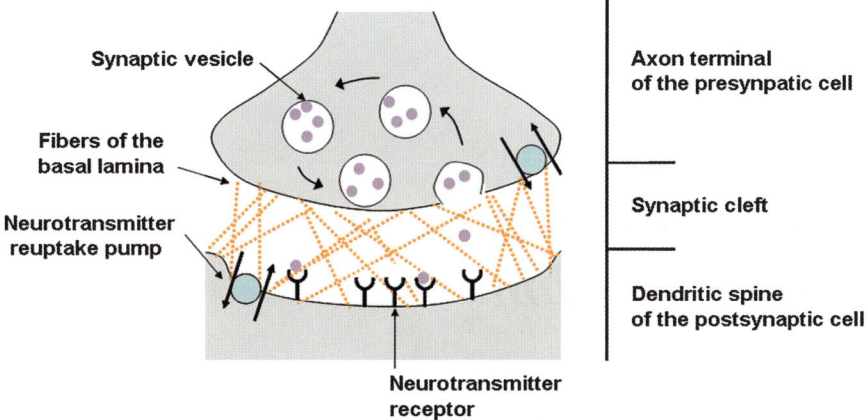

Synaptic vesicle

Fibers of the
basal lamina

Neurotransmitter
reuptake pump

Axon terminal
of the presynpatic cell

Synaptic cleft

Dendritic spine
of the postsynaptic cell

Neurotransmitter
receptor

Figure 3.8   Structure of a chemical synapse between two neurons. Chemical synapses are specialized junctions through which cells of the nervous system signal to one another. The electrical impulses arriving at the axon terminal trigger the release of neurotransmitters stored in vesicles. Neurotransmitters diffuse across the synaptic cleft to receptors located on the adjacent dendrite. This will affect the likelihood that an electrical impulse will be triggered in the latter neuron. Once released, the neurotransmitter is rapidly metabolized or is pumped back into a neuron.

concentrations which, combined with their receptors, are able to induce the transcription of the T3 receptor and other genes.[13]

References

1. C. Nielsen, *Evol. Devel.*, 2008, **10**, 241.
2. C. Mieko Mizutani and E. Bier, *Nat. Rev. Genet.*, 2008, **9**, 663.
3. F. Cebria, *Dev. Genes Evol.*, 2007, **217**, 733.
4. E. R. Kandel, J. H. Schwartz and T. M. Jessel, in *Principles of Neural Science*, ed. E. R. Kandel, J. H. Schwartz and T. M. Jessel, Appleton & Lange, Norwalk, Connecticut, USA, 3rd edn., 1991, pp. 271–295.
5. S. M. Abmayr, M. S. Erickson and B. A. Bour, *Trends Genet.*, 1995, **4**, 153.
6. J. E. Lai-Cheong, K. Arita and J. A. McGrath, *J. Invest. Dermatol.*, 2007, **127**, 2713.
7. M. -S. Yoon, L. Puelles and C. Redies, *J. Comp. Neurol.*, 2000, **421**, 451.
8. M. W. Bennett, L. C. Barrio, T. A. Bargiello, D. C. Spray, E. Hertzberg and J. C. Sáez, *Neuron*, 1991, **6**, 305.
9. S. C. Guthrie and N. B. Gilula, *Trends Neurosci.*, 1989, **12**, 12.
10. J. Briscoe, L. Sussel, P. Serup, D. Hartignan-O'Connor, T. Jessell, J. L. Rubenstein and J. Ericson, *Nature*, 1999, **398**, 622.
11. J. Briscoe, A. Pierrani, T. Jessell and J. Ericson, *Cell*, 2000, **101**, 435.
12. T. Jessell, *Nat. Rev. Genet.*, 2000, **1**, 20.
13. M. Paris and V. Laudet, *Genesis*, 2008, **46**, 657.

CHAPTER 4

Understanding Cellular Differentiation

TANIA VITALIS

CNRS-UMR 7637, Laboratoire de Neurobiologie et Diversité Cellulaire, ESPCI, 10 rue Vauquelin, 75005, Paris, France

Abstract

Developmental biology is making tremendous progress in describing the mechanisms that coordinate developmental programs and lead to the specification and differentiation of the correct cell at the correct position with appropriate synaptic or cell–cell contacts. In parallel, progress in microscale and nanoscale technologies and microfluidics is revealing new insights into single cell development in a specific environment. With these approaches it becomes possible to pinpoint the master genes that control the specification and differentiation of a single cell. In addition, these new technologies are also providing means to control microenvironments and possibly direct cell differentiation. They could also be valuable in understanding the role of a cell environment versus the genetic determinants. To revue what micro and nanotools could offer to understand development, we will focus on the development of the nervous system.

4.1 Development of the Cerebral Cortex

In adult mammals, the cerebral cortex is organized in a laminar structure comprising six layers. The layering is produced by variation in the type and density of cell bodies through the cortical depth. Each layer contains a

RSC Nanoscience & Nanotechnology No. 15
Unravelling Single Cell Genomics: Micro and Nanotools
Edited by Nathalie Bontoux, Luce Dauphinot and Marie-Claude Potier
© Royal Society of Chemistry 2010
Published by the Royal Society of Chemistry, www.rsc.org

complement of pyramidal, so-called for the shape of their soma, and non-pyramidal neurons. Pyramidal neurons, which make up approximately 75% of all neurons in the adult cortex, are projection neurons that send axons to other areas of the cortex and to distant parts of the brain. They utilize the excitatory amino acid glutamate as a neurotransmitter. Non-pyramidal cells are inter-neurons, which only make local connections. They all contain the inhibitory neurotransmitter γ-aminobutyric acid (GABA) and also one or more neuro-peptides and/or calcium-binding proteins.[1,2]

The cerebral cortex develops from the rostral part of the neural tube, named the telencephalic pallium. The wall of the pallium is initially formed of neuro-epithelial germinal cells whose continued proliferation causes the outward bulging of the pallial walls to form the cerebral vesicles. In rodents, pyramidal and non-pyramidal neurons originate from different regions of the tele-ncephalon, and pyramidal neurons are generated in the cortical ventricular zone (VZ), whereas non-pyramidal neurons are generated in the ganglionic eminence of the basal telencephalon (reviewed by Corbin *et al.*[3] and Parnavelas *et al.*[4]). Pyramidal neurons follow a "conventional scheme of cortical forma-tion", early postmitotic neurons migrating radially away from the VZ towards the surface of the cerebral vesicles to form the primordial plexiform layer or preplate (PP)[5–7] (Figure 4.1a). The later-generated neurons migrate to form a layer within the PP, the so-called cortical plate (CP), thus splitting it into a superficial marginal zone (MZ; layer I) and a deep subplate (SP). The neurons of the CP assemble into layers II–VI in an "inside-out" sequence; the deepest cellular layers are assembled first and those closest to the surface are assembled last (Figure 4.1a). In rodents, non-pyramidal neurons originate from the basal telencephalon, mainly from the ganglionic eminences (GE) and migrate tangentially following three parallel migratory streams, in SVZ, IZ, and MZ, and progressively incorporate in the developing cortical plate.[3,4] Glial cells are generated postnatally and a fraction of them are derived from the differentia-tion of radial glial cells into astrocytes.

4.2 Neuronal Differentiation

While building the cerebral cortex several key steps can be identified: (1) the cell cycle exit, (2) the regulation of genetic programs determining neuronal specification, (3) neuronal commitment, and (4) neuronal differentiation *per se.*

The cell cycle exit is an essential step in the progression of neuronal lineages, and is tightly coupled to specification of progenitors to a particular neuronal identity. This commitment usually occurs during the last cycle and the differ-entiation program usually starts after cell cycle exit.[8,9] Proneural genes con-taining the basic helix–loop–helix (bHLH) transcriptional activator are both implicated in cell cycle exit and neuronal differentiation. They are able to regulate the activity of cyclin-dependent kinase inhibitors. Regulation of cell cycle progression and cell cycle length is a crucial process in corticogenesis and

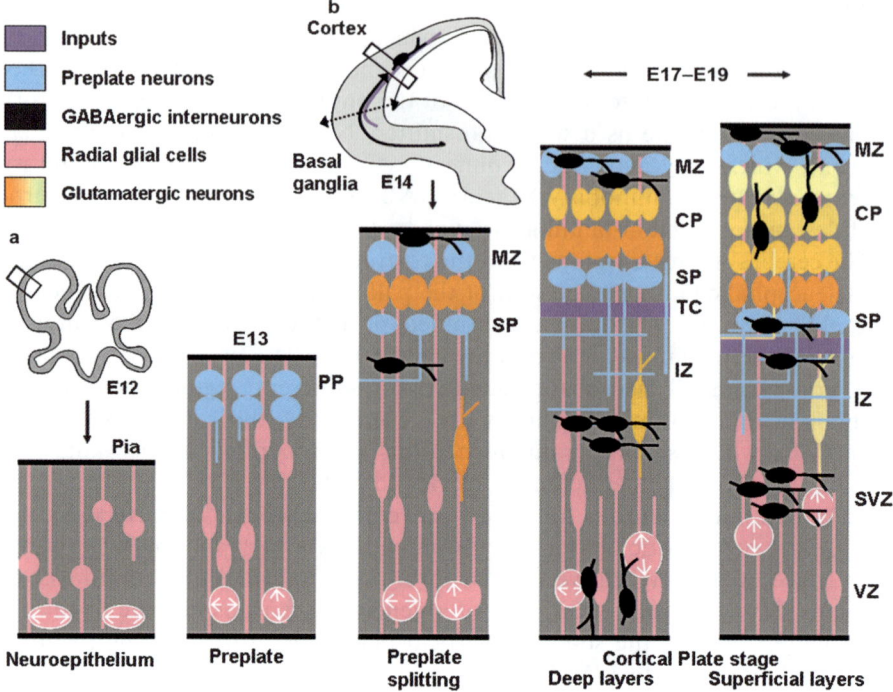

Figure 4.1   Cortical development. Panels a and b are schematic drawings of coronal sections of the brain of mouse embryos illustrating the development of thalamocortical axons (input violet) and the migration of telencephalic interneurons from the basal telencephalon. Enlargement of the boxes drawn in a and b are represented below; they depict the stratification of the early developing cortical anlagen. Neurogenesis in the rat neocortex occurs from embryonic day E12 (left) to E17 (right). Cortical development begins with the appearance of a population of cells along the lateral ventricle, known as the ventricular zone (VZ). This population of cells gives rise to most of the neurons and glial cells of the cerebral cortex. Once generated, neurons migrate towards the pial surface and complete their differentiation in the cortical plate. Neurons destined to populate the deeper layers of the cortex are generated and then migrate from the VZ earlier than the neurons destined for progressively more superficial layers. On E13 the cerebral wall is bilaminar consisting of the VZ and overlying primitive preplate. By E17–E19 the thickness of the overlying intermediate zone/with matter and developing cortical plate are at their maximum widths, with all neuronal cells having exited the cell cycle and migrated to their final laminar distribution within the developing cortex. CP, cortical plate; IZ, intermediate zone; MZ, marginal zone; PP, preplate; SP, sub-plate; SVZ, subventricular zone.

determines the total number of cortical neurons produced as well as the relative ratio of neurons in different cortical layers and their thickness.[10,11]

Proneural genes are both necessary and sufficient to promote the differentiation of postmitotic neurons. Proneural genes regulate key steps of commitment of stem cells to neuronal fate. In the cerebral cortex, neurogenin 1 and 2 (*Ngn 1* and *Ngn 2*) and *Mash1* have been identified to achieve such role.[12,13] *Ngn2* positively regulates *Ngn1* and negatively regulates *Mash1*. In the cerebral cortex, modifications of proneural gene expression alter neuronal commitment by a misregulation of several gene expression, such as *Hes5*, *NeuroD* (differentiation genes) and of Notch signalling. Loss of *Ngn2* and *Mash1* result in a dramatic decrease of the cortical plate size, a reduction in neurogenesis and an increase in astrocytogenesis.[12,13]

Differentiation *per se* requires the sequential expression of additional transcription factors of the bHLH family, including *NeuroD1*, *NeuroD2*, *Math2*, *Math3*, and *Nscl1*.[14] In addition, other regulatory genes expressed sequentially in cortical neurons are involved in the differentiation process such as the T-box gene *Tbr1*[15] (Figure 4.2).

4.3 Single Cell Analysis in Differentiation Processes

The extraordinary complexity and heterogeneity of the mammalian brain significantly limits the power of gene microarray for gene expression analysis. Because many neuronal and non-neuronal cell populations are highly intermixed, microarray analyses of a given region either during development, at the mature stage, or after behavioral or pharmacological manipulation, or subsequent to a disease process, only provides a composite view of gene expression.

For example, if a gene is expressed at low or moderate levels in a small group of cells of a given sample a change of expression will not be detected by microarray analysis. In addition, if a highly expressed gene is only expressed by a minority of cells, its expression may be completely diluted and masked in the microarray.

Analysis of the transcriptome of single cells represents a valuable tool to apprehend the transcriptome of specific cells in a specific state. Recently, this approach has been successfully used in several studies that aimed to gain insight into the early steps of organ development,[16,17] in several aspects of central nervous system development,[18–20] or to analyze specific cell populations at the mature stage.[21,22] In particular, Kawaguchi *et al.*[20] have used single cell microarrays for analyzing the transcriptome of developing neocortical progenitors at mid-embryonic stages. Cells were obtained from small pieces of tissues that were chemically dissociated. In their study they wanted to determine the level of heterogeneity within the neocortical progenitors. They were able to identify different clusters of cells behaving similarly: the undifferentiated progenitors located in the ventricular zone (also known as self-renewable apical progenitors), basal progenitors located in the subventricular zone (also known as intermediate progenitors), young neurons and mature neurons mainly located in the cortical plate (Figure 4.1). They further demonstrate that progenitor cell heterogeneity was encoded by both the time of genesis and the place they are occupying at a given time. In addition, they also demonstrate that the cluster of undifferentiated progenitors could be further subdivided according to

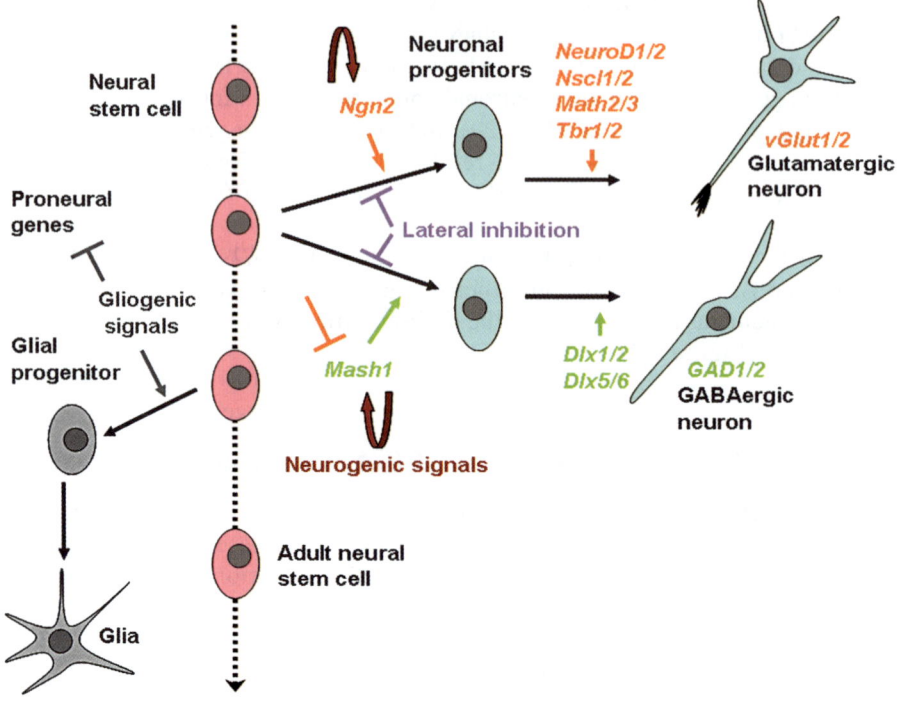

Figure 4.2 Model of the genetic programs controlling neuronal specification in the mouse telencephalon. Neural stem cells are multipotent and can generate all neurons and astrocytes and oligodendrocytes. Generally, neural stem cells generates neurons and later produce glial cells the switch from neurogenesis to gliogenesis is the result of changes in stem-cell properties that are controlled by both extrinsic and intrinsic cues. During the first period of differentiation neurogenic signals (such as secreted morphogenes; see Chapter 3) induce proneural gene expression. The proneural genes are involved in the specification of cortical (glutamatergic neurons) neurons and of basal ganglia neurons (GABAergic interneurons) by coordinately regulating several aspects of their identity, including their regional identity and neurotransmission phenotype. Some of the genes involved in region-specific differentiation and neurotransmission that are regulated by proneural genes are listed. At the same time, through lateral inhibition, Notch signaling down-regulates and/or inhibits proneural genes in other cells preventing them from being differentiated into neurons.

the cell-cycle phase they were in. Their analysis further showed that numerous genes are differentially expressed between apical progenitors and basal progenitors. Some of them involve the Notch pathway and the fibroblast growth factor (FGF) and sonic hedgehog pathways, consistent with previous findings[23] (see also Chapter 3). In another study the group of Tiejen *et al.*[18] has used a laser microdissector (see Chapter 3) to harvest the cytoplasm of individual cells in association with single cell microarrays to analyze the development of the olfactory system. Laser dissection has already been successfully used to determine the microarray of individual cells[18] although, in most studies, several cells thought to belong to the same subtype of cells are pooled. Dulac's group[18] has analysed the development of the olfactory system, which includes highly heterogeneous pollution of olfactory sensory neurons in the main olfactory epithelium and their progenitors of olfactory sensory neurons target cells (the mitral cells). The neuronal diversity of olfactory sensory neurons is best exemplified by the fact that a single olfactory sensory neuron expresses a unique olfactory receptor gene (belonging to a large family of a thousand genes). Their study allowed them to pinpoint new genes expressed by discrete olfactory sensory neurons that had not previously been reported and to discover multiple regulatory pathways underlying olfactory neuronal diversity. However, they point to the fact that larger amounts of a single cell microarray needs to be obtained to provide a clear picture since there is a large amount of hetero-geneity within olfactory sensory neurons. Similarly to the previous study it is clear that the phase of the cell-cycle largely contributes to determining the set of genes a cell will express. In another promising study performed by the group of Esumi *et al.*,[19] a single cell microarray has been used to determine the expression profile of young GABAergic neurons that started to invade

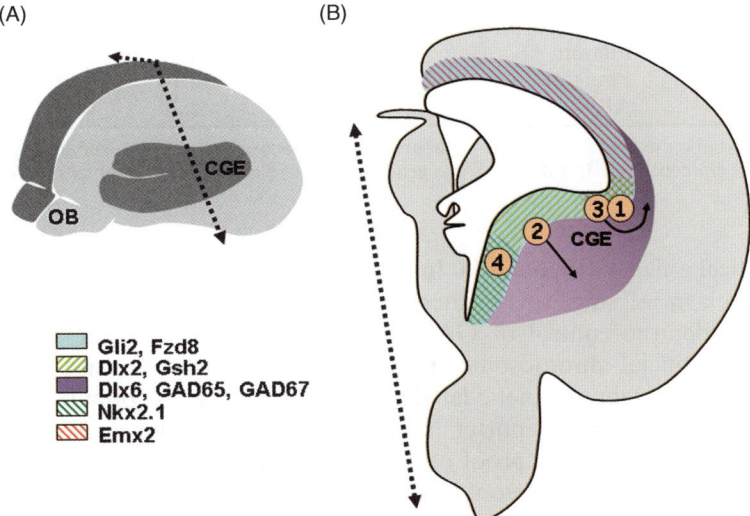

Figure 4.3 Microarray analysis of four cells extracted from the mouse caudal gang-lionic eminence allow their place of origin and maturation state to be determined. (A) Drawing of a whole mouse brain at embryonic day 14. Note the location of the caudal ganglionic eminence (CGE) in the basal telencephalon. The olfactory bulb is indicated (OB). (B) Drawing of a coronal section of an E14 embryo at the level of the caudal ganglionic eminence. Regions of transcription factor expression that are known to be restricted to specific GABAergic interneurons or to specific differentiation states are drawn. Note that the four cells could be clearly positioned.

Table 4.1   Markers and level of gene expression detected in four cells after microfluidic reverse transcription.

Abbreviated name		Cell 1	Cell 2	Cell 3	Cell 4	CGE
Neuron markers						
	Tuj-1 (all neurons)	65.25	61.25	42.19	106.97	30.80
	Dlx2 (all GABA neurons)	12.5	49.95	9.28	23.02	22.34
Transcription factors used for cell positioning						
	Emx2	15.83	5.87	8.55	2.62	2.80
	Fzd8	61.32	58.12	86.46	115.83	10.03
	Gli2	7.73	25.84	5.5	5.95	6.70
	Gsh2	7.09	12.76	4.43	8.34	6.09
	Nkx2.1	1.54	2.58	1.68	9.69	6.59
Markers expressed in individual cells						
	Calbindin	12.6	9.15	3.35	6.19	5.58
	Calretinin	7.14	21.8	5.97	3.46	3.98
	Neuropeptide Y	27.61	3.99	50.68	9.59	5.47
Example of newly identified transcription factor						
	CoupTF2	1.61	58.70	1.57	3.30	21.14
Markers implicated in human pathology due to impaired migration						
	Doublecortin	40.06	37.63	26.05	5.49	29.7
	Filamin 1	27.17	17.6	34.75	3.91	10.75
	Pomgntl1	49.2	65.5	38.14	30.38	4.79
	ARX	3.73	3.17	6.28	5.93	5.49

Note that marker expression has been compared with microarray results obtained from full caudal ganglionic eminence (CGE) mRNA. Some genes are more selectively expressed by one of the four cells.

the cortical anlagen (Figure 4.1). In this study the authors selectively picked up the cytoplasm of single dissociated cells expressing green fluorescent protein (GFP) under the control of GAD67 promoter. Indeed, recent advances in bacterial artificial chromosome (BAC)-mediated transgenic mice allow relatively high throughput genetic labeling of distinct neuronal populations in the brain using fluorescent reporter protein (Gensat, Rockefeller project). However, the study was just a proof of concept to show that a "simplified method for single-cell microarray" could be used. Indeed, in their study they have only fully analyzed eight GAD67–GFP-positive cells with special attention for genes known to characterize different populations of mature GABAergic neurons or glutamatergic neurons. More, the authors have reported several limitations of this technique mainly due to non-unbiased amplification of the messengers. One exciting future goal of several laboratories aiming to combine molecular biology, electrophysiology, and morphology is to develop strategies that will allow the analysis of different cell parameters and their transcriptome in an unbiased manner. Association between reverse transcription (RT) in micro-fluidic devices and polymerase chain reaction (PCR) in a tube have also recently been reported to provide unbiased amplification of the transcriptome of a single cell.[24] The comparison of four cells derived from the same developmental region (the basal telencephalon) has revealed that the state of maturation and crude position of genesis could be extracted from microarray analysis (Figure 4.3 and Table 4.1). From these data it is also clear that some genes will be more specifically expressed in one cell type (*i.e.* Doublecortin, filamin 1, CoupTF2). This also suggests that these cells would be differentially affected by the deletion of specific genes, such as genes implicated in migratory processes and correct positioning within the cortical circuits. For instance, it is clear that the transcription factor CoupTf2 detected in caudal ganglionic eminence (CGE) explants is enriched in cell 2 that expresses calretinin, a cell population known to be mostly generated in the caudal ganglionic eminence.[24] CoupTfs are important transcription factors that have been shown to be involved in targeting of interneurons.[25]

References

1. J. G. Parnavelas, A. Dinopoulos and S.W. Davies, in *Handbook of Chemical Neuroanatomy.* ed. A. Björklund, T. Hökfelt and L.W. Swanson, Elsevier, Amsterdam, Vol. 7, 1989, Integrated Systems of the CNS, Part II, pp. 1–164.
2. J. DeFelipe, *Cereb. Cortex*, 1993, **3**, 273–289.
3. J. G. Corbin, S. Nery and G. Fishell, *Nat. Neurosci. Rev.*, 2001, **4**(Suppl), 1177–1182.
4. J. G. Parnavelas, P. Alifragis and B. Nadarajah, *Prog. Brain. Res.*, 2002, **136**, 73–80.
5. P. Rakic, *J. Comp. Neurol.*, 1972, **145**, 61–83.
6. P. Rakic, in *The Organization of the Cerebral Cortex*, ed. F. O. Schmitt, MIT Press, Cambridge, MA, 1981, pp. 7–28.
7. H. B. M. Uylings, C. G. Van Eden, J. G. Parnavelas and A. Kalsbeek, in *The Cerebral Cortex of the Rat*, ed. B. Kolb and R. C. Tees, MIT Press, Cambridge, MA, 1990, pp. 35–76.
8. S. K. McConnell, *Neuron*, 1995, **15**, 761–768.
9. T. Edlund and T. M. Jessel, *Cell*, 1999, **96**, 211–224.
10. V. S. Caviness Jr, T. Goto, T. Tarui, T. Takahashi, P. G. Bhide and R. S. Nowakowski, *Cereb. Cortex*, 2003, **13**, 592–608.
11. F. Polleux, C. Dehay, B. Moraillon and H. Kennedy, *J. Neurosci.*, 1997, **17**, 7763–7783.
12. C. Fode, Q. Ma, S. Cassarosa, S. L. Ang, D. J. Anderson and F. Guillemot, *Genes Dev.*, 2000, **14**, 67–80.
13. M. Nieto, C. Schuurmans, O. Britz and F. Guillemot, *Neuron*, 2001, **29**, 401–413.
14. C. Schuurmans, O. Armant, M. Nieto, J. M. Stenman, O. Britz, N. Klenin, J. Seibt, C. Brown, H. Tang, J. M. Cunningham, R. Dyck, C. Walsh,

K. Campbell, F. Polleux and F. Guillemot, *EMBO J.*, 2004, **23**, 2892–2902.

15. R. F. Hevener, L. Shi, N. Justice, Y. Hsueh, M. Sheng, S. Smiga, A. Bulfone, A. M. Goffinet, A. T. Campagnoni and J. L. Rubenstein, *Neuron*, 2001, **29**, 353–366.
16. K. Kurimoto, Y. Yabuta, Y. Ohinata, Y. Ono, K. D. Uno, R. G. Yamada, H. Ueda and M. Saitou, *Nucleic Acids Res.*, 2006, **34**(5), C42.
17. Y. Yabuta, K. Kurimoto, Y. Ohinata, Y. Seki and M. Saitou, *Biol. Reprod.*, 2006, **75**, 705–716.
18. I. Tiejen, J. M. Rihel, Y. Cao, G. Koentges, L. Zakhary and C. Dulac, *Neuron*, 2003, **38**, 161–175.
19. S. Esumi, S.-X. Wu, Y. Yanagawa, K. Obata, Y. Sugimoto and N. Tamamaki, *Neurosci. Res.*, 2008, **60**, 439–451.
20. A. Kawaguchi, T. Ikawa, T. Kasukawa, H. Ueda, K. Kurimoto, M. Saitou and F. Matsuzaki, *Development*, 2008, **135**, 3113–3124.
21. S. Gustincich, M. Contini, M. Gariboldi, M. Puopolo, K. Kadota, H. Bono, J. LeMieux, P. Walsh, P. Carninci, Y. Hayashizaki, Y. Okazaki and E. Raviola, *Genome Res.*, 2003, **13**, 1395–1401.
22. F. Kamme, R. Salunga, J. Yu, D.-H. Tran, J. Zhu, A. Bittner, H.-Q. Guo, N. Miller, J. Wan and M. Erlander, *J. Neurosci.*, 2003, **23**, 3607–3615.
23. F. Guillemot, *Curr. Opin. Cell Biol.*, 2005, 17, 639–647.
24. N. Bontoux, L. Dauphinot, T. Vitalis, V. Studer, Y. Chen, J. Rossier and M. C. Potier, *Lab Chip*, 2008, **8**, 443–450.
25. M. Tripodi, A. Filosa, M. Armentano and M. Studer, *Development*, 2004, **131**, 6119–6129.

CHAPTER 5

Realistic Models of Neurons Require Quantitative Information at the Single-cell Level

NICOLAS LE NOVÈRE

EMBL-EBI, Wellcome-Trust Genome Campus, Hinxton CB10 1SD, UK

Abstract

Detailed modelling of neurons is now a recognised sub-field of neurobiology. Such models rely on accurate and quantitative experimental measurements. For instance, modelling electrophysiology requires morphological reconstructions of identified neurons. Similarly, understanding the biochemical basis of neurotransmission becomes possible if we know about the molecular composition of the connected neurons. In this chapter we will describe the pitfalls of generic models that seek to reproduce common features of groups of neurons, and in particular, the artifacts generated by an excessive abstraction. Instead, we advocate the development of typological models, seeking to describe accurately a given neuron, generic inferences being derived afterwards.

5.1 Introduction

Understanding how our brain works, and trying to improve its performance, has always been very high on life science research agenda, but now it is a concern of everyone. From psychoanalysis to pills improving cognitive

RSC Nanoscience & Nanotechnology No. 15
Unravelling Single Cell Genomics: Micro and Nanotools
Edited by Nathalie Bontoux, Luce Dauphinot and Marie-Claude Potier
© Royal Society of Chemistry 2010
Published by the Royal Society of Chemistry, www.rsc.org

function, everyone has a miracle solution, based on some sort of "scientific" understanding. Moreover, with the general improvement of life conditions and of life expectancy in developed countries, the impact of neuropathologies on society becomes heavier. Neurodegenerative disorders such as Parkinson's disease or Alzheimer's disease are touching an ever larger share of the population. Mental illnesses are becoming significant public health concerns, schizophrenia, for example, having an incidence approaching 0.5–1% of the population.[1] Finally, drug addiction remains the most widespread mental disorder and the associated direct or indirect mortality a major worldwide societal problem. Neurosciences have a long and successful tradition of quantitative modeling, where theory and experiment have always formed a happy couple. The work of Warren Sturgis McCulloch and Walter Pitts on formal neural networks[2] gave rise to one of the best examples of cross-fertilizing scientific fields, which resulted in many advances both in information technology and cognitive science. Almost as soon as digital computers became available, they were used by neuroscientists to quantitatively test their theories. One of the first numerical simulations in biology was the famous model by Alan Lloyd Hodgkin and Andrew Huxley,[3] which explained the propagation of action potentials along axons. Quantitatively describing a cellular behavior emerging from the interaction between two different molecular components, a potassium channel and a sodium channel, the model of Hodgkin–Huxley can be seen as the beginning of computational systems biology.

Models are used in biology as simplified representations, summarising our knowledge in order to explain the workings of a system, or a population of systems. We can divide models in two different kinds: **generic** and **typological** (Figure 5.1). Generic models aim to represent populations of systems by abstracting their properties. In order to build a generic model, one needs to identify a set of characteristics common to all the biological systems belonging to a certain population. Models can then be built, in which behavior will exhibit those characteristics while overlooking the particularities of each individual system. Although the vast majority of formal models in science belong to this category, this approach suffers from a series of pitfalls that are not alien to the fact that theoretical biology has been considered with contempt by most of the biologists until the recent rise of systems biology. The first problem is the **analogy** problem. Although, some systems may exhibit similar or identical behaviors, this does not mean that the behaviors reflect the same properties in the different systems. Analogy in turn often leads to the **tautology** problem. Once an abstract model is capable of reproducing a common characteristic observed in different systems, can we get more insight? This problem was recognized by one the fathers of both computing and theoretical biology, Alan Turing, when he wrote[4] (about the development from one developmental pattern to another, which is a nonlinear process):

One would like to be able to follow this more general process mathematically also. The difficulties are, however, such that one cannot hope to have any very

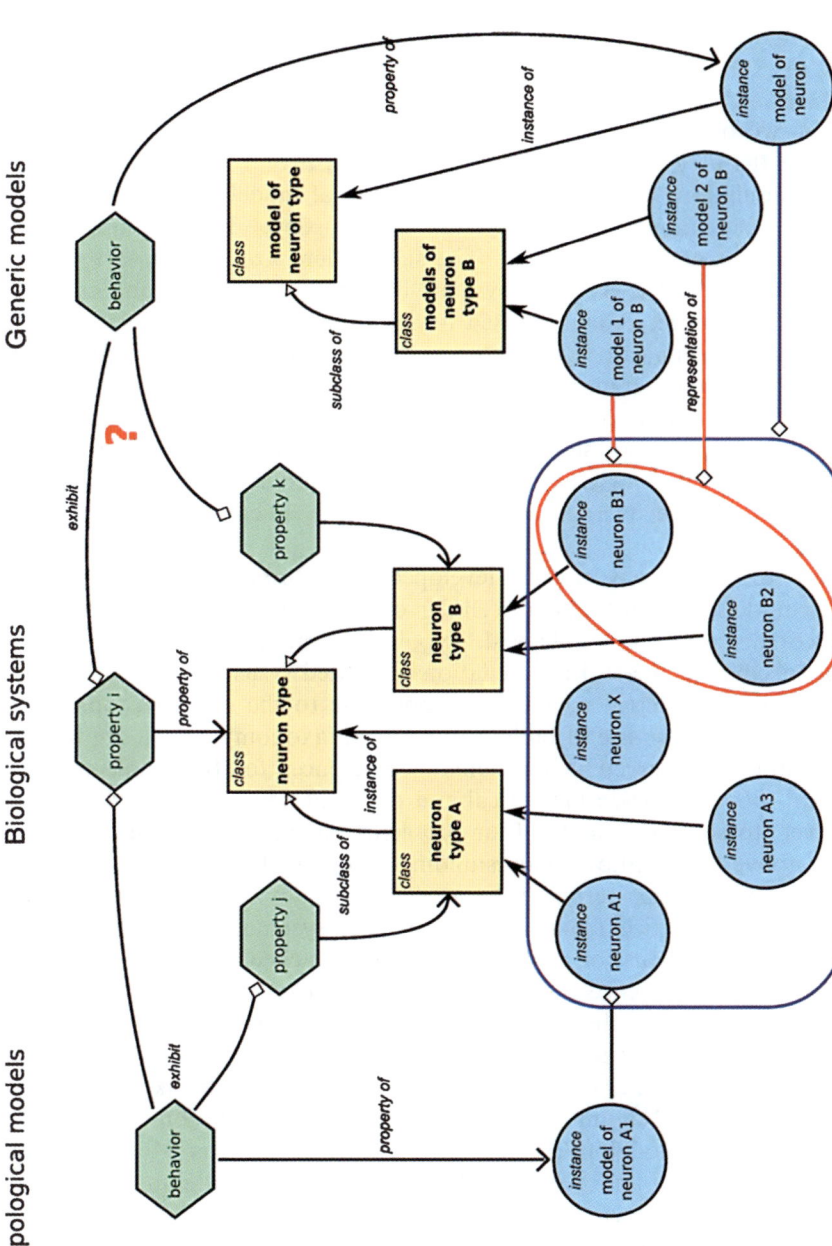

Figure 5.1 Diagram representing the relationships between neurons, classes of neurons, typological, and generic models. A generic model able to a represent the properties displayed by all neurons may not necessarily exhibit all the desirable properties of a given group of neurons (*e.g.* neuron B).

embracing theory of such processes, beyond the statement of the equations. It might be possible, however, to treat a few particular cases in detail with the aid of a digital computer. This method has the advantage that it is not so necessary to make simplifying assumptions as it is when doing a more theoretical type of analysis.

In other words, very often, mathematical modeling only gets out what it puts in in the first place. Of course, Turing's wish of a solution involving numerical simulations of example cases (typological models, see below), was granted the very same year by Hodgkin and Huxley.[3] The last problems are somehow mirrors of each other. The **subtraction** problem appears when generic descriptions of a group of systems become too abstract, and cannot reproduce the common behaviors used to define the group in the first place. The analogy problem can lead to the subtraction problem. For instance, if the oscillations of different systems are due to different underlying mechanisms, the resulting common subset may not generate oscillations. Finally, the **spherical animal** problem results in abstract systems that produce behaviors lacking in any of the original systems (if one assumes for the sake of simplification that the average animal is spherical, it can roll on a plane surface).

Contrary to the formal models developed by theoreticians, most of the biological models, whether animal, cellular or molecular, or of physiological processes or diseases, are typological. Typological models aim to accurately represent specific members of a population of related systems, and from their particular behaviors derive conclusions applicable to the whole group. It is similar in essence to the use of biological types in taxonomy, which are particular individuals conserved in museums as references for the whole taxon. Typological models provide theoretical and practical advantages. First of all, they answer, to some extent, the four problems plaguing generic models, as described above. In addition, understanding in detail the relation structure/function of given systems permits rules applicable to any related systems to be generated. On the practical front, focusing on given systems permits the critical mass of available information and expertise to be reached.

Modeling in neuroscience has long used the generic modeling approach, simplifying the neuron to the ultimate abstraction of a mathematical function with McCulloch–Pitts-like artificial neurons. Since then, a whole domain has developed, based on the use of phenomenological models like Integrate and Fire, FitzHugh–Nagumo, Hindmarsh–Rose, or based on models of ionic channels like Hodgkin–Huxley, or even both, such as Morris–Lecar (reviewed by Gerstner and Kistler[5]). In fact, although Hodgkin–Huxley is often considered as an abstract model, it was based on precise measurements of biochemical parameters, and led the way to a completely different type of neuronal model. While such approximate models may be useful to comprehend the function of neural networks on the large scale, they are completely inappropriate if the purpose is to understand how the neuron itself functions. The molecular and cellular events mediating neuronal transmission span several spatial and temporal scales. While the signal received from a glutamatergic terminal is decoded by a dendritic spine of width 500 nm,[6] the resulting action potential can be propagated along axons up to 1 m in length. Understanding synaptic function also means deciphering the effect of conformational transitions of ion channels, taking place on the microsecond range, onto long-term synaptic modifications lasting several weeks. Moreover, most assumptions used to simplify modeling in other fields of cell biology, such as homogenous concentrations and spatial isotropy are inappropriate. The geometry of subcellular compartments strongly affects their functions,[7] as does the relative location of molecular partners and their diffusion. Finally, the morphology of neurons changes over time, and itself depends on the activity of the neurons.[8,9]

5.2 The Importance of Precise Neuronal Morphology

Ever since the first stunningly precise images generated by Ramon Y. Cajal using the method of Camillo Golgi, it was clear that the shapes of the cells, soon to be named "neurons" by Heinrich Wilhelm Gottfried von Waldeyer-Hartz, were incredibly diverse. Additionally, the reproducibility of the neuronal shape within a given cerebral structure suggested a strong correlation with the function. Understanding how the properties of a cerebellar Purkinje neuron, whether in terms of electrical behavior or integrative function, differ from those of a medium-spiny neuron (MSN) of the striatum or a pyramidal neuron of the cortex, require morphological parameters to be taken into account. Indeed, experiment and modeling showed that neurons with different morphology and, in particular, dendritic structure exhibited different properties,[10,11] as in the case of dendritic spines.[12]

With the cable approximation, Wilfrid Rall opened the way to realistic multi-compartment electrical models.[13] This approach assimilates a portion of dendrite to a simple electrical circuit that can then be assembled serially. These models quickly spanned several scales, encompassing synaptic contacts between neurons,[14] models of multicellular structures,[15] and even of several coupled brain structures,[16] culminating with modeling entire cortex with millions of semi-realistic neurons.[17] The availability of powerful and easy-to-use simulators to develop such multi-compartment models, like NEURON[18] and GENESIS,[19] allowed the construction of extremely detailed models of neurons. Those neurons based on recorded morphologies can reproduce neuronal behaviors very adequately.[20] Advanced computing facilities now permit the development of large heterogeneous neuronal assemblies, where each neuron possesses a realistic geometry and specific electrophysiological properties determined by a given set of ion channels. The most ambitious project in this domain may be the Blue Brain Project,[21] which aims to simulate a whole mammalian cerebral cortex using a super-computer. As a proof of concept, simulations of a neocortical column containing 10 000 neurons have been run. Each neuron is reconstructed from experimental measurements, and their realistic morphology contains hundreds of compartments. Multi-compartment models include not only electrical behavior, but also ion diffusion,[22] another process strongly affected by cell morphology.[23] Elucidation of the relationships between morphology, connectivity, and function requires modeling of the three aspects for given neurons, and cannot be achieved with abstract models that would be the "average neuron", or even the average pyramidal neuron of CA3. Recent techniques such as the **brainbow**[24] resulted in a leap in monitoring the morphological diversity of "identical" neurons in a given neuronal structure. Furthermore, one can access the precise morphology of individual neurons in which activity is recorded in parallel,[25–28] and this allows building more realistic models.

5.3 Each Neuron has a Unique Neurochemistry

While the intrinsic behavior of a neuron largely depends on its morphology, the widespread belief that neuronal cells are just complex geometric transistors is particularly misleading. In particular, their activity depends not just on the excitation and inhibition by main synaptic inputs (whether excitatory or inhibitory), but also on the effect of many neuromodulators, either on synaptic plasticity or intrinsic plasticity. As misleading is the idea that all neurons of the same "type" in a neural structure are neurochemically identical. For instance, it was shown that the nicotinic receptors expressed in the catecholaminergic neurons of the locus ceruleus defined several classes of cells,[29] with different physical properties. The same situation was observed in the dopaminergic neurons of the mesencephalon[30] and the neurons of the habenular–interpeduncular pathway.[31] Moreover, in each of these cases, the separation in different families was not clear cut, and virtually all combinations of nicotinic subunits could be found.

One of the best examples of such a neurochemical diversity can be observed in the striatum. Ninety percent of the neurons of the striatum project GABAergic (MSNs). Classically, those cells are classified into striatonigral neurons ("direct" path) and striatopallidal neurons ("indirect" path). The former express substance P and dopamine D1 receptors, while the latter express preproenkephalin and dopamine D2 receptors. However, this division is not that absolute; 4% of the MSNs express preproenkephalin and D1 receptors, 4% express SP and D2 receptors, and 3% express SP and preproenkephalin.[32] This diversity is further increased by the differential expression of receptors for other modulator systems, such as (but not limited to) adenosine receptors,[33] muscarinic receptors,[34] cannabinoid receptors.[35] Finally, the MSNs of the striatum are segregated in the striosomes and the matrix, neurons located in the former substructures expressing the opioid mu-receptor.[36] Therefore, a population of neurons that looked fairly homogeneous in terms of morphology and connectivity is found to exhibit dozens if not hundreds of different neurochemical phenotypes.

Therefore, if we wish to accurately model a neuron, we need to analyze precisely the transcripts expressed by an identified cell, if possible after having recorded its electrical behavior (and, ideally, also its morphology; cf. above). Good examples are the classification of cortical interneurons[37] or the experimental characterization of neurons that informed the Blue Brain Project.[38]

5.4　Conclusions

To accurately model neuronal function presents many challenges, and stretches the techniques and resources of computational biology to their limits. The molecular and cellular events mediating neuronal transmission span several spatial and temporal scales. While the signal received from a glutamatergic terminal is decoded by a dendritic spine of 500 nm width, the resulting action potential can be propagated along axons up to 1 m in length. Understanding synaptic function also means deciphering the effect of conformational transitions of ion channels, taking place on the microsecond range, onto long-term synaptic modifications lasting several weeks. Moreover, most assumptions used to simplify modeling in other fields of cell biology, such as homogenous concentrations and spatial isotropy, are inappropriate. The geometry of subcellular compartments strongly affects their functions, as does the relative location of molecular partners and their diffusion. Finally, the morphology of neurons changes over time, and itself depends on the activity of the neurons.

Although computational systems neurobiology is still far ahead of other fields when it comes to multi-scale, multi-algorithms modeling, the coupling of signaling pathways, electrical dynamics, and ionic diffusion is still infrequent. Even more serious is the fact that some crucial cellular functions or behaviors are barely considered at all when it comes to quantitative modeling. Modifications of gene expression and protein translation have been largely studied in synaptic function and plasticity, or in the symptomatology of neuronal diseases. Due to the different time scales involved, and the difficulty of building hybrid models able to provide continuous descriptions of electrical, metabolic, and signaling events, together with stochastic or even logical descriptions of gene regulatory networks, those aspects of neuronal physiology are mainly considered separately. The recent availability of new types of quantitative data should help to expand the models in new directions. However, the "omics" approaches are currently most often applied to whole brains or, at best, entire brain structures. Those techniques will have to be more targeted if we want to build realistic models of given neurons. Other cutting-edge technologies, like single-particle tracking in living cells,[39] will allow the development of more detailed models, and will enable the investigation of the role of micro-domains and supra-macromolecular complexes.

References

1. E. Goldner, L. Hsu, P. Waraich and J. Somers, *Can. J. Psychiatry*, 2002, **47**, 833–843.

2. W. McCulloch and W. Pitts, *Bull. Math. Biophys.*, 1942, **5**, 115–133.
3. A. L. Hodgkin and A. F. Huxley, *J. Physiol.*, 1952, **117**, 500–544.
4. A. M. Turing, *Phil. Trans. R. Soc. London*, 1952, **B237**, 37–72.
5. W. Gerstner and W. Kistler, *Spiking Neuron Models. Single Neurons, Populations, Plasticity,* Cambridge University Press, 2002.
6. M. B. Kennedy, H. C. Beale, H. J. Carlisle and L. R. Washburn, *Nat. Rev. Neurosci.*, 2005, **6**, 423–434.
7. R. Araya, J. Jiang, K. B. Eisenthal and R. Yuste, *Proc. Natl. Acad. Sci. U.S.A.*, 2006, **103**, 17961–17966.
8. E. Nimchinsky, B. Sabatini and K. Svoboda, *Annu. Rev. Physiol.*, 2002, **64**, 313–353.
9. T. Tada and M. Sheng, *Curr. Opin. Neurobiol.*, 2006, **16**, 95–101.
10. Z. F. Mainen and T. J. Sejnowski, *Nature*, 1996, **382**, 363–366.
11. P. Vetter, A. Roth and M. Häusser, *J. Neurophysiol.*, 2001, **85**, 926–937.
12. R. Araya, J. Jiang, K. B. Eisenthal and R. Yuste, *Proc. Natl. Acad. Sci. U.S.A.*, 2006, **103**, 17961–17966.
13. W. Rall, *Exp. Neurol.*, 1959, **1**, 491–527.
14. G. Shepherd and R. Brayton, *Brain Res.*, 1979, **175**, 377–382.
15. R. Traub and R. S. K. Wong, *Science*, 1982, **216**, 745–747.
16. R. Traub, D. Contreras, M. Cunningham, H. Murray, F. LeBeau, A. Roopun, A. Bibbig, W. Wilent, M. Higley and M. Whittington, *J. Neurophysiol.*, 2005, **93**, 2194–2232.
17. E. M. Izhikevich and G. M. Edelman, *Proc. Natl. Acad. Sci. U.S.A.*, 2008, **105**, 3593–3598.
18. M. Hines, *Int. J. Biomed. Comput.*, 1989, **24**, 55–68.
19. M. A. Wilson, U. S. Bhalla, J. D. Uhley and J. M. Bower, in *Advances in Neural Information Processing Systems*, ed. G. Touretzky, 1989, 485–492.
20. E. De Schutter and J. M. Bower, *J. Neurophysiol.*, 1994, **71**, 375–400.
21. H. Markram, *Nat. Rev. Neurosci.*, 2006, **7**, 153–160.
22. E. De Schutter and P. Smolen, in *Methods in Neuronal Modeling: From Synapses to Networks*, ed. C. Koch and I. Segev, MIT Press, Cambridge, MA, USA, 1998, pp. 211–250.
23. K. Holthoff, D. Tsay and R. Yuste, *Neuron*, 2002, **33**, 425–437.
24. J. Livet, T. A. Weissman, H. Kang, R. W. Draft, J. Lu, R. A. Bennis, J. R. Sanes and J. W. Lichtman, *Nature*, 2007, **450**, 56–62.
25. A. Larkman and A. Mason, *J. Neurosci.*, 1990, **10**, 1407–1414.
26. A. Mason and A. Larkman, *J. Neurosci.*, 1990, **10**, 1415–1428.
27. T. J. Grudt and E. R. Perl, *J. Physiol.*, 2002, **540**, 189–207.
28. J. V. Le Bé, G. Silberberg, Y. Wang and H. Markram, *Cereb. Cortex*, 2007, **17**, 2204–2213.
29. C. Léna, A. de Kerchove d'Exaerde, M. Cordero-Erausquin, N. Le Novère, M. M. Arroyo-Jimenez and J.-P. Changeux, *Proc. Natl. Acad. Sci. U.S.A.*, 1999, **96**, 12127–12131.
30. M. Zoli, M. Moretti, A. Zanardi, J. M. McIntosh, F. Clementi and C. Gotti, *J. Neurosci.*, 2002, **22**, 8785–8789.

31. S. R. Grady, M. Moretti, M. Zoli, M. J. Marks, A. Zanardi, L. Pucci, F. Clementi and C. Gotti, *J. Neurosci.*, 2009, **29**, 2272–2282.
32. C. Le Moine and B. Bloch, *J. Comp. Neurol.*, 1995, **355**, 418–426.
33. P. Svenningsson, C. Le Moine, B. Kull, R. Sunahara, B. Bloch and B. B. Fredholm, *Neuroscience*, 1997, **80**, 1171–1185.
34. Z. Yan, J. Flores-Hernandez and D. J. Surmeier, *Neuroscience*, 2001, **103**, 1017–1024.
35. H. Hermann, G. Marsicano and B. Lutz, *Neuroscience*, 2002, **109**, 451–460.
36. M. Herkenham and C. B. Pert, *J. Neurosci.*, 1982, **2**, 1129–1149.
37. B. Cauli, J. T. Porter, K. Tsuzuki, B. Lambolez, J. Rossier, B. Quenet and E. Audinat, *Proc. Natl. Acad. Sci. U.S.A.*, 2000, **97**, 6144–6149.
38. M. Toledo-Rodriguez, B. Blumenfeld, C. Wu, J. Luo, B. Attali, P. Goodman and H. Markram, *Cereb. Cortex*, 2004, **14**, 1310–1327.
39. L. Groc, M. Heine, L. Cognet, K. Brickley, F. Stephenson, B. Lounis and D. Choquet, *Nat. Neurosci.*, 2004, **7**, 695–696.

CHAPTER 6

Application to Cancerogenesis: Towards Targeted Cancer Therapies?

BERNHARD POLZER AND CHRISTOPH A. KLEIN

Division of Oncogenomics, University of Regensburg, Franz-Josef-Strauss-Allee 11, 93053 Regensburg, Germany

Abstract

Although molecular diagnosis in cancer has advanced over the last decades translation into successful therapies of solid cancers is lagging behind. In this chapter we will provide insights into the molecular analysis of single disseminated tumour cells (DTCs) – a cell population that comprises the founder cells of lethal metastasis. Here, recent data challenge the traditional approach of using the primary tumour as the surrogate marker for the selection of adjuvant therapies and emphasize the need to understand the biology of systemic cancer progression for personalized targeted therapies.

6.1 Molecular Diagnosis in Cancer

Cancer is a leading cause of death in the European Union, with about one million people dying every year.[1] Over the last decades, molecular understanding and diagnosis of cancer has greatly advanced and this knowledge is now being translated into therapeutic success. In cases of chronic myeloid leukemia, for example, cytogenetic detection of a common translocation between chromosomes 9 and 22 ("Philadelphia chromosome") or the

RSC Nanoscience & Nanotechnology No. 15
Unravelling Single Cell Genomics: Micro and Nanotools
Edited by Nathalie Bontoux, Luce Dauphinot and Marie-Claude Potier
© Royal Society of Chemistry 2010
Published by the Royal Society of Chemistry, www.rsc.org

PCR-based molecular diagnosis of the resulting fusion gene *bcr-abl* makes patients eligible to receive a specific therapy against the activated ABL1 tyrosine kinase with the drug imatinib, and the first results on the long-term survival of imatinib-treated patients are very encouraging.[2] In breast cancer, the analysis of hormone receptor status and the detection of an overexpressed *Her2* oncogene are important for the assessment of prognosis and the selection of therapy. On the other hand, some "breakthroughs" in molecular diagnosis are still controversially discussed. For example, the excessive screening for prostate specific antigen (PSA) in the serum of men as a marker for prostate cancer in the early 1990s resulted in enhanced early detection of prostate cancer.[3] Although this phenomenon was accompanied by a decrease in the incidence of high-grade cancer and mortality rates, current status is that to avoid one death of prostate cancer one would need to screen 1410 men and to treat 48 additional cases of prostate cancer.[4] Also, while the importance of targeted therapies for hematologic cancer is undeniable, the progress for solid tumors has been much slower. This indicates that cancer disease is still poorly understood, which might be due to the outstanding cellular heterogeneity of solid tumors. Convincing data suggest that there exist no two cancer cells in a patient that are identical.[5] This challenges the traditional approach to study cancer biology with samples from pooled cells of the primary tumor and advocates addressing the process of selection and mutation by analyzing individual cancer cells.[6]

This perspective is particularly justified when we consider the most fatal process of the disease, metastasis, which accounts for the majority of cancer-related deaths. Important steps during the metastatic process include dissemination of cells from the primary tumor, intravasation, survival in the circulation, extravasation, resistance to anoikis at the distant site, and colonization of the ectopic organ (Figure 6.1). Although recent studies imply that most of these obstacles can be overcome by a few specific genetic changes that induce epithelial–mesenchymal transition in tumor cells,[7] the outgrowth to macrometastasis in breast cancer may take decades. Currently, it is unknown which mechanisms govern such long latency periods and which eventually trigger metastatic colonization. Molecular characterization of disseminated tumor cells (DTCs) therefore could shed light into this dark stage of cancer progression and result in novel therapy strategies.

6.2 Detection and Malignant Origin of Disseminated Cancer Cells

Latent systemic disease, and therefore single disseminated tumor cells (DTCs), are the main target for most adjuvant therapies. In epithelial cancers, early detection of DTCs is mostly based on the expression of epithelial markers such as cytokeratins or the epithelial cell adhesion molecule (EpCAM), which are specific markers in mesenchymal organs like blood, lymph node, or bone marrow.[8,9] For 20–50% of cancer patients without manifest metastasis, these

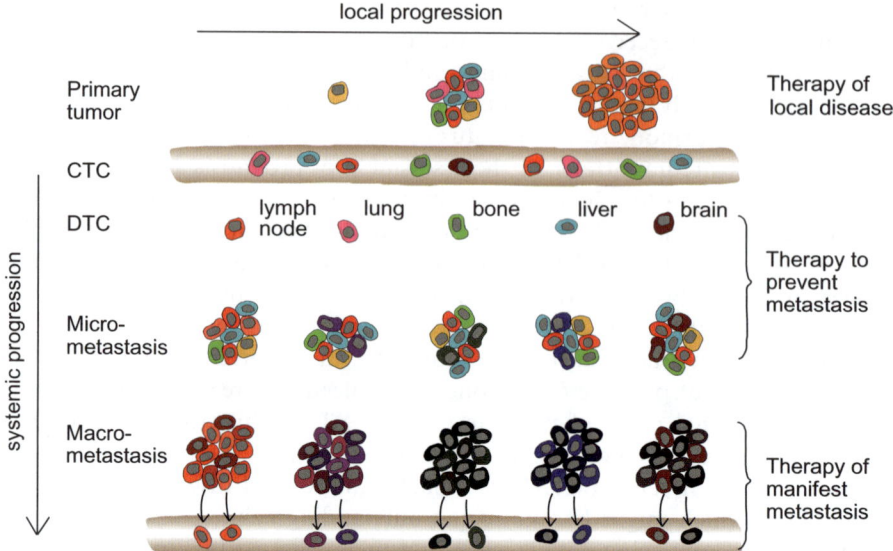

Figure 6.1   Local and systemic cancer progression: implications for therapy. Mutation and selection of cancer cells results in accumulation of genetic and
epigenetic alterations, both at the primary site and at distant sites.
Advantageous changes are then clonally expanded. Currently, therapy
decisions are based on the analysis of primary tumors. However, molecular and clinical data suggest that dissemination takes place early, leading
to diversification of clones at the primary and at different distant sites.
After homing to distant sites, circulating tumor cells (CTCs) become
disseminated tumour cells (DTCs) and continue to evolve. This cellular
variation in time and space (with different rates of acquiring changes in
different organs, as indicated by the different colors) is complicating
therapy development.

marker-positive cells are found at a frequency of one to two DTCs per million
bone marrow cells. Importantly, the detection of cytokeratin-positive cells in
bone marrow or EpCAM-positive cells in lymph nodes of patients suffering
from epithelial cancers has been demonstrated to be an important risk factor
associated with reduced overall and disease-free survival in a variety of cancers,[10,11] the most convincing data being available for breast cancer.[12] Since the
1980s, accumulating clinical evidence has made these marker-positive DTCs
prime candidates for direct metastatic precursor cells.

The first clinical studies investigating the prognostic impact of DTCs
in patients without manifest metastasis became available in the early 1990s.
As only one or two cells per sample were usually detected for analysis, the
molecular characterization was hampered for many years. Indeed, the only
techniques that could be applied used a marker antibody to detect the tumor
cells plus one additional molecule, like an antigen of interest or specific DNA
probes. These studies, using double-labeling approaches, demonstrated either
the malignant[13] or the histogenetic origin[10,14] of cytokeratin-positive cells from
bone marrow. It should also be noted that until today, the only cell line derived

from DTCs without further genetic manipulation was established from micrometastatic cells of an esophageal cancer patient without clinical evidence of nodal metastasis. A lymph node taken into culture and the esophageal cancer cell line derived thereof (LN1590) formed tumors in immuno-compromised mice.[15]

New methods for the analysis of single cells were thus needed for further insight into the molecular biology of these rare metastatic precursor cells, PCR techniques that enabled reliable whole genome amplification of minute DNA amounts allowed comprehensive genetic information from disseminated cancer cells to be obtained (*cf.* Chapter 8).

6.3 Genomic Studies of Single Disseminated Cancer Cells

The traditional view of metastasis holds that tumor cells accumulate genetic and epigenetic changes, expand clonally at the primary site, and reach a fully malignant state. Advantageous genetic and epigenetic alterations finally enable few variant tumor cells (or all, as some authors believe) to disseminate and establish distant metastases. Regardless of details, the traditional concept holds that metastatic dissemination generally is a late event in the history of a tumor. Based on this model it could be expected that several individual DTCs from the same patient should be genomically similar, as a result of their clonal descent, to the most advanced predominant clone within the primary tumor.

In one study, we screened over 500 bone marrow and lymph node samples of various carcinomas and detected more than one DTC in 71 samples (14%). After global amplification of the DNA of single DTCs and comparative genomic hybridization, we found that several cytokeratin-positive cells isolated from the bone marrow of one individual indeed displayed a high degree of similarity in their chromosomal imbalances, but, only when the patient was in the stage of manifest metastases (stage M1). In contrast, when patients were clinically free of metastases (*i.e.* in the clinical stage M0) we observed unexpected genomic heterogeneity of tumor cells from the same individual, and this in most patients.[16] These data suggest that during disease progression from the M0 to M1 stage, very aggressive cancer cells are selected from a high number of variant cells, which clonally expand at a later time point at the ectopic site. In a study on patients with breast cancer, genetically advanced cytokeratin-positive cells from stage M1 patients displayed characteristic chromosomal changes, such as amplification of 8q and 17q and loss of 17p and 18q, which are only rarely detected in DTCs isolated from M0 patients. Interestingly, these aberrations, together with changes on 12q, provide information as to whether the patient has manifest metastasis or not, with 85% accuracy.[17] Such selection of specific comparative genomic hybridization (CGH) alterations in stage M1 DTCs was also found in prostate cancer; again amplification on chromosome 8q was among the most prominent aberrations in single DTCs[18] as well as in pools of 10–20 DTCs.[19] The repeated observation of 8q gains suggests some functional role of this alteration for metastatic colony formation in breast and prostate cancers.

While it is interesting that disease stage can be deduced from the genomic profile of a single cytokeratin-positive cell from the bone marrow of a patient with breast cancer, the observation is clinically irrelevant as manifest metastases are diagnosed by modern imaging techniques. However, it is remarkable that the genome of a single DTC contains information that cannot be obtained by CGH analysis of the primary tumors. Likewise, although the genetic heterogeneity prior to metastasis partially explains the complexity and phenotypic heterogeneity of solid tumors, it complicates the search for novel drug targets. Indeed, as oncogene dependence is a prerequisite for therapy response in many instances, therapy target structures that are genetically activated in all cancer cells would be needed.

Since genetic analysis of DTCs has demonstrated mostly the genetic disparity between primary tumors and DTCs as well as between several DTCs from the same individual in early stage of disease, there is a need to increase the resolution of analyses. This may uncover changes that are clonal. Most likely, such changes – if they exist – are generated early in the tumor progression, *i.e.* during the transformation of epithelial tissue, and therefore could be transmitted to the progeny more frequently than changes that arise late in the process of clonal diversification. In this context, it is particularly interesting that 60% of breast cancer and 30% of prostate cancer DTCs of stage M0 patients do not display chromosomal aberrations at the resolution of metaphase CGH (which detects gains and losses comprising 10–20 Mb).[17,18] This finding was unexpected, as karyotypic abnormalities are present in primary tumors early on,[20] and it prompted us to investigate polymorphic markers for the detection of allelic losses in breast cancer DTCs by loss of heterozygosity (LOH) analysis. These small losses were abundantly detected in cytokeratin-positive cells that displayed normal karyotypes, proving there malignant origin. Additionally, some of the changes were shared with the matched primary tumors and therefore most likely identified genetic lesions that were present before dissemination. This finding of subchromosomal aberrations in disseminated cancer cells with a normal karyotype points to an early time point of dissemination, before the onset of chromosomal instability.[21] Indeed, we recently confirmed the early time point of dissemination in mouse models and breast cancer patients with ductal carcinoma *in situ* (DCIS).[22]

6.4 Oncogene Dependence and Tumor Suppressor Sensitivity in Metastasis Founder Cells

Until today, molecular analysis of cancers in clinical research and molecular diagnostics has been confined to pools of tumor cells taken from the primary tumor. At best, adjuvant therapies are then administered according to molecular traits of these cells, based on the assumption that metastatic cells will display the same characteristics. Recent data challenge this simplistic view of linear cancer progression and question the hypothesis that therapy decisions can be reliably based on molecular data of the primary tumor. Since systemically spread tumor cells are genetically different from the primary tumor, targeting them requires their direct analysis. For example, comparison of chromosomal aberrations found in DTCs and the corresponding primary tumor of breast cancer patients showed that a high similarity of CGH aberrations could be detected in only two out of 14 cases.[17] This weak concordance was also shown for esophageal cancer[23] and for M0 patients with prostate cancer, where 10 of 11 matched samples harbored not a single identical metaphase CGH aberration. Furthermore, the genetic similarity of M1 DTCs and corresponding primary tumor was weak.[18] Together with the accumulation of characteristic aberrations in DTCs of patients with metastatic prostate cancer these observations suggest parallel genetic progression in local and systemic cancer.[24] This is strengthened by the growth kinetics of primary carcinomas and metastases[25] and highlights the need for molecular characterization of single DTCs to stratify patients for adjuvant therapies.

Of clinical relevance, in about 20% of early disseminated breast cancer cells we found a gain of the well-known oncogene *HER2*, which renders the patient eligible for therapy with trastuzumab (Herceptin®), an antibody directed against the oncogene. This finding was surprising because most of the primary tumors of these patients did not display a *HER2* amplification[21] and would therefore not receive adjuvant *HER2* therapy. Clinical studies that address the question whether or not patients with *HER2*-amplified disseminated cancer cells (in the absence of a *Her2* amplification in the primary tumor) would benefit from trastuzumab treatment are currently under way.

Likewise, in patients with esophageal cancer we found that *Her2* amplifications in disseminated tumor cells but not in the primary tumor were prognostically relevant. In DTCs from patients with esophageal adenocarcinoma, amplifications of the *Her2* gene were even more frequent than in DTCs of patients with breast cancer. In future therapeutic trials these patients should be stratified by *Her2* gene amplifications of their disseminated tumor cells.[23]

Parallel progression of DTCs will likely occur at different ectopic sites. Tumor cells that are able to home to bone marrow or lung might be different in their genetic aberrations from those that disseminate to the brain or to other sites. Furthermore, to grow out at the ectopic site, those clones might need further characteristic changes, suggesting that genetic progression occurs over time at different sites in parallel. Thus, single DTCs might be subjected to different mechanisms than small colonies or manifest metastasis, which would complicate targeted therapies (Figure 6.1). While DTCs from bone marrow and lymph nodes – although rare and difficult to detect – can be accessed rather easily by bone marrow aspiration or lymphadenectomy at primary surgery, the isolation of DTCs from other sites of distant metastasis (as, for example, lung or liver) poses a greater obstacle for clinicians.

Yet, to understand the biology of systemic cancer and especially the progression to metastases it will be important to look at the single disseminated cell. Improvements in the detection, isolation, and molecular analysis of these single cells will hopefully lead the way to new and better therapies for cancer patients.

References

1. P. Boyle, A. d'Onofrio and P. Maisonneuve, *et al., Ann. Oncol.*, 2003, **14**, 1312–1325.
2. B. J. Druker, F. Guilhot and S. G. O'Brien, *et al., N. Engl. J. Med.*, 2006, **355**, 2408–2417.
3. B. F. Hankey, E. J. Feuer and L. X. Clegg, *et al., J. Natl. Cancer Inst.*, 1999, **91**, 1017–1024.
4. F. H. Schroder, J. Hugosson and M. J. Roobol, *et al., N. Engl. J. Med.*, 2009, **360**, 1320–1328.
5. J. H. Bielas, K. R. Loeb, B. P. Rubin, L. D. True and L. A. Loeb, *Proc. Natl. Acad. Sci. U. S. A.*, 2006, **103**, 18238–18242.
6. C. A. Klein, *Adv. Cancer Res.*, 2003, **89**, 35–67.
7. K. Polyak and R. A. Weinberg, *Nat. Rev. Cancer*, 2009, **9**, 265–273.
8. B. Kubuschok, B. Passlick, J. R. Izbicki, O. Thetter and K. Pantel, *J. Clin. Oncol.*, 1999, **17**, 19–24.
9. G. Schlimok, I. Funke and B. Holzmann, *et al., Proc. Natl. Acad. Sci. U. S. A.*, 1987, **84**, 8672–8676.
10. K. Pantel, R. J. Cote and O. Fodstad, *J. Natl. Cancer Inst.*, 1999, **91**, 1113–1124.
11. S. Riethdorf, H. Wikman and K. Pantel, *Int. J. Cancer*, 2008, **123**, 1991–2006.
12. S. Braun, F. D. Vogl and B. Naume, *et al., N. Engl. J. Med.*, 2005, **353**, 793–802.
13. P. Mueller, P. Carroll and E. Bowers, *et al., Cancer*, 1998, **83**, 538–546.
14. R. Oberneder, R. Riesenberg and M. Kriegmair, *et al., Urol. Res.*, 1994, **22**, 3–8.
15. S. Hosch, J. Kraus and P. Scheunemann, *et al., Cancer Res.*, 2000, **60**, 6836–6840.
16. C. A. Klein, T. J. Blankenstein and O. Schmidt-Kittler, *et al., Lancet*, 2002, **360**, 683–689.
17. O. Schmidt-Kittler, T. Ragg and A. Daskalakis, *et al., Proc. Natl. Acad. Sci. U. S. A.*, 2003, **100**, 7737–7742.
18. D. Weckermann, B. Polzer and T. Ragg, *et al., J. Clin. Oncol.*, 2009, **27**, 1549–1556.
19. I. N. Holcomb, D. I. Grove and M. Kinnunen, *et al., Cancer Res.*, 2008, **68**, 5599–5608.
20. R. A. DePinho and K. Polyak, *Nat. Genet.*, 2004, **36**, 932–934.
21. J. A. Schardt, M. Meyer and C. H. Hartmann, *et al., Cancer Cell*, 2005, **8**, 227–239.
22. Y. Husemann, J. B. Geigl and F. Schubert, *et al., Cancer Cell*, 2008, **13**, 58–68.
23. N. H. Stoecklein, S. B. Hosch and M. Bezler, *et al., Cancer Cell*, 2008, **13**, 441–453.
24. C. A. Klein, *Nat. Rev. Cancer*, 2009, **9**, 302–312.
25. C. A. Klein and D. Holzel, *Cell Cycle*, 2006, **5**, 1788–1798.

CHAPTER 7
Capturing a Single Cell

CATHERINE REY,[a, b] ANNE WIERINCKX,[a, c] SÉVERINE CROZE,[a] CATHERINE LEGRAS-LACHUER[a, d] AND JOEL LACHUER[a, c]

[a] Genomic and microgenomic core facility: ProfileXpert platform (Federative Institute of Neurosciences), 16 rue du Doyen Lépine, 69676 Bron cedex, France; [b] Institut Fédératif des Neurosciences, IFNL, F-69676 Bron, France; [c] Université de Lyon 1, Université Lyon 1, Inserm U842, F-69372 Lyon, France; [d] Université de Lyon, INRA, UMR754, Université Lyon 1, Ecole Nationale Vétérinaire de Lyon, Ecole Pratique des Hautes Etudes, IFR 128, 50 avenue Tony Garnier, F-69366 Lyon, cedex 07, France

Abstract

A major problem encountered in genomic and proteomic studies arises from the heterogeneous nature of different tissue. Analysis of a pure cell population is essential for correlating relevant molecular signatures in diseased and disease-free cells. During the last 30 years this challenge has led to the development of different technologies able to isolate cells of interest. Laser capture micro-dissection (LCM) is the last available technology using the precision of a laser beam to isolate single cells from complex tissue. In this chapter we will review the different technologies available and some applications.

7.1   Introduction

The development of technologies for whole genome analysis of the transcriptome and proteome has allowed spectacular advancement in numerous biological fields. However, in many cases, the biological significance of results is tightly

RSC Nanoscience & Nanotechnology No. 15
Unravelling Single Cell Genomics: Micro and Nanotools
Edited by Nathalie Bontoux, Luce Dauphinot and Marie-Claude Potier
© Royal Society of Chemistry 2010
Published by the Royal Society of Chemistry, www.rsc.org

associated with the cellular composition of analyzed tissue. For example, genomic and proteomic analysis in the context of cancer investigation is susceptible to contamination by non-neoplastic cells (as inflammatory cells and vascular cells) which can mask tumor cell specific alterations. The degree of masking will depend on the percentage of tumor cells versus non-tumor cells located in the tissue analyzed.[1,2] The same problem of signal dilution is observed in all heterogeneous tissue and more particularly in complex tissue such as brain.

The development of technologies allowing the isolation of cells from heterogeneous tissue and methods to amplify small quantity of nucleic acids (DNA and RNA) answers this problem.[3] In this chapter, we will review these methods.

7.2   Overview of Cell Sorting Technologies

A first batch of approaches available to concentrate and purify cells of interest is to use cell sorting techniques such as density gradients,[4] fluorescence-activated cell sorting,[5] antibody-labeled magnetic beads.[6,7]

- The *density gradients* procedure is based on the possibility of producing a linear gradient density with different media as sucrose, metrizamide, Ficoll™ and Percoll and to separate cells on the basis of their density following centrifugation or sedimentation. This procedure is extensively used to separate monocytes from whole blood.
- *Fluorescence-activated cell sorting* is a specialized type of flow cytometry. It provides a method for isolating cell populations upon specific fluorescent labeling of cells of interest. The procedure uses several steps: (1) fluorescent labeling of intact cells, (2) separation of cells in individual liquid droplets, (3) measuring the fluorescence with a specific laser, (4) placement of electric charges in fluorescence-positive cells, and (5) separation of charged cells in an electric field.
- The method using *antibody-labeled magnetic beads* is based on the property of uniform polystyrene spherical beads to be magnetizable. The attachment of target-specific antibodies to the surface of the beads allows the capture and isolation of intact cells directly from a complex suspension of cells. This procedure is accomplished without intervention of column and centrifugation steps. Positive or negative cell isolation can be performed regardless of the availability of the antibody able to recognize specific cell surface markers. This technology is particularly used to isolate CD34+ stem cells.

These methods need the creation of a suspension of individual cells, which could be rarely applicable for solid tissue without disturbing cell phenotypes. Consequently, a chemical cell-dissociation step such as trypsinization is applied. The other approaches developed allow selective isolation of cells from their non-disturbed biological environment. These techniques – ranging from

Table 7.1 Cell isolation methodologies.

Methodology	Principle	Applications
Density gradients	Cell sorting using media forming a gradient	Cell suspension
Fluorescence activated cell sorting (FACS)	Cell sorting using immunolabeling	Cell suspension
Magnetic beads (Dyna-beads, Dynal Biotech) (www.invitrogen.com)	Cell isolation with magnetic beads coated with an antibody	Cell suspension
Needle attached to a micromanipulator	Mechanical isolation	Frozen and fixed paraffin-embedded section
Laser microdissection	Cell cutting or capture	Frozen and fixed paraffin-embedded section
	Mechanical cell recovery	Living cell culture, smears, cytospin

the lowest to the highest degree of resolution – include manual microdissection with a razor blade or needle attached to a micromanipulator,[8] and more recently technologies based on laser microdissection under microscopy cell visualization. The different technologies are presented in Table 7.1.

7.3 Laser Capture Microdissection Technologies

Laser capture microdissection (LCM) appeared in the 1970s but really became available during the mid 1990s with the work by the Emmert-Buck team at the National Institutes of Health, Bethesda, USA.[9] There are two general classes of laser microdissection systems: infrared (IR) capture systems and ultraviolet (UV) cutting systems. The fundamental features of the laser microdissection process are: (1) visualization of the cells *via* microscopy, (2) transfer of laser energy to a thermoplastic film (IR system) or cutting surrounding a selected area (UV system), and (3) removal of the cells of interest from the heterogeneous tissue section by different procedures. The different characteristics are summarized in Table 7.2.

7.3.1 Infrared Laser Capture Systems

LCM technology (the Arcturus system, Figure 7.1) uses a low-power infrared laser to melt a special thermoplastic film over the cell(s) of interest (Figure 7.2). CapSure® HS or CapSure® Macro Caps, which are coated with this thermoplastic film and are especially developed for this technology, are placed on the tissue section or cytology sample. The PixCell® IIe LCM instrument is then used to direct the laser through the cap to melt the film onto the cells of interest. In addition, the CapSure LCM Caps work with ExtracSure™ devices that minimize the dilution of biological molecules extracted from captured cells. This optimizes the recovery of these molecules for downstream molecular analysis.

Table 7.2 Features of the equipment from different manufacturers.

System	Laser	Resolution of dissection	Cell collection process	Specimen source
Arcturus Pixcell II (www.moleculardevices.com)	IR (810 nm)	5–7.5 μm to 30 μm	Plastic melting and mechanical removal	Frozen and FFPE tissues, smears, cytospin
Biorad Clonis (www.microscopy.bio-rad.com)	IR (780 nm)	2 μm	Capture on a non-toxic absorbing film	Living cells and tissues
PALM (www.zeis.com)	UVa (337 nm)	0.5–1 μm	Pressure catapulting system	Frozen and FFPE tissue, smears, cytospin, living cells, chromosome spreads
Leica system (LMD 6000) (www.leica-microsystem.com)	UV (355 nm)	4–5 μm	Gravity	Frozen and FFPE tissue, smears, cytospin, living cells, chromosome spreads
Molecular Machines and Industries (MMI) (www.molecular-machines.com)	UV (350 nm)	1 μm	Adhesive Eppendorf cap	Frozen and FFPE tissue, smears, cytospin, living cells
Veritas, XT (Arcturus) (www.moleculardevices.com)	UV/IR (810/355 nm)	5/0.2 μm	Adhesive collection	Frozen and FFPE tissue, smears, cytospin, living cells, chromosome spreads

FFPE, formalin-fixed paraffin-embedded.

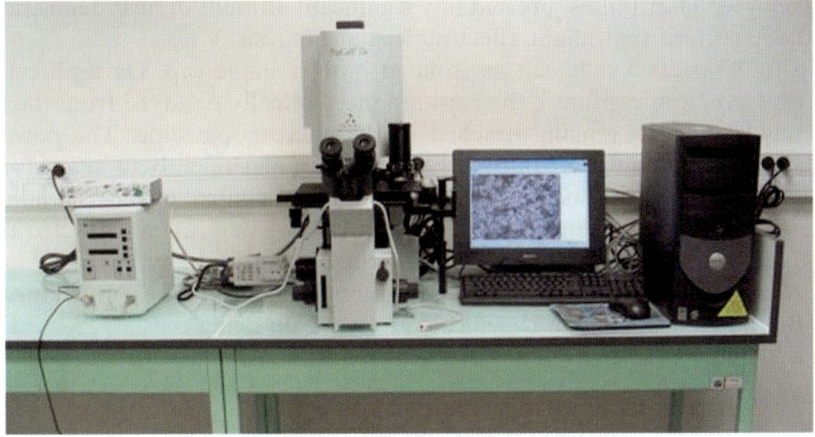

Figure 7.1   Laser capture microdissection (LCM) apparatus (Pixcell II from Arcturus).

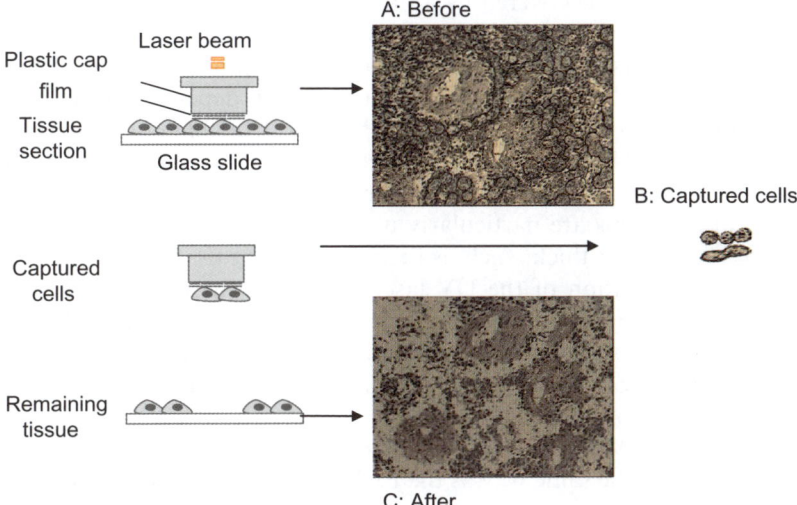

Figure 7.2   The principle of laser capture microdissection (LCM).

The cap acts as an optic for focusing the laser in the same plane as the tissue section. The polymer melts only in the vicinity of the laser pulse, forming a polymer–cell composite. A dye incorporated into the polymer serves two purposes:

- Absorption of laser energy, preventing damage to the cellular constituent.
- Visualization of melted polymer areas.

The combination of (1) the short laser pulse duration used, (2) a low laser power level, (3) absorption of the laser pulse by the polymer, and (4) the long time between laser pulses, prevent any significant amount of heat deposition at the tissue surface that might affect molecular analysis.

Up to 3000–5000 cells can be isolated onto a single cap. Once all cells of interest have been captured, the caps are mechanically removed from the slide and unwanted cells remain attached to the microscopic slide. The power of the laser and its diameter can both be adjusted independently to adapt for the dissection of different kinds of tissues. After visual control of the cap, adhered target cells are subsequently lysed and DNA/RNA or protein extracted using extraction and purification methods with adequate sensitivity.

An automated version is available (AutoPix, Veritas and XT from Arcturus).[10] The Biorad clonis platform is dedicated to work on living cells.

7.3.2 Ultraviolet Cutting Systems

The shorter pulse duration in combination with the optimal optical transmission at 337–355 nm enables fast and highly precise cutting. However, these systems require special slides for the work. Tissue sections are first transferred onto a microscope slide covered by a thin polyethylene membrane. Single cells or group of cells are excised by circumcision with a high-energy focused laser beam. The laser cuts a beam spot size of less than 1 μm in diameter. For a UV system, unlike an IR system, a precise cutting line depends on the objective magnification. With LPC[pat] technology (PALM),[11] the high photonic pressure force of the focused laser beam ejects selected sample from the object plane and catapults it directly into the cap of a microcentrifuge tube.[12]

UV cutting systems are particularly useful for the microdissection of tissue sections up to 200 μm thick, such as sections of plant tissue.

A potential limitation of the UV laser systems is the putative UV-induced damage in the final cell population. The UV system may be used to ablate unwanted tissue. Laser microbeam microdissection systems use a much thinner laser beam diameter in contrast to the IR-LCM and enable clear cut separation from the neighboring tissue.

The Molecular Machines and Industries (MMI) CellCut and SmartCut systems operate in the same way as the PALM. The Leica system uses the same cutting procedure but target cells are collected by placing an adhesive cap (of a microcentrifuge tube) onto the cut area.

The ArcturusXT™ and Veritas systems are a unique microdissection instruments that combine laser capture microdissection (LCM) and ultraviolet (UV) laser cutting in one platform.

7.4 Protocols Before Laser Microdissection (Tissue Sampling and Preparation)

To obtain relevant results from tissue isolated cells, the procedure involves the optimization of critical preparation steps:[13] (1) maintaining the tissue morphology to allow good identification of the cells of interest; and (2) maintain the integrity and a high yield for DNA/RNA and protein recovery after microdissection. These two critical steps are largely dependent on the tissue preparation (fresh frozen (FF) tissue and formalin-fixed paraffin-embedded (FFPE) tissue).

It was found that the main factors influencing tissue morphology, LCM capture success and acid nucleic/protein integrity are: (1) slide temperature for collecting tissue, (2) method and temperature of fixation, and (3) temperature and nature of staining and dehydration solution.

For the LCM system, dehydration steps are essential for capture success. The dehydration allows a decrease of the RNase or protease activity during microdissection.

7.4.1 Dissection from Fresh Frozen Tissue

Frozen sections are highly recommended to maximize quantity and quality of RNA/DNA and protein recovery.[14] In Figure 7.3 are presented two electropherograms (Bioanalyser 2100, Agilent) showing the very similar quality of two RNA preparations coming from fresh tissue or microdissected cells from the same tissue (data from our group). Different cryo-protection methods facilitate cutting frozen sections without freezing artifact. Optimal Cutting Temperature (OCT™) is a useful widely used product to preserve tissue morphology but this compound can inhibit polymerase chain reaction.[15]

However, in frozen section it is often difficult to recognize histological details after routine staining, such as with hematoxylin and eosin (H&E), because in this case laser microdissection requires desiccated sections without cover slips. To circumvent this limitation, specialized staining methods have been developed for distinguishing cells of interest from surrounding the stroma, including Nissl's stain (NS), immunofluorescence (IF), and immunohistochemistry (IHC). In most cases, cresyl violet staining will be sufficient for cell identification, a basic dye that stains negatively charged nucleic acids in the nucleus of a cell with a dark blue color. It also stains the rough endoplasmic reticulum of neurons and preserves RNA/DNA and proteins.

In the case of the tissue morphology damages, cresyl violet might not clearly show cell distribution. In such cases, the acridine orange stain may provide better contrast between clusters of cells.

The stain used before microdissection has to be compatible with a good preservation of molecular targets and must be checked in pilot experiments.

7.4.2 Dissection from Formalin-fixed Paraffin-embedded Tissue

In clinical research, it is often the case that morphology of frozen tissue is not sufficient for complete identification of cells of interest and the use of formalin-fixed paraffin-embedded tissues section is preferred. Formalin fixation operates by creating extensive cross-links within and between proteins. Disruption of the cross-links produces peptide and protein fragments rather than intact proteins.

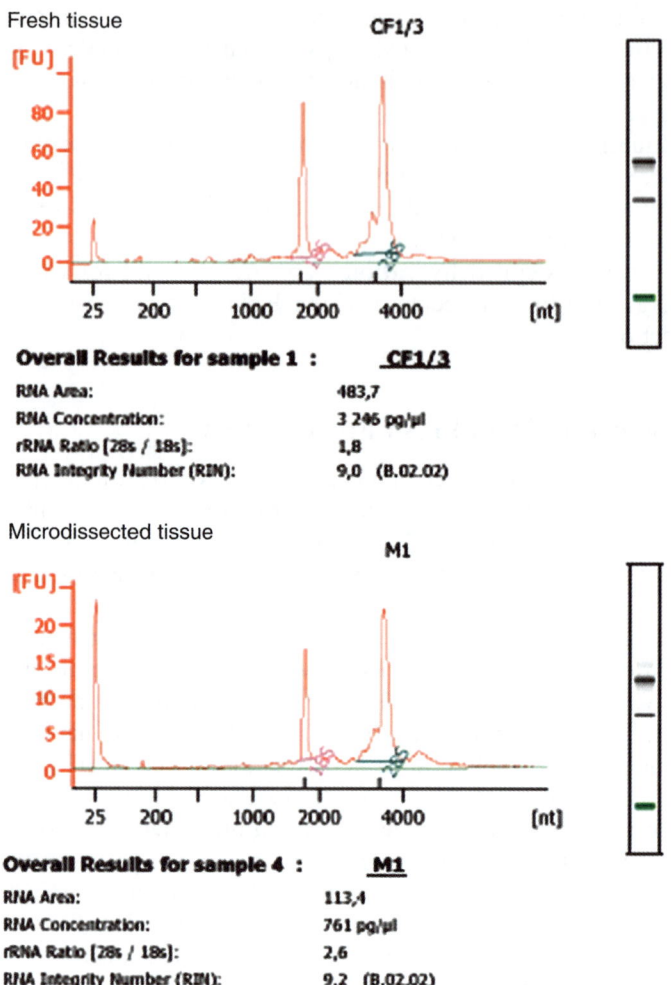

Figure 7.3 Comparison of RNA quality. Total RNA extracted from fresh tissues or the same microdissected tissue is analyzed using a 2100 Bioanalyzer (Agilent). The RNA integrity number (RIN) proves the good preservation of RNA quality during the whole procedure of IR microdissection (data from our group).

Despite the development of new technologies for reversing cross-linking, the yield and quality of protein/RNA/DNA remain low.[16,17] Different reversible cross-linkers, such as dithiobis(succinimidyl) propionate (DSP), have been developed and used successfully.[18] While the results of some studies are discouraging, archived FFPE samples have been successfully used in some cases to identify prognostic and diagnostic gene signatures for numerous diseases.[19,20] Paraffin-embedded sections can be used with conventional staining techniques, including immunostaining.

7.4.3 Immuno Laser Capture Microdissection

Immuno-LCM, which involves immunohistochemical staining of tissue before laser microdissection, enhances the ability to identify cells of interest in complex tissue by combining morphology and immunophenotype. Immunostaining

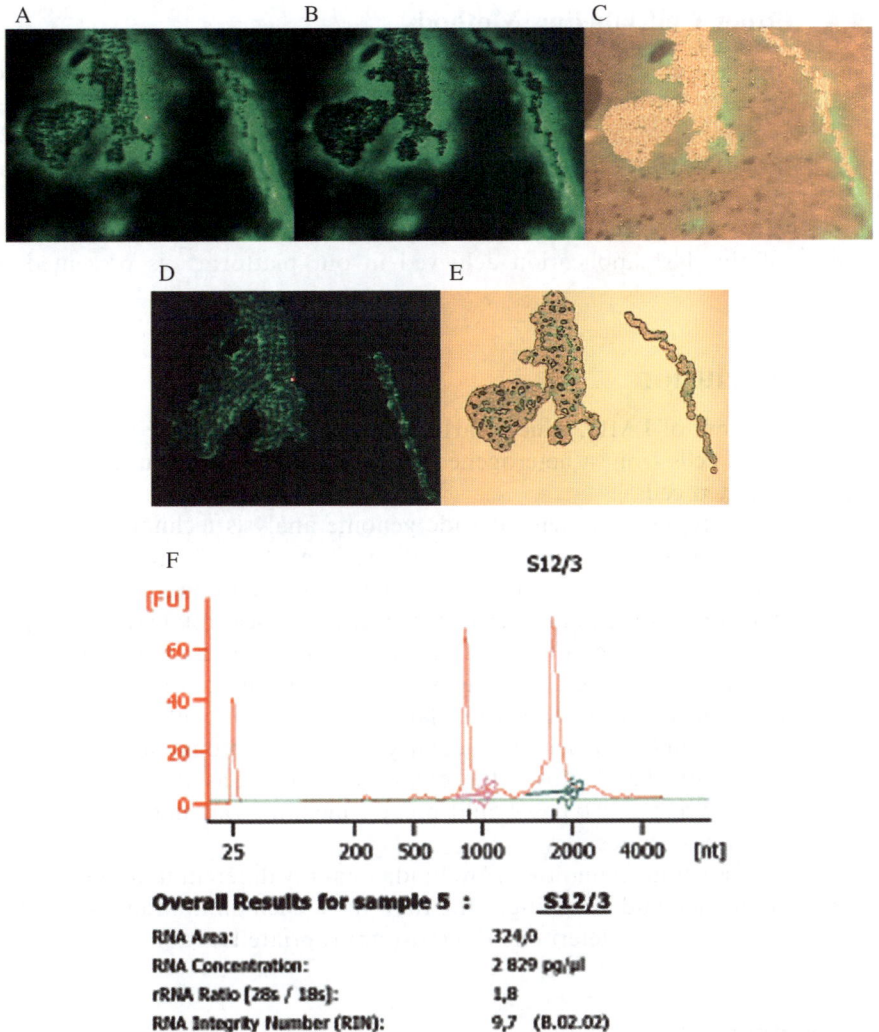

Figure 7.4 Laser capture microdissection of fluorescent cells. Laser microdissection has been performed on rat brain expressing GFP-siRNA directed against the Huntington gene. Microdissection is performed using Pixcell II from Arcturus under UV exposure. (A) Before capture; (B) and (C) after capture; (D) and (E) isolated cell on caps; (F) quality of total RNA obtained from microdissected cells.

protocols are available for frozen tissue or FFPE tissue. In this strategy, the optimal immuno-LCM procedure, in which RNA is not degraded during the staining process, requires the use of high-affinity antibodies at high concentration with a short overall staining period.[21,22] It is worth mentioning that DNA *in situ* hybridization has also been successfully combined with laser microdissection to analyze cells based on their genotype.[23]

7.4.4   Other Cell-labeling Methods

Other strategies can be used to avoid pre-treatment of cells which could lead to nucleic acid and protein degradation. These methods include: (1) the pre-labeling of cells *in vivo* in animal models by injection of a fluoro-gold label;[24] (2) the use of lectin probes for labeling brain microvessels; [25] and (3) the use of transgenic animals expressing the fluorescent transgene as green fluorescent protein (GFP) in specific cells is easily detectable under UV exposure.[26] An example of this last application achieved in our platform[27] is presented in Figure 7.4.

7.5   Conclusion

The development of LMD, which started in the 1990s, has made it possible to isolate defined cells from a heterogeneous cell population without contamination of unwanted cells.

Simultaneously, development of wide genome analysis technologies (genomic, transcriptomic, and proteomic) allow, "in theory", the single cell machinery to be deciphered. It may be argued that examining the expression profile of the complete cellular microenvironment of disease state (for example, in oncology) is more representative of the ongoing progress than the separation of individual cells. However, previous studies clearly showed that analysis of bulk tissue (tumor) gives few relevant data in comparison with the huge data collection and sample analysis (particularly for transcriptomic analysis). Studies that gradually break down the complex interactions that exist *in vivo* between neighboring cell types would greatly facilitate our understanding of normal and disease states.

All microdissection technologies are fundamentally different in terms of their physical principles and handling, and they have their limitations. Only the specific application can determine the most appropriate technology to use.

References

1. D. C. Sgroi, S. Teng, G. Robinson, R. LeVangie, J. R. Hudson Jr. and A. G. Elkahloun, *Cancer Res.*, 1999, **59**, 5656–5661.
2. B. Domazet, G. T. Maclennan, A. Lopez-Beltran, R. Montironi and L. Cheng, *Int. J. Clin. Exp. Pathol.*, 2008, **1**, 475–488.

3. F. Ducray, A. Wierinckx, C. Rey, C. Legras, M. F. Belin, J. Honnorat and J. Lachuer, *Discovery Matters*, 2006, **3**, 16–17.

4. R. J. Colello and C. Sato-Bigbee, *Curr. Protoc. Neurosci.*, 2001, **3**, 3–12.

5. L. A. Herzenberg, D. Parks, B. Sahaf, O. Perez, M. Roederer and L. A. Herzenberg, *Clin. Chem.*, 2002, **48**, 1819–1827.

6. A. A. Neurauter, M. Bonyhadi, E. Lien, L. Nøkleby, E. Ruud, S. Camacho and T. Aarvak, *Adv. Biochem. Eng. Biotechnol.*, 2007, **106**, 41–73.

7. K. Pike-Overzet, D. de Ridder, T. Schonewille and F. J. Staal, *Methods Mol. Biol.*, 2009, **506**, 403–421.

8. S. Hernández and J. Lloreta, *Ultrastruct. Pathol.*, 2006, **30**, 221–228.

9. M. R. Emmert-Buck, R. F. Bonner, P. D. Smith, R. F. Chuaqui, Z. Zhuang, S. R. Goldstein, R. A. Weiss and L. A. Liotta, *Science*, 1996, **274**, 998–1001.

10. Arcturus Biosciences, Inc., User guide for the Veritas microdissection instrument, Version A, Molecular Devices, Sunnyvale, CA, USA.

11. P. Micke, A. Ostman, J. Lundeberg and F. Ponten, *Methods Mol. Biol.*, 2005, **293**, 151–156.

12. D. E. Kuhn, S. Roy, J. Radtke, S. Khanna and C. K. Sen, *Am. J. Physiol. Heart Circ. Physiol.*, 2007, **292**, H1245–H1253.

13. V. Espina, J. D. Wulfkuhle, V. S. Calvert, A. VanMeter, W. Zhou, G. Coukos, D. H. Geho, E. F. Petricoin 3rd and L. A. Liotta, *Nat. Protoc.*, 2006, **1**(2), 586–603.

14. H. Wang, J. D. Owens, J. H. Shih, M. C. Li, R. F. Bonner and J. F. Mushinski, *BMC Genomics*, 2006, **27**, 97.

15. G. R. Turbett and L. N. Sellner, *Diagn. Mol. Pathol.*, 1997, **6**, 298–303.

16. S. R. Shi, C. Liu, B. M. Balgley, C. Lee and C. R. Taylor, *J. Histochem. Cytochem.*, 2006, **54**, 739–743.

17. L. Gianni, M. Zambetti, K. Clark, J. Baker, M. Cronin, J. Wu, G. Mariani, J. Rodriguez, M. Carcangiu, D. Watson, P. Valagussa, R. Rouzier, W. F. Symmans, J. S. Ross, G. N. Hortobagyi, L. Pusztai and S. Shak, *J. Clin. Oncol.*, 2005, **23**, 7265–7277.

18. C. C. Xiang, E. Mezey, M. Chen, S. Key, L. Ma and M. J. Brownstein, *Nucleic Acids Res.*, 2004, **32**, e185.

19. K. Specht, T. Richter, U. Müller, A. Walch, M. Werner and H. Höfler, *Am. J. Pathol.*, 2001, **158**, 419–429.

20. G. Fedorowicz, S. Guerrero, T. D. Wu and Z. Modrusan, *BMC Med. Genomics*, 2009, **2**, 23.

21. K. Kinnecom and J. S. Pachter, *Brain Res. Protoc.*, 2005, **16**, 1–9.

22. C. Rupp, H. Dolznig, C. Puri, N. Schweifer, W. Sommergruber, N. Kraut, W. J. Rettig, D. Kerjaschki and P. Garin-Chesa, *Diagn. Mol. Pathol.*, 2006, **1**, 35–42.

23. L. M. Gjerdrum and S. Hamilton-Dutoit, *Methods Mol. Biol.*, 2005, **293**, 139–149.

24. F. Yao, F. Yu, L. Gong, D. Taube, D. D. Rao and R. G. MacKenzie, *J. Neurosci. Methods*, 2005, **143**, 95–106.

25. J. Mojsilovic-Petrovic, M. Nesic, A. Pen, W. Zhang and D. Stanimirovic, *J. Neurosci. Methods*, 2004, **133**(1–2), 39–48.
26. V. Bhattacherjee, P. Mukhopadhyay, S. Singh, E. A. Roberts, R. C. Hackmiller, R. M. Greene and M. M. Pisano, *Genesis*, 2004, **39**, 58–64.
27. V. Drouet, V. Perrin, R. Hassig, N. Dufour, G. Auregan, S. Alves, G. Bonvento, E. Brouillet, R. Luthi-Carter, P. Hantraye and N. Déglon, *Ann Neurol.*, 2009, **65**, 276–285.

CHAPTER 8

Looking at the DNA of a Single Cell

BERNHARD POLZER AND CHRISTOPH A. KLEIN

Division of Oncogenomics, University of Regensburg, Franz-Josef-Strauss-Allee 11, 93053 Regensburg, Germany

Abstract

Small amounts of genomic DNA frequently impede genetic analyses in many fields of research, including forensic research, prenatal diagnosis or, as discussed earlier in chapter 6, oncology. To overcome these limitations amplification of genomic DNA may provide quantities of DNA needed for downstream applications. This chapter will describe methods currently available to analyse the DNA of single cells. A focus will be on whole genome amplification (WGA) of single cells and the advantages and drawbacks of available protocols will be discussed.

8.1 Challenges of Single Cell DNA Amplification

With the development of methods to detect and isolate few or even single cells, such as fluorescence activated cell sorting (FACS), laser capture microdissection, capillary micromanipulation or optical tweezers, as well as with the advent of stem cell biology and the diversification of cell lineages and cell types, the study of single cells is of increasing interest. In oncology, genetic diversity and cellular heterogeneity – while being a hallmark of cancer – has been addressed only by a limited numbers of methods. For example, fluorescence *in situ* hybridization (FISH) has been used to determine genetic differences between

RSC Nanoscience & Nanotechnology No. 15
Unravelling Single Cell Genomics: Micro and Nanotools
Edited by Nathalie Bontoux, Luce Dauphinot and Marie-Claude Potier
© Royal Society of Chemistry 2010
Published by the Royal Society of Chemistry, www.rsc.org

cells and to assess clonal expansion of specific changes; likewise, single cell PCR has been used to determine clonality in a variety of lymphomas by analysis of clonal T-cell receptor or antibody rearrangements. In both cases, only selected chromosomal loci were analyzed, while the low amount of DNA (6–7 pg) had prevented genome-wide studies of DNA changes.[1] To overcome these limitations, the DNA has to be amplified after isolation and before molecular genetic analyses.

Whole genome amplification of a single cell poses three major problems. First, the human genome is complex and is comprised of 3×10^9 nucleotides, such that full genomic coverage is difficult to achieve. Second, single-copy sequences must be amplified and risk losing one or both copies in a diploid genome as artefact of the method is substantial. This phenomenon is called allelic drop out (ADO) and is thought to be caused by varying distributions of reagents at the target sites such that not all components necessary for the PCR reaction are available.[2] Third, for whole genome analysis of copy number changes, by comparative genomic hybridization (CGH) on metaphase chromosomes or array-CGH or single nucleotide polymorphism (SNP) arrays, homogenous amplification of single cell DNA has to be achieved in order retrieve the correct karyotype of a cell.

8.2   Methods for Amplifying Genomic DNA of Single Cells

Several protocols for DNA amplification of few or even single cells have been published from the early 1990s, as summarized in Table 8.1. The first widely used method uses degenerate oligonucleotide primers (DOPs) for PCR. It relies on a mixture of six random core nucleotides that are flanked by defined 5′ and 3′ sequences (CGACTCGAGNNNNNNATGTGG), which enable annealing of the primers at low temperature at approximately one million sites throughout the DNA template.[3] DOP PCR has been applied to the analysis of single or small pools of cells and in pre-implantation genetic diagnosis,[4,5] although with low input of genomic DNA (less than 1 ng, *i.e.* fewer than 100 cells) the fidelity and the genome coverage decrease and the likelihood of ADO increases.[6] Another Taq polymerase-based approach is primer extension pre-amplification (PEP) PCR, which uses highly degenerate oligonucleotides consisting of 15 random bases to prime template DNA.[7–9] All methods employing degenerate oligonucleotides suffer, however, from the limitation that not all genomic regions will be equally primed and amplification conditions for the 4096 DOP or 10^9 PEP primers within the reaction vary considerably. Yet, an improved procedure (I-PEP) was successfully used for single cell analysis of defined genomic loci.[10]

Some problems inherent to DOP or PEP protocols were overcome by the development of multiple strand displacement amplification (MDA),[11] an isothermal amplification method. Here, the reaction is catalyzed by the large fragment of the Bst polymerase or the highly processive Phi29 DNA

Table 8.1 Methods for whole genome amplification of single cells.

Characteristic	Method					
	DOP-PCR	PEP-PCR	Combination of DOP and PEP	MDA	SCOMP	TLAD
Pattern	Thermal cycling with semi-random primers	Thermal cycling with random 15mer primers	Thermal cycling (GenomePlex by Sigma Aldrich)	Isothermal amplification	Thermal cycling with one universal primer	T7-based thermal cycling with reverse transcription
Whole genome representation of single cells	No reliable metaphase CGH	Not reported	Smallest aberration in array CGH 8.3 Mb	Smallest aberration in array CGH 34 Mb	Smallest aberration in array CGH 1–2 Mb; best representation of the original genomic DNA	Not reported
Allelic drop out rate of single cells	Not reported	Not reported	25–33%	10–31%	5–10%	Not reported
Level of feasibility	Easy	Easy	Easy but expensive with kit	Easy	Difficult and time-consuming	Difficult and expensive

CGH, comparative genomic hybridization; DOP, degenerate oligonucleotide primers; MDA, multiple strand displacement amplification; PCR, polymerase chain reaction; PEP, primer extension pre-amplification; SCOMP, single cell comparative hybridization; TLAD, T7-based linear amplification of DNA.

polymerase, which possesses proofreading activity, resulting in error rates 100 times lower than the Taq polymerase.[12,13] MDA is based on the annealing of random hexamers to denatured DNA, followed by strand-displacement synthesis at a constant temperature.[14] The result is a network of relatively long DNA fragments. This technique has successfully been used for amplifying diluted DNA samples[15] as well as single cells.[16,17] Unfortunately, it has been shown that MDA, when used on complex genomes, results in a significant rate of ADO.[18]

While all the methods mentioned so far are suited for the analysis of pre-defined genetic loci, they failed to convincingly demonstrate a homogenous amplification of a diploid single-cell genome. As a consequence, comparative genomic hybridization (CGH), a technique that enables genome-wide screening of DNA copy number variations,[19,20] cannot be applied to single cells. To overcome this, we developed an adapter–linker PCR strategy that homogeneously amplifies the whole genome[21] and, for the first time, enabled reproducible application of metaphase comparative genomic hybridization (CGH). We therefore called the procedure SCOMP (for single cell comparative hybridization). The resulting genomic representation of single cells can not only be used for genome-wide screens but also as a template for multiple sequence-specific PCR.

Recently, based on the protocol originally designed by Phillips and Eberwine to amplify mRNA,[22] a T7-based linear amplification of DNA (TLAD) has been adopted for WGA.[23] Here, a polyT tail is added on the 3′ terminus by a terminal transferase after an Alu I restriction endonuclease digestion. The amplification is driven by a 5′ T7 promotor and a 3′ polyA tract, as well as Taq polymerase to synthesize the second strand followed by *in vitro* transcription and reverse transcription. TLAD is resistant to sequence- and length-dependent bias but has not been widely used up to now, as it is rather time-consuming.[24]

8.3   Array Comparative Genomic Hybridization of Single Cells

As mentioned before, all methods of WGA studies were demonstrated to be suitable for multiple analyses of specific genomic loci and some of the methods were also suitable for metaphase CGH. However, there is a need to apply genome-wide techniques that provide higher resolution than metaphase CGH, such as array CGH. Array CGH uses oligonucleotides or cloned genomic DNA (*e.g.* bacterial artificial chromosomes (BACs)) as hybridization matrix instead of metaphase chromosomes (as in metaphase CGH). So far, several hundred nanograms of cellular input DNA are needed. In a comparative study that evaluated the suitability of WGA protocols to amplify 10 ng of genomic DNA – approximately 1000-fold more DNA than in a single cell – prior to array CGH, SCOMP showed superior representation of the original DNA over

MDA and DOP-PCR,[25] a finding that corroborates the lower prevalence of ADO by SCOMP compared to the other methods (Table 8.1).

Using DOP-, PEP- or MDA-based techniques for array-CGH of single cells has not yet resulted in an increase of resolution beyond metaphase CGH as the signals of several BAC clones had to be integrated for noise reduction.[26,27] Single cell array CGH using DOP- or MDA-based amplification protocols reached a resolution of 34–60 Mb.[27] Recently, a method combining DOP and PEP PCR for single cell genome amplification (GenomePlex®, Single Cell Whole Genome Amplification Kit) was tested on a genome-wide array comprising more than 30 000 BAC clones. The authors analyzed single epithelial cancer cells from cell lines and uncovered an amplification of 8.3 Mb.[28] While changes of this size may be detected in high-quality metaphase CGH experiments, it was clearly demonstrated that single cell array CGH is able to resolve aberrations smaller than chromosome arms. The failure to increase resolution further likely results from uneven amplification of single cell DNA. Possibly, the ADO of 25–30% of the GenomePlex® kit is too high to achieve further improvements. In this respect it is noteworthy that SCOMP has an allelic dropout of only 5–10%.[29] Indeed, first data on our customized SCOMP array CGH increased the resolution of metaphase CGH about ten-fold and was able to detect genomic aberrations down to a resolution of 1–2 Mb in single T-47D breast cancer.[30] A definite answer on the maximal resolution of this amplification method awaits the application of high-density BAC arrays, as in contrast to the GenomePlex amplified products, which were hybridized on a 30 K array, we used so far BAC arrays with a spacing of 1 Mb and more (comprising 3000 clones). It should also be noted that hybridization of SCOMP amplicons onto an oligonucleotide array comprising up 244 K probes did not result in higher resolution when compared to the 3 K BAC array.[30] The reason for this observation is still unclear but its clarification might guide future development towards improved resolution.

8.4   Combined Genome and Transcriptome Analysis of Single Cells

Changes in genomic DNA may also translate in deregulated RNA and protein expression and may then alter cell function. In cancer, progression from local to metastatic disease is driven by rare disseminated tumor cells that are left behind in the patient after removal of the primary tumor (see Chapter 6). In this context, it is noteworthy that after isolating single viable cells of interest, SCOMP can be combined with single cell analysis of gene expression (SCAGE) on gene expression microarrays.[31,32] In this case, solid phase capturing of the mRNA and ethanol precipitation of the nuclear DNA and subsequent parallel isolation and amplification of the nucleic acids are performed allowing for gene expression and genetic alterations analyses of the same single cell. For a limited number of antigens, multicolor immunofluorescence staining can help further

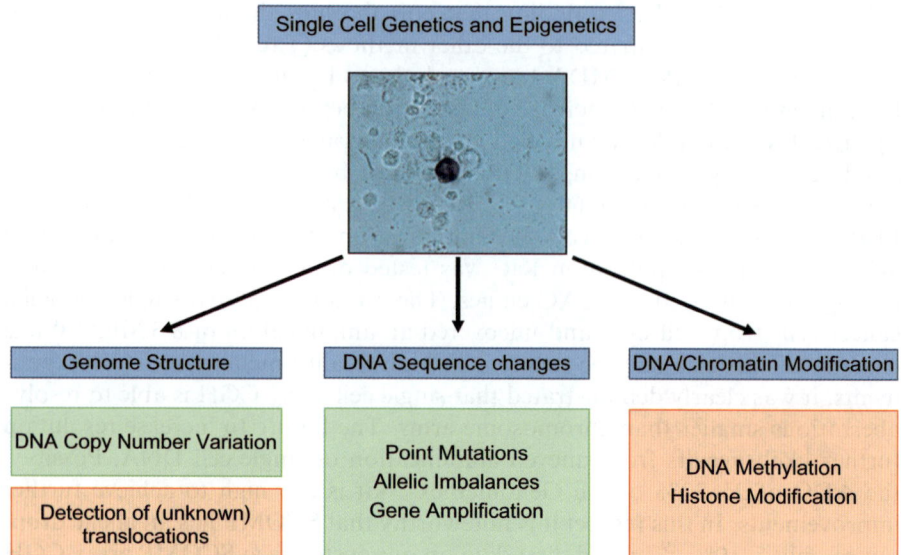

Figure 8.1   Single cell genetics and epigenetics. New technologies to analyze the genome of single cells now enable detailed molecular analyses. The analysis of specific DNA sequence variations as well as the global analysis of chromosomal gains and losses of single cells (both depicted in a green box) are well established. On the other hand, methods for identifying unknown chromosomal rearrangements as well as epigenetic DNA and chromatin modifications (both depicted in a red box) still need to be developed.

specify and characterize individual cells, whenever the marker of interest is expressed on the cell surface.

8.5   Perspective on Single Cell DNA Analysis

Although, as discussed in this chapter, single cell DNA analysis has already opened new opportunities for researchers, several molecular analyses still have to be established for single cells. Indeed, it is still not possible to screen for unknown balanced translocations in single cells. Additionally, the high-throughput methods that allow "deep sequencing" and therefore enable sequencing of whole genomes in a short time have to be adapted to current WGA protocols. Finally, epigenetics, such as the analysis of the methylation status of individual cells or of histone modifications, awaits inclusion into single cell research (Figure 8.1).

References

1. T. Araki, A. Yamamoto and M. Yamada, *Histochemistry*, 1987, **87**, 331–338.

2. I. Findlay, P. Ray, P. Quirke, A. Rutherford and R. Lilford, *Hum. Reprod.*, 1995, **10**, 1609–1618.

3. H. Telenius, N. P. Carter, C. E. Bebb, M. Nordenskjold, B. A. Ponder and A. Tunnacliffe, *Genomics*, 1992, **13**, 718–725.

4. E. Fragouli, D. Wells and A. Thornhill, *et al., Hum. Reprod.*, 2006, **21**, 2319–2328.

5. D. Wells and J. D. Delhanty, *Mol. Hum. Reprod.*, 2000, **6**, 1055–1062.

6. R. Kittler, M. Stoneking and M. Kayser, *Anal. Biochem.*, 2002, **300**, 237–244.

7. M. T. Barrett, B. J. Reid and G. Joslyn, *Nucleic Acids Res.*, 1995, **23**, 3488–3492.

8. M. C. Snabes, S. S. Chong, S. B. Subramanian, K. Kristjansson, D. DiSepio and M. R. Hughes, *Proc. Natl. Acad. Sci. U.S.A.*, 1994, **91**, 6181–6185.

9. L. Zhang, X. Cui, K. Schmitt, R. Hubert, W. Navidi and N. Arnheim, *Proc. Natl. Acad. Sci. U.S.A.*, 1992, **89**, 5847–5851.

10. W. Dietmaier, A. Hartmann and S. Wallinger, *et al., Am. J. Pathol.*, 1999, **154**, 83–95.

11. F. B. Dean, J. R. Nelson, T. L. Giesler and R. S. Lasken, *Genome Res.*, 2001, **11**, 1095–1099.

12. K. A. Eckert and T. A. Kunkel, *PCR Methods Appl.*, 1991, **1**, 17–24.

13. J. A. Esteban, M. Salas and and L. Blanco, *J. Biol. Chem.*, 1993, **268**, 2719–2726.

14. L. Blanco, A. Bernad, J. M. Lazaro, G. Martin, C. Garmendia and M. Salas, *J. Biol. Chem.*, 1989, **264**, 8935–8940.

15. P. Paul and J. Apgar, *Biotechniques*, 2005, **38**, 553–554.

16. A. H. Handyside, M. D. Robinson and R. J. Simpson, *et al., Mol. Hum. Reprod.*, 2004, **10**, 767–772.

17. A. Hellani, S. Coskun and M. Benkhalifa, *et al., Mol. Hum. Reprod.*, 2004, **10**, 847–852.

18. K. K. Murthy, V. S. Mahboubi and A. Santiago, *et al., Hum. Mutat.*, 2005, **26**, 145–152.

19. S. du Manoir, M. R. Speicher and S. Joos, *et al., Hum. Genet.*, 1993, **90**, 590–610.

20. A. Kallioniemi, O. P. Kallioniemi and D. Sudar, *et al., Science*, 1992, **258**, 818–821.

21. C. A. Klein, O. Schmidt-Kittler, J. A. Schardt, K. Pantel, M. R. Speicher and G. Riethmuller, *Proc. Natl. Acad. Sci. U. S. A.*, 1999, **96**, 4494–4499.

22. J. Phillips and J. H. Eberwine, *Methods*, 1996, **10**, 283–288.

23. C. L. Liu, S. L. Schreiber and B. E. Bernstein, *BMC Genomics*, 2003, **4**, 19.

24. S. Hughes, N. Arneson, S. Done and J. Squire, *Prog. Biophys. Mol. Biol.*, 2005, **88**, 173–189.

25. Y. S. Lee, C. N. Tsai and C. L. Tsai, *et al., Taiwan J. Obstet. Gynecol.*, 2008, **47**, 32–41.

26. D. G. Hu, G. Webb and N. Hussey, *Mol. Hum. Reprod.*, 2004, **10**, 283–289.

27. C. Le Caignec, C. Spits and K. Sermon, *et al., Nucleic Acids Res.*, 2006, **34**, e68.

28. H. Fiegler, J. B. Geigl and S. Langer, *et al., Nucleic Acids Res.*, 2007, **35**, e15.
29. J. A. Schardt, M. Meyer and C. H. Hartmann, *et al., Cancer Cell*, 2005, **8**, 227–239.
30. C. Fuhrmann, O. Schmidt-Kittler and N. H. Stoecklein, *et al., Nucleic Acids Res.*, 2008, **36**, e39.
31. C. H. Hartmann and C. A. Klein, *Nucleic Acids Res.*, 2006, **34**, e143.
32. C. A. Klein, T. J. Blankenstein and O. Schmidt-Kittler, *et al., Lancet*, 2002, **360**, 683–689.

CHAPTER 9
Gene Analysis of Single Cells

BRUNO CAULI AND BERTRAND LAMBOLEZ

Université Pierre et Marie Curie Paris 6, CNRS UMR7102, Neurobiologie des Processus Adaptatif, 9 quai saint Bernard, 75005 Paris, France

Abstract

Shortly after the initial demonstration of the feasibility of gene analysis at the single cell level,[1] the two major technical trends enabling gene expression analyses of single cells developed in the early 90s. These developments were aimed either at whole transcriptome analysis based on RNA amplification,[2,3] or at limited gene expression profiling using RT-PCR for correlating molecular and functional properties.[4] The brain complexity and cellular diversity has been a strong incentive for the development of these tools at a time when many of the major constituents of neurotransmission had been cloned. Both techniques initially relied on the use of the patch-clamp technique[5] to harvest selectively the cell's mRNAs. In this chapter we will detail the key steps, which assessed the reliability and functional relevance of the "single cell RT-PCR after patch-clamp" technique (scPCR, Lambolez *et al.*, 1922), and describe its evolutions. We will also share our observations on the design and interpretation of scPCR experiments and discuss the limits of this approach.

9.1 Single Cell RT-PCR After Patch Clamp

Figure 9.1 outlines the general protocol for scPCR. After recording of a cell in the whole-cell configuration of the patch-clamp technique, the cell's content is collected into the electrode and expelled into a test tube. Reagents are then added to perform first strand cDNA synthesis from the cell's mRNAs.

RSC Nanoscience & Nanotechnology No. 15
Unravelling Single Cell Genomics: Micro and Nanotools
Edited by Nathalie Bontoux, Luce Dauphinot and Marie-Claude Potier
© Royal Society of Chemistry 2010
Published by the Royal Society of Chemistry, www.rsc.org

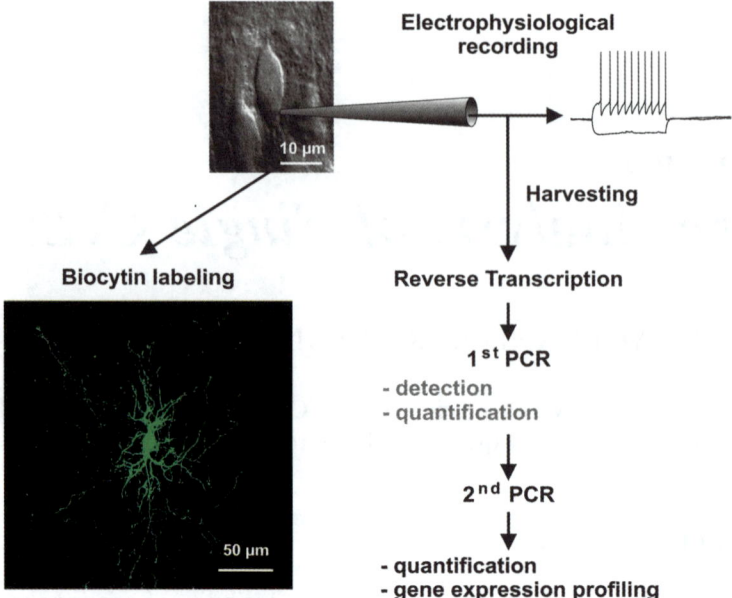

Figure 9.1 General principle of scPCR. A neuron is first recorded in whole-cell
configuration. At the end of the recording the cell's content is harvested
into the recording pipette, expelled into a test tube and submitted to a
reverse transcription step. Next, cDNAs are amplified during a first PCR
round. The first PCR product can then be detected, and for some applications used for quantitative analyses of expression levels. A refined
analysis is generally performed following re-amplification through a second PCR. During whole-cell recording, the neuronal tracer biocytin
contained in the intracellular solution diffuses into the neuritic processes
of the cell, hence allowing morphological analyses after slice fixation and
histochemistry.

Following overnight incubation of the RT reaction, a PCR is set in the same
tube to amplify one or multiple cDNAs of interest. Hence, the procedure is fast
and simple because one tube corresponds to one cell and few manipulations are
required. The first PCR product can then be detected and, for most applications, re-amplified through a second PCR aimed at more refined analyses of
molecular complexity.[1] In multiplex scPCR conditions,[2,3] the presence of
multiple primer pairs in the first PCR decreases the amplification efficiency,
making second, gene-specific, PCRs necessary to achieve consistent detection
by agarose gel electrophoresis.

This basic scPCR scheme allowed multiple developments and applications,
some of which are described in this chapter. Besides flexibility and robustness, a
major advantage of scPCR resides in its speed and ease of application, which
allows rapid feedback between functional and molecular analyses. These
properties proved essential to establishing correlations between molecular and

functional properties at the single cell level, beyond the large cellular diversity of brain tissues.

9.2 Correlating mRNA Expression and Functional Properties of Single Cells

The scPCR technique was initially designed to establish correlates between the functional properties of native glutamate receptor channels of the α-amino-3-hydroxy-5-methyl-4-isoxazolepropionic acid (AMPA) subtype and their mRNA expression.[1] Subsequent studies of native AMPA receptors illustrate the power of this approach, but also the limits of correlations established by comparing different cell types, whose multiple differences extend beyond the transcriptome level. AMPA receptors are multimeric assemblies of four different subunits GluR1–4 and mediate fast excitatory synaptic transmission in the central nervous system (CNS). Further diversity is generated by alternative splicing and mRNA editing.

The initial scPCR report established the reliability and selectivity of this technique by showing that single cerebellar Purkinje neurons expressed a relatively constant set of AMPA receptor subunit mRNAs, which differed from that found in cerebellar granule cells. AMPA receptor mRNAs were not found in glial cells unresponsive to glutamatergic agonists. Conversely, expression of glial fibrillary acidic protein was observed in glial cells but not in Purkinje cells.[1]

The functional relevance of scPCR analyses was next assessed by demonstrating that the calcium permeability of AMPA receptor is determined by transcriptional control of the expression level of the GluR2 subunit. This was established by showing that the abundance of the GluR2 mRNA relative to that of other subunits correlates with the channel properties of native AMPA receptors in different cell types.[4,5] Besides its scientific importance, the study by Jonas *et al.*[5] was the first to report scPCR from acute brain slices, which preserve the cyto-architecture and connectivity of brain tissues.

In contrast with channel properties, similar experiments investigating the molecular determinants of AMPA receptor kinetics gave rise to divergent interpretations. Upon examination of a limited number of cell types, slow and fast kinetics were correlated with either the abundance of GluR2 and GluR4 mRNAs,[6] respectively, or to that flip and flop splice variants.[7] However, a subsequent study exploring a larger neuronal diversity showed that these correlations were not linked to the molecular composition of native AMPA receptors, but to other, unidentified properties of the cell types examined.[8] The more recently described interactions of AMPA receptors with transmembrane AMPA receptor regulatory proteins (TARPs) and Cornichon regulatory proteins,[9,10] which influence AMPA receptor kinetics, may provide the key to this unresolved issue.

9.3   Quantitative Analyses by scPCR

Studies of AMPA receptors relied on relative quantification between AMPA receptor subunits mRNAs and their splice and editing variants expressed by single neurons. This was made possible by the fact that a single primer pair co-amplified the four subunits cDNAs, and that their original mRNA ratios were maintained throughout the whole RT-PCR process.[1,11] This relative quantification was performed by cloning PCR products[5] or using subunit-specific restriction enzymes following amplification with radiolabeled primers,[7] or fluorescently labeled primers.[8]

The issue of comparing mRNA expression levels between different cells has been elegantly addressed by Liss *et al.*[12] using real-time quantitative PCR following patch-clamp harvesting of the cell's content and RT reaction as described above. Using this technique to investigate the variability of dopaminergic neurons pacemaker frequency, the authors demonstrated a linear correlation between Kv4α and Kv4β potassium channel mRNA abundance, the intensity of the corresponding A-type potassium current, and the pacemaker frequencies of single neurons. It is noteworthy that the protocol by Liss *et al.*[12] required precipitation of the minute amount of cDNA obtained from single cells to achieve real-time quantitative PCR, presumably to eliminate the inhibitory effect of RT enzymes on Taq polymerase.[13]

Finally, absolute quantification of AMPA receptor subunits mRNAs expressed by single neurons was performed using radiolabeled primers and an internal standard consisting in a full-length GluR2 RNA bearing a point mutation.[11] This allowed determination of the number of AMPA receptor mRNA copies harvested by patch-clamp from two different types of neurons in culture, namely pyramidal cells and interneurons. Comparison with mRNA copy numbers determined from whole culture RNA purification after cell counting further allowed estimating the proportion of AMPA receptor mRNA harvested by patch-clamp from single neurons, hence their respective distribution in somatic versus neuritic compartments. It should be stressed that absolute quantification of mRNA using scPCR requires careful design, quantification, and validation of the internal standard RNA in cell-free assays, as well as adjustment of internal standard to amounts comparable to those of the mRNA of interest in single cells.[11]

9.4   Molecular and Functional Phenotyping of Neuronal Types

Multiplex scPCR was introduced to correlate functional properties and subunit composition of native glutamate receptors of the kainate subtype,[2] which required several primer pairs because of the low sequence homology between constituent subunits. Its field of application was extended to the detection of multiple unrelated genes in single cells, aimed at establishing detailed electrophysiological, gene expression, and morphological phenotypes of single neurons.[3] In the multiplex protocol (Figure 9.2), a limited set of cDNA species is amplified by two PCR rounds using a specific primer pair for each gene. In the first PCR, cDNAs of interest are simultaneously amplified by mixing all different primer pairs. The purpose of the first PCR round is to increase the numbers of cDNA copies to ensure a reliable aliquoting for subsequent reamplification. Indeed, primer dimers and non-specific primer–amplicon interactions substantially decrease multiplex PCR amplification efficiency. The second PCR rounds, in which each gene is individually amplified from the first PCR product, achieve high efficiency allowing amplicon detection by agarose gel analysis. The use of internal (nested) primers during the second PCR rounds,[14] which enhances amplification specificity and minimizes the carry-over of primer dimers from the first PCR, and improves amplification efficiency and specificity. We routinely use this protocol to probe expression of 30 or more different genes from single neurons (*e.g.* Cauli *et al.*[15])

Multiplex scPCR is an ideal tool to establish a detailed molecular, physiological, and morphological phenotype of single cells.[3] This approach has

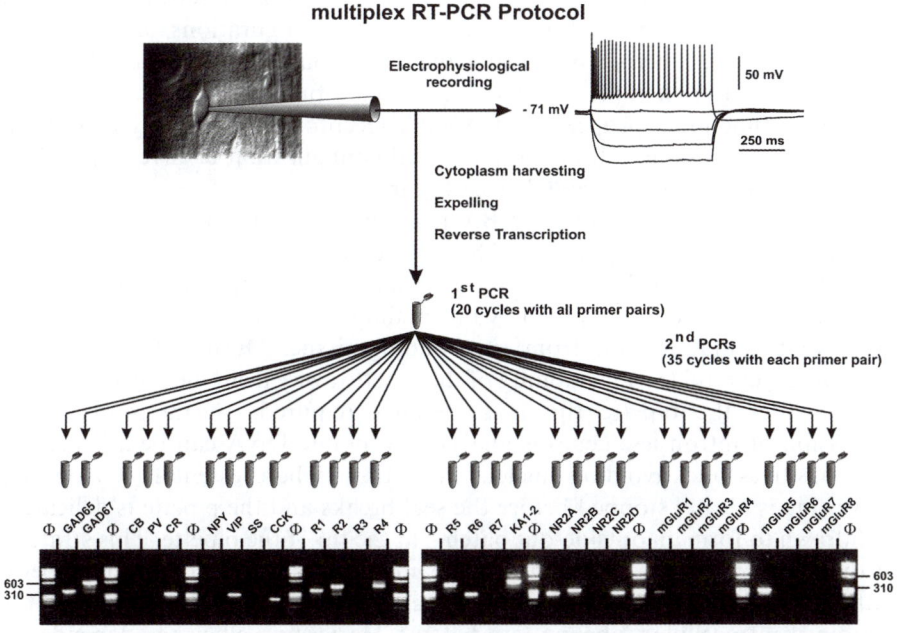

Figure 9.2   Gene expression profiling by multiplex scPCR. Procedures for cytoplasm harvesting, expelling and reverse transcription are identical to those described in Figure 9.1. In the first amplification step, all cDNAs of interest are co-amplified by mixing their specific PCR primers together. PCR products are then individually re-amplified using their specific PCR primers and analyzed by agarose gel electrophoresis. Note that expression of 16 genes was detected in this cortical GABAergic neuron among the 30 genes probed, helping to determine cell-type specific expression profiles.

significantly contributed to establishing the distinctive molecular, functional, and morphological phenotypes of the diverse neuronal types in the cerebral cortex.[3,16–19] This approach also disclosed the existence of tight correlation between molecular, functional, and morphological features. For instance, the expression profile of voltage-dependent ion channels was shown to be a good predictor of the electrophysiological behavior of neurons.[20] Finally, these multivariate phenotypes have been used to classify cortical neuronal types based on unsupervised multiparametric clustering analyses,[15,19] hence establishing their statistical significance on a large array of features.

9.5   Patch-clamp Harvesting of Single Cells

A key issue in single cell molecular biology, especially in tissue slices, is to harvest selectively the cellular content without contamination from surrounding cells. The whole-cell configuration of the patch-clamp technique[21] confers a tightly insulated physical and electrical access to a cell's cytoplasm. It thus provides an ideal means of harvesting in complex tissue following electrophysiological characterization. It is noteworthy that electrophysiological characterization can be performed using other configurations of the patch-clamp technique,[21] before harvesting the cell's content under the whole-cell configuration. For example, Angulo *et al.*[8] successfully used excised patches to investigate functional properties of AMPA receptors, followed by cytoplasm harvesting of the same cell under whole-cell configuration to investigate their molecular composition by scPCR (see Figure 9.3).

Patch-clamp harvesting of a GABAergic neuron from a slice of rat cerebral cortex is illustrated in Figure 9.4. The cell's content is harvested under visual control to ensure an effective collection of materials by applying a gentle suction through the recording pipette while maintaining the giga-ohmic tight seal to prevent contamination from surrounding tissue. During the harvesting procedure a careful (re-)positioning of the pipette away from the nucleus, which can obstruct the pipette tip, is sometimes required. Furthermore, when expression of intron-less genes is analyzed, genomic DNA can be a source of false positives and avoiding nucleus collection is here essential. Cytoplasm harvesting is ideally stopped before the seal breaks and the pipette is delicately withdrawn to form an outside-out patch[21] at the tip of the pipette. This strategy further guarantees that no surrounding material is collected during pipette withdrawal from the slice, and favors closure of the cell membrane required for combined intracellular labeling (see below). Harvesting efficiency depends on the internal diameter of the patch pipette tip, which must be as large as possible while matching cell size. We found that pipettes with a 1–2 µm open tip diameter are suitable for most neuronal types. Virtually any type of patch-clamp intracellular solution can be used, as long as it does not interfere with RT-PCR efficiency in cell-free assays. In our experience, intracellular solutions containing K^+ or Cs^+ cations, and Cl^-, gluconate or methylsulfate anions, are equally suitable to scPCR. Furthermore, an intracellular label such as biocytin

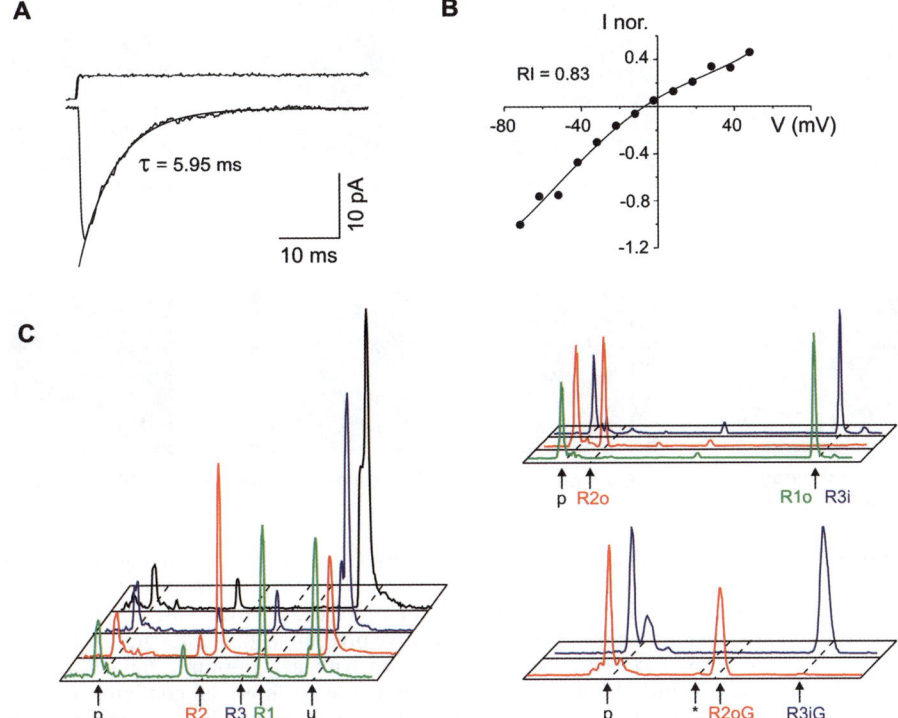

Figure 9.3 Functional and molecular analysis of AMPA receptors on a single cortical neuron. Functional properties were investigated on an excised patch. Subsequently, a second whole-cell recording was obtained on the same cell, and the cellular content was harvested to investigate the GluR1-4 combination expressed. (A) Averaged response (bottom trace) to 100 ms step applications (top trace) of 10 mM glutamate. The current decay was fit by a single exponential function. (B) I–V curve of glutamate-induced peak currents at different potentials. (C) Quantification of GluR1–4, flip/flop splicing, and R/G editing proportions. GluR1–4 cDNA fragments were amplified with a fluorescent primer, cut with restriction enzymes, and submitted to capillary electrophoresis. Time of migration is represented horizontally (fragment size increasing from left to right), and fluorescence intensity is represented vertically. (p) unincorporated primers. Left, GluR1–4 proportions. GluR1–4 cDNA fragments were co-amplified and cut with subunit-specific enzymes. Positions of uncut (u; GluR1–4 cDNAs resistant to the specific enzyme) and cut cDNA fragments are indicated. No GluR4 was found in this cell. The small peak present on all profiles at the same position (left of R2 peak) is a PCR artifact uncut by the enzymes. Top right panel, Flip/flop proportions of GluR1–4 subunits expressed. Subunit-specific amplifications were followed by restriction with splice variant-specific enzymes. GluR1 and GluR3 were uncut (flop and flip, respectively), whereas GluR2 was cut (flop). Bottom right panel, R/G edition proportions. Products of subunit-specific amplification were cut with an editing variant-specific enzyme. Positions of edited flip (R3iG), edited flop (R2oG), and unedited (asterisk; not present but found in other analyzed cells) peaks are indicated. In this cell GluR3 flip and GluR2 flop were edited. Reproduced, with permission, from Angulo *et al.*[8] © 1997 Society for Neuroscience.

Identification   **Approach**   **Cell attached**   **Whole cell**

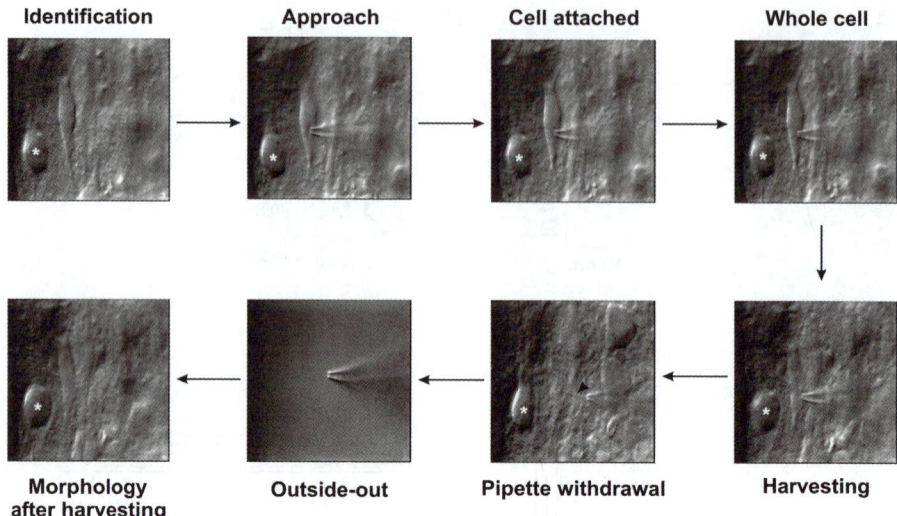

Morphology   **Outside-out**   **Pipette withdrawal**   **Harvesting**
after harvesting

Figure 9.4   Patch-clamp harvesting in brain slices. Once a cell is visually identified, the recording pipette is approached with a positive pressure in order to avoid cellular debris contamination at the tip of the pipette. Note the dimple on the membrane of this neuron. Positive pressure is then interrupted in order to form a cell-attached configuration with a giga-ohmic tight seal, and subsequently brief suctions allow going into whole-cell configuration. At the end of the recording the cytoplasm is harvested by applying a gentle negative pressure into the pipette while the tight seal is maintained. Note the shrinkage of the cell body during the harvesting procedure. The recording pipette is then gently withdrawn to form an outside-out patch, which favors cell membrane closure for subsequent biocytin revelation, and preservation of harvested material in the patch pipette.

can be included in the intracellular solution for combined electrophysiological, molecular, and detailed morphological investigations of single cells[3] and of their anatomical relationships with other cells or cellular processes.[22,23] In our hands, exploitable labeling is achieved in only 50% of molecularly and physiologically characterized cells,[19] which is not surprising considering that the extensive cytoplasm harvesting required for scPCR often affects morphological preservation.

Laser microdissection is an alternative to patch-clamp for harvesting single cells in complex tissues. Laser microdissection offers the advantage of rapidly collecting large samples of single cells. However, this approach does not allow investigating the physiological and morphological properties of the cells, and thus critically relies on prior selective labeling of cells to be collected. Provided that labeling specificity is adequate, laser microdissection further allows overcoming the variability due to detection limits of scPCR by pooling cells of the same type (*e.g.* Liss *et al.*[24]).

9.6   Sensitivity Limits

There is currently no evidence of mRNAs detected by scPCR in post-natal neural cells in the absence of the corresponding protein. Conversely, many observations clearly demonstrate a low sensitivity of scPCR as compared to electro-physiological recording. Our studies of neuropeptide receptors provide striking examples of the sensitivity limits of scPCR. While virtually all cortical pyramidal neurons showed electrophysiological responses to the three neuropeptides neurokinin B, cholecystokinin, and corticotropin-releasing factor, expression of their cognate receptors mRNAs was only detected in 25–30% of these neurons by scPCR.[25] Similarly, ì-opioid receptor expression was only detected in 60% of ì-opioid agonist responsive cortical interneurons.[26] This presumably relates to the general scarcity of G protein-coupled receptor mRNA expression, since expression of the serotonin 5-HT3 receptor-channel mRNA in the same inter-neuron type was reliably detected in cells responsive to a 5-HT3 agonist.[23]

The low scPCR efficiency is due in part to the low efficiency of reverse transcriptases, especially at low RNA copy number.[27] Indeed, Tsuzuki *et al.*[11] showed by serial dilution of AMPA receptor RNA that RT-PCR detection becomes unreliable at 10 copies of RNA molecule, while PCR reliably detects single DNA molecules. Another limiting factor is the proportion of cellular mRNA content harvested, essentially from the soma. Indeed, while the detection of the medium–high abundance AMPA receptor mRNA[15] by scPCR was close to 100% in cultured neurons (*e.g.* Lambolez *et al.*[1]), it only reached 80–90% in acute brain slices depending on the neuronal type (*e.g.* Lambolez *et al.*[7] and Angulo *et al.*[8]), presumably due to a lesser harvesting of the cells soma in slices to avoid contamination (see above). Yet another factor is the duration of whole-cell recordings. Indeed, although efficient detection of medium–high abundance AMPA receptor mRNAs is achieved after 30 min of whole-cell recording,[11] we observed that the detection rate of low abundance mRNAs decreases after 10 min (*e.g.* G protein-coupled receptor mRNAs[25,26]). Finally, scPCR sensitivity also depends on the size of the cell, which appears to determine in part the amount of mRNA it expresses. For instance, the number of AMPA receptor mRNA molecules per hippocampal cell in culture was ∼2400 in large pyramidal neurons, ∼1800 in medium-size interneurons, and ∼600 in small interneurons.[11] The cell size dependence of scPCR is illustrated by our studies of glutamate receptors of the AMPA and N-methyl-D-aspartic acid (NMDA) subtypes in cerebellar granule cells in culture,[1,28] which are among the smallest neurons in the brain (soma diameter, 6–9 μm). Although all cells expressed functional glutamate receptors, scPCR detection of their cognate mRNAs was only achieved in ∼70% of the cells.

9.7   Controls

Contamination of scPCR experiments can stem from several sources. Contamination from laboratory plasmids, RNA preparations, or PCR products is easily identified and easily avoided by taking standard precautions. Contamination from the biological sample is more difficult to tackle. Although we avoid harvesting the cell's nucleus in brain slices (see above), genomic DNA contamination can still occur at a low frequency,[29] potentially confounding the detection of intron-less mRNA expression. We addressed this issue by including a control amplifying an intronic DNA sequence in the scPCR reaction, and found that this control efficiently detected genomic DNA contamination.[29] Harvesting mRNA from cells surrounding the target cell or released from dying cells is yet another possible source of contamination. For each new primer pair or combination of primers, we test mRNA contamination from surrounding tissue by placing a patch-clamp pipette into the slice without establishing a seal. Following removal of the pipette its content is then processed by RT-PCR. In our hands, this control has always yielded negative results. It must be noted, however, that a better negative control consists in scPCR of cells devoid of the mRNA of interest, from the same biological preparation. Unfortunately, the absence of an mRNA species in a given cell is generally difficult to presume in heterogeneous tissues, unless the corresponding protein can be probed *in situ* with sufficient sensitivity (*e.g.* electrophysiologically). For this reason, we consider the comparison of different cell types yielding different scPCR expression profiles as a valuable control of general applicability for scPCR experiments.

9.8   Interpretation of scPCR Results

ScPCR relies on comparison between single cells to correlate mRNA expression with other phenotypic properties of the cells. It can help establish a causal link in simple cases where a given cellular property is determined by transcriptional control of single or few molecular species. This is exemplified by studies of AMPA receptors, which also illustrate the limits of such correlations in predicting causality in the complex environment of a cell (see above). As in any scientific study, it is therefore advisable to examine the validity of the correlation across a large cellular diversity before inferring any causal relationship.

Not surprisingly considering its generally higher sensitivity, analyses of different cell types by scPCR provide less contrasted molecular expression patterns than with most other molecular detection techniques. Furthermore, scPCR often reveals cell-to-cell molecular variability within cell types, which may sometimes appear as a surprise although such variability is well established at the electrophysiological and morphological levels. These issues are addressed by the use of internal controls ranging from other, cell-type specific mRNAs to functional and anatomical phenotyping, which greatly help cell-type identification (see above), and thus interpretation of the data. We consistently found that the combination with electrophysiology, which is able to provide sensitive and quantitative analyses of multiple functional properties, is of great value in the interpretation of molecular scPCR results. As a consequence of this combined phenotypic approach of cellular diversity, our recent studies often rely on multifactorial, unsupervised statistical analyses to interpret the scPCR data.[15]

Finally, the above considerations and the demonstrated occurrence of false negatives in the current state-of-the-art of scPCR detection (see above) raise an important question: how many genes can be examined to infer meaningful correlates without increasing enormously the size of the sample of single cells? This question, which presumably applies to other techniques of single cell RNA expression analysis and obviously depends on the correlation between genes expressions, *i.e.* on their cell type specificity, must be carefully considered in the design, but also in the interpretation of the experiments.

Conclusion

Throughout the years following its first report,[4] scPCR has established itself as a biologically meaningful technique, which, in spite or perhaps because of its simplicity, has greatly contributed to the development of single cell molecular biology. The efficiency of scPCR relies primarily on the ability of gene-specific PCR to detect single molecules reliably. Ongoing and future developments will certainly enhance the efficiency of the harvesting and RT steps, and of large-scale nucleic acid amplification and detection to enable whole transcriptome analyses of single cells. It is likely that considerable efforts will be required to establish the biological relevance of these complex data, which may in turn provide completely new insights into the basic rules governing cellular diversity.

Acknowledgement

BC is supported by a Human Frontier Science Program grant (HFSP, RGY0070/2007) and by a CNRS grant ("Nitrex").

References

1. B. Lambolez, E. Audinat, P. Bochet, F. Crepel and J. Rossier, *Neuron*, 1992, **9**, 247.
2. D. Ruano, B. Lambolez, J. Rossier, A. V. Paternain and J. Lerma, *Neuron*, 1995, **14**, 1009.
3. B. Cauli, E. Audinat, B. Lambolez, M. C. Angulo, N. Ropert, K. Tsuzuki, S. Hestrin and J. Rossier, *J. Neurosci.*, 1997, **17**, 3894.
4. P. Bochet, E. Audinat, B. Lambolez, F. Crepel, J. Rossier, M. Iino, K. Tsuzuki and S. Ozawa, *Neuron*, 1994, **12**, 383.
5. P. Jonas, C. Racca, B. Sakmann, P. H. Seeburg and H. Monyer, *Neuron*, 1994, **12**, 1281.
6. J. R. Geiger, T. Melcher, D. S. Koh, B. Sakmann, P. H. Seeburg, P. Jonas and H. Monyer, *Neuron*, 1995, **15**, 193.

7. B. Lambolez, N. Ropert, D. Perrais, J. Rossier and S. Hestrin, *Proc. Natl. Acad. Sci. U. S. A.*, 1996, **93**, 1797.
8. M. C. Angulo, B. Lambolez, E. Audinat, S. Hestrin and J. Rossier, *J. Neurosci.*, 1997, **17**, 6685.
9. R. A. Nicoll, S. Tomita and D. S. Bredt, *Science*, 2006, **311**, 1253.
10. J. Schwenk, N. Harmel, G. Zolles, W. Bildl, A. Kulik, B. Heimrich, O. Chisaka, P. Jonas, U. Schulte, B. Fakler and N. Klocker, *Science*, 2009, **323**, 1313.
11. K. Tsuzuki, B. Lambolez, J. Rossier and S. Ozawa, *J. Neurochem.*, 2001, **77**, 1650.
12. B. Liss, O. Franz, S. Sewing, R. Bruns, H. Neuhoff and J. Roeper, *EMBO J.*, 2001, **20**, 5715.
13. L. N. Sellner, R. J. Coelen and J. S. Mackenzie, *Nucleic Acids Res.*, 1992, **20**, 1487.
14. B. Liss, R. Bruns and J. Roeper, *EMBO J.*, 1999, **18**, 833.
15. B. Cauli, J. T. Porter, K. Tsuzuki, B. Lambolez, J. Rossier, B. Quenet and E. Audinat, *Proc. Natl. Acad. Sci. U. S. A.*, 2000, **97**, 6144.
16. Y. Wang, A. Gupta, M. Toledo-Rodriguez, C. Z. Wu and H. Markram, *Cereb. Cortex*, 2002, **12**, 395.
17. Y. Wang, M. Toledo-Rodriguez, A. Gupta, C. Wu, G. Silberberg, J. Luo and H. Markram, *J. Physiol*, 2004, **561**, 65.
18. C. J. Price, B. Cauli, E. R. Kovacs, A. Kulik, B. Lambolez, R. Shigemoto and M. Capogna, *J. Neurosci.*, 2005, **25**, 6775.
19. A. Karagiannis, T. Gallopin, C. David, D. Battaglia, H. Geoffroy, J. Rossier, E. M. Hillman, J. F. Staiger and B. Cauli, *J. Neurosci.*, 2009, **29**, 3642.
20. M. Toledo-Rodriguez, B. Blumenfeld, C. Wu, J. Luo, B. Attali, P. Goodman and H. Markram, *Cereb. Cortex*, 2004, **14**, 1310.
21. O. P. Hamill, A. Marty, E. Neher, B. Sakmann and F. J. Sigworth, *Pflugers Arch.*, 1981, **391**, 85.
22. B. Cauli, X. K. Tong, A. Rancillac, N. Serluca, B. Lambolez, J. Rossier and E. Hamel, *J. Neurosci.*, 2004, **24**, 8940.
23. I. Férézou, B. Cauli, E. L. Hill, J. Rossier, E. Hamel and B. Lambolez, *J. Neurosci.*, 2002, **22**, 7389.
24. B. Liss, O. Haeckel, J. Wildmann, T. Miki, S. Seino and J. Roeper, *Nat. Neurosci.*, 2005, **8**, 1742.
25. T. Gallopin, H. Geoffroy, J. Rossier and B. Lambolez, *Cereb. Cortex*, 2006, **16**, 1440.
26. I. Férézou, E. L. Hill, B. Cauli, N. Gibelin, T. Kaneko, J. Rossier and B. Lambolez, *Cereb. Cortex*, 2007, **17**, 1948.
27. G. F. Gerard and J. M. D'Alessio, *Methods Mol. Biol.*, 1993, **16**, 73.
28. E. Audinat, B. Lambolez, J. Rossier and F. Crepel, *Eur. J. Neurosci.*, 1994, **6**, 1792.
29. E. L. Hill, T. Gallopin, I. Férézou, B. Cauli, J. Rossier, P. Schweitzer and B. Lambolez, *J. Neurophysiol.*, 2007, **97**, 2580.

CHAPTER 10
Proteomics

ANNE-MARIE HESSE AND JOËLLE VINH

Biological Mass Spectrometry and Proteomics, USR 3149 CNRS/ESPCI ParisTech, Ecole Superieure de Physique et de Chimie Industrielles, 10 rue Vauquelin, 75231 Paris cedex 05, France

Abstract

Understanding the functioning of a living cell supposes to decipher the complex molecular mechanisms which underlie the various cellular activities. All the genes of an organism, or its genome, constitute a static and specific databank of this living being. From a unique genome, every cellular type of an organism is going to express a set of proteins, or proteome, which is going to vary according to the environment of the cells. As proteins constitute the final actors of the biological processes, their study may offer the most relevant vision of the functioning of a living cell.

In this part the specificity of proteomic approaches are discussed. Analytical strategies involving biological mass spectrometry are presented in association with liquid chromatography. Multidimensional analyses are introduced for complex mixture analysis and quantification. In conclusion specific technological developments towards the miniaturization, the integration and the automation of the analysis of samples available in very low amounts are introduced as the next step towards single cell proteomics.

10.1   Motivation to Study Proteins at the Single Cell Level

Understanding the functioning of a living cell requires deciphering the complex molecular mechanisms which underlie the various cellular activities. All the

RSC Nanoscience & Nanotechnology No. 15
Unravelling Single Cell Genomics: Micro and Nanotools
Edited by Nathalie Bontoux, Luce Dauphinot and Marie-Claude Potier
© Royal Society of Chemistry 2010
Published by the Royal Society of Chemistry, www.rsc.org

genes of an organism, or its genome, constitute a static and specific databank of this living being. From a unique genome, every cellular type of an organism will express a set of proteins, or proteome, which will vary according to the environment of the cells. As proteins constitute the final actors of the biological processes, their study may offer the most relevant vision of the functioning of a living cell.

The concept of proteomic analysis was introduced in 1998 by Anderson and Anderson, who defined it as the quantitative analysis of the levels of expression of proteins, to decipher the mechanisms of control of genomic expression.[1] The proteomics studies evolved from the general overview of the protein contents to the specific detection of a biomarker. They could be: (1) exhaustive, to detect as many proteins as possible in a complex sample; (2) differential, to compare protein extracts from different cellular origin and bring to light for example markers of pathologies; or (3) targeted, to highlight proteins involved in a given cellular function.

What are the challenges in such a "dream"?

10.1.1 Proteins, mRNAs and DNA

The sequencing of whole genomes of prokaryotes and eukaryotes and the completion of the Human Genome Project in 2003 aroused a renewed interest for proteomics. The study of all the proteins produced by living cells involves the identification of proteins of the organism and the determination of their roles in physiological and pathological functions. The term proteome includes all the proteins expressed by a given genome in a given environment, at a given time. Whereas a genome remains globally unchanged, every cell expresses its particular proteins set in a different way according to the stimulation of its environment.[1] As far as proteins are directly involved at the molecular scale in most of normal and pathological biochemical processes, a more complete understanding of the diseases can be obtained by a direct study of proteins from a diseased cell or tissue. By definition, the proteome is dynamic, leading some researchers to use the term "functional proteomics" to describe the study of the expression level modulation of the proteins expressed by a specific cell in a given time. While the human genome contains approximately 22 000 genes coding potentially for 40 000 different proteins, the alternative splicing of RNA and post-translational modifications increase the number of proteins or fragments of proteins up to 2 000 000.[2] As a consequence, the proteome is much more complex than the genome. The proteome is a dynamic entity which depends on the studied cellular type and on its environment; the genome is a static entity which depends on the organism studied only.

10.1.1.1 No Amplification Method

Genomics brings a large quantity of information linking the activity of a gene to a disease, but it does not predict the maturation processes and post-translational modifications carried by numerous proteins. The sequence of a gene and its profile of activity are not directly correlated to the complete and precise image of the abundance of a protein, neither to its final molecular structure nor to its state of activity. The transcript of a gene can be spliced in various ways before translation in proteins.

Working with extremely small amounts of sample has been made possible in genomics thanks to the polymerase chain reaction technique,[3] which amplifies the signal resulting from a DNA. This amplification strategy has no analogue for protein study. Moreover, the microarray methods are still under development for proteomics and lack robustness, reproducibility, and sensitivity.

10.1.1.2   Protein Dynamics and Post-translational Modifications

After ribosomal synthesis, proteins are cleaved to eliminate the initiation sequences, then transfer sequences and signal sequences undergo post-translational modifications (PTMs), like sugar addition or phosphorylations. Such modifications play a vital role in the modulation of the proteins function but are not directly encoded at the genes level. PTMs are numerous (more than 300 were counted). They can be static or dynamic. As a consequence, a unique gene can encode up to about 50 protein species. Also, the quantification of RNA does not still reflect the corresponding protein expression levels.[4] Beyond the genomic aspect, it is clear that the information brought by the genome cannot give a precise profile of the proteins abundance, structure or activity.

The genomics and the proteomics are complementary domains. Thus, genomics and proteomics could significantly influence the future of medicine and biotechnology, but there is still much work to do in order to understand their specific inputs and to switch to clinical applications.

10.1.2   Sample Preparation

Constitutively, the samples of interest could be very heterogeneous from the protein's point of view at the microscopic level. A first study pioneered detection of protein by mass spectrometry on single cell with the detection of human α and β hemoglobin from a single erythrocyte. The study used a combination of very sensitive techniques (capillary zone electrophoresis and Fourier transform ion cyclotron resonance mass spectrometry) for the detection of a major protein (estimated at 450 amol/cell).[5] Whereas manual microdissection has been successfully reported for single cell analysis by different groups,[6,7] a huge effort has been made for sample preparation even before protein extraction. Working from whole tissues, laser-captured tissue microdissection, combined with optimized coloration methods and histopathology approaches, has allowed miniaturization of sample handling. Collection of a single cell from a preparation of dissociated cells was made possible using optical tweezers[8] as applied to fungi and yeast,[9,10] immunomagnetic cell separation[11] or dielectrophoresis-activated cell sorting (DACS).[12,13] Cell separation approaches have been reviewed by Dainiak *et al.*[14]

The extraction of protein itself is a challenge when working with very small samples. The chemical or mechanical extraction of proteins for "micro-proteomics", their solubilization and the modes of separation (electrophoresis or chromatography) have been carefully reviewed by Gutstein *et al.*[15] The major idea that should be recalled is that even if protein diversity decreases in small samples, the quantity of each species also decreases and only the major species will be above the limits of detection. Moreover, if the amount of starting material decreases, non-specific loss is not correlated to this amount and could be a major limitation.

10.1.3   Sub-proteome Analysis

Sub-proteome analysis has been developed to reduce sample complexity.[16] This fractionation approach was widely applied either to study a specific organ (cardiac[17] or placental proteome,[18] for example) or a specific group of protein (membrane proteins,[19] kinases[20] or phosphoproteins,[21–26] for example). The study of subgroups of proteins gives a more detailed description of the proteins contents. Some species, which would have remained below the limit of detection in a complex proteome, are enriched to be detected. Some species expressed at very low relative concentration, which would have been hidden by major signals coming from constitutive and major proteins, are not suppressed anymore in the mass spectra. However, due to the lack of sensitivity of current techniques, this approach does not seem to be suited for single cell analysis. The ionization process itself can be regarded as a fractionation step, because of the suppression effect. Most single cell applications have focused on subsets of specific molecules, whose ionization efficiency has been optimized with this aim. A first publication has pioneered this strategy with the work from van der Greef and co-workers on giant neurons from *Lymnea stagnalis*[27] and Sweedler's group has focused on the preparation, detection,[28] and quantification of signaling neuropeptides in single neurons from *Aplysia californica* or tissue sample.[29] Protocols have been published extending the application to smaller mammalian cells.[30] A peptidomic approach was also applied to *Drosophila* single neurons for neuropeptide mapping.[31] Single cell detection is required for the study of cell expression specificity. The cellular expression of the two pre-cursors of the hyperglycemic hormone (cHH) was analyzed in neurosecretory cells (30 μm diameter) from the crayfish *Orconectes limosus.*[7] In this early work the combination of dual characterization was successfully scaled down to the level of a single crayfish neurosecretory cell. MALDI-TOF MS (see section 10.2.1 for details on mass spectrometry), which is considered to be a highly powerful tool for peptide profiling of complex samples such as organs or cells,[32] was associated with immunological tools such as enzyme-linked immunosor-bent assay (ELISA) and immunocytochemistry to probe for specific peptides in biological extracts or cells. Direct peptide profiling by MALDI-TOF MS on

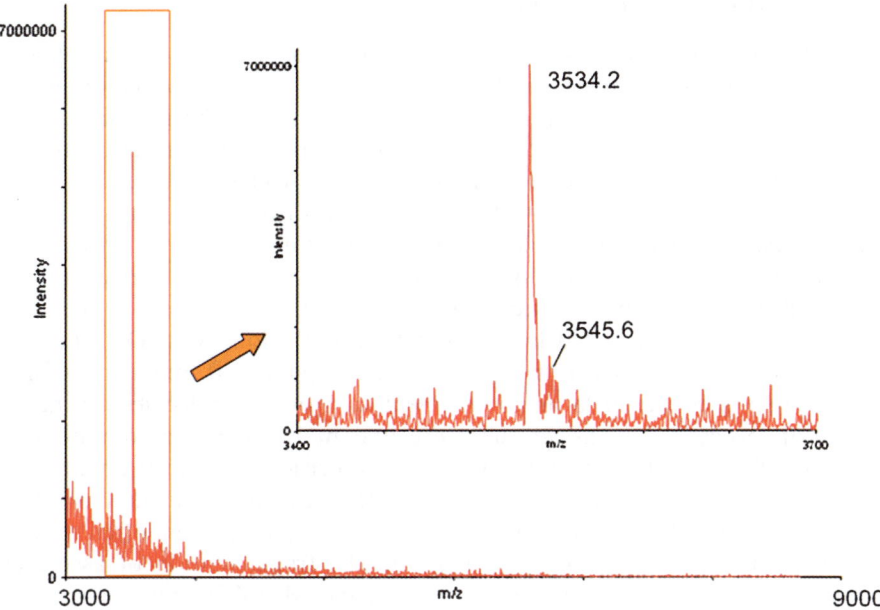

Figure 10.1 Single cell MALDI-TOF MS of one single neurosecretory cell from *Astacu leptodactylus*. MALDI preparation was optimized to enhance desorption of the species of interest. The enlarged figure shows that the cHH isoform at *m/z* 3534.2 seems more abundant. Additional data using MS/MS for sequence analysis is needed to confirm peptide identification.

a single cHH-producing cell previously identified by immunocytochemistry demonstrated that both procHH isoforms were expressed in each cell analyzed. This work was applied to other species such as *Astacus leptodactylus*. The spectrum (Figure 10.1) illustrates single cell MALDI analysis.

10.2 Analytical Strategies

The proteomic analysis is an integrative science. Its intrinsic analytical challenge relies on the incredible chemical variety of proteins and the very wide range of concentrations in which they are expressed. To answer this challenge, powerful techniques of separation have been developed. Liquid chromatography is generally associated with mass spectrometry, which helps in the identification of the proteins being studied. In particular, reversed phase liquid chromatography is required for the analysis of complex mixtures of proteins or proteolytic peptides. However, it is generally accepted that no monodimensional chromatographic separation could offer sufficient peak capacity to resolve a complex mixture of peptides resulting from the digestion of a complex proteome. That is why development efforts are focused on increasing the peak capacity of chromatographic separations.

10.2.1 Mass Spectrometry

To study the protein profiles expressed in various physiological or pathological conditions, analysis by mass spectrometry (MS) is usually used as a powerful tool for proteomics. A mass spectrometer records mass-to-charge ratio (m/z) of proteins (or other analytes). Molecules to be identified are ionized in the gas phase and the ions are sent towards the analyzer which separates these ions according to their m/z. The detector sends the information towards the computer to be analyzed. The biological mass spectrometry approaches rapidly established themselves as sensitive and universal characterization techniques for proteins, since the introduction at the end of the 1980s of the two ionization source types: matrix-assisted laser desorption ionization (MALDI) and electrospray ionization (ESI). Both types are considered as soft ionization sources because they generate little or no fragmentation of the molecules during the ionization process. To answer the more and more important requirements in terms of sensitivity, dynamic range, and precision, new instruments such as Fourier transform mass spectrometry (FTMS) instruments have been interfaced with MALDI and ESI. The identification of a protein often requires partial sequence information obtained with the fragmentation of associated proteolytic peptides by tandem mass spectrometry (MS/MS). Powerful software applications have been developed, allowing the automatic interpretation of MS/MS spectra, in order to associate amino acid sequences with the fragment ions observed.

10.2.1.1 Ionization Modes

10.2.1.1.1 Electrospray Ionization. Electrospray ionization (ESI) was developed by Fenn and his co-workers, and described for the first time in 1985[33] and was applied to the study of biomacromolecules 3 years later.[34] Since then, ESI has been very widely used in proteomics. ESI implies the production of gaseous ions *via* the application of an electric potential in a solution of molecules of interest in movement. It results in the formation of a spray of small droplets containing solvent and analytes. The solvent evaporates from droplets. The size of the droplets decreases more and more until they become unstable (limit of Rayleigh), what leads to the dissociation in even smaller droplets. Finally, the electrostatic repulsion becomes strong enough to induce desorption of analyte ions into the gas phase, which are then sent towards the mass spectrometer. The ions generated in ESI are generally multiply charged and are of type $(M + nH)^{n+}$.

The ESI mode allows a high flexibility for choosing the type of mass spectrometer, because of the small m/z ratios generated by the formation of multiply charged ions. This makes possible the use of ion traps and quadrupoles, Fourier transform (FT) ion cell (cyclotron resonance or orbital resonance) and time-of-flight (TOF) analyzers. On the other hand, the formation of complex mass spectra as a result of the multiple ion species generated from the same analyte is a major drawback. The important consumption of sample was relatively corrected with miniaturization in nanospray.[35] Classical ESI cannot be used for molecular imaging. Recently, desorption electrospray ionization (DESI) was presented by Cooks and co-workers.[36] So, the ESI mode has become compatible with the analysis of solid samples for molecular imaging but, at the moment, with spatial resolution and ionization efficiency much lower than those obtained in MALDI.

10.2.1.1.2 MALDI. In 1988, Hillenkamp and Karas discovered that molecular ions of large proteins can be generated by laser desorption without fragmentation when these biomolecules are co-crystallized with a matrix.[37] Matrix preparation presents a crystalline structure of small organic chromophore molecules. The wavelength of the laser used for the ionization has to be in the range of absorption of the matrix crystals. After the absorption of the energy carried by photons of a pulsed laser, small molecules of matrix are vaporized with molecules of proteins/peptides. However, the theory of the MALDI process is not completely understood. A better understanding of the ionization mechanisms could become essential in the improvement of the detection by mass spectrometry. Today, MALDI is very effective for the detection of peptides; some tens of attomoles to some femtomoles only are necessary by sample spot preparation.[38] On the other hand, the detection of larger biomolecules, in particular masses greater than 100 000 Da is more difficult. Ionized proteins (or peptides) are accelerated by an electric field towards the mass analyzer. MALDI is often coupled to a TOF analyzer. Accelerated by a constant potential, the ions reach the detector after a time of flight proportional to $(m/z)^{1/2}$. According to the required sensitivity and resolution, the TOF instrumentation in linear and reflectron modes is widely used for the analyses of biomolecules. The resolution of a MALDI-TOF in linear mode can reach 5000; in reflectron mode it can reach 20 000. However, it is difficult to apply a reflectron mode for molecules with $m/z = 20\,000$ Da.

Mass spectra are simple: most of the peptide ions are detected as singly charged species. MALDI is compatible with molecular imaging on tissues and thus proteomic imaging is feasible because the laser beam can be focused on a very small spot area (less of 10 µm diameter). On the other hand, MALDI presents some drawbacks such as a weak reproducibility due to the difficult control of the crystallization process and of the ionization efficiency (which depends on the fluence of the laser, *i.e.* the energy per unit area). The resolution for the very large molecules is weak due to the wide distribution of kinetic energy of ionized biomolecules, due to the possible formation of matrix adducts, and due to metastable fragmentation of large proteins. Matrix ions interfere in the low mass range and prevent the detection of small molecules. The direct ionization on silica or on metal was developed to reduce this problem.[39,40] However, MALDI remains a tool of choice for biomarker screening, in particular for the imaging of large biomolecules.[41]

The MALDI mode is preferably used for single cell analysis because it provides high sensitivity, simple MS profiles (one molecular species, one major ionized species), low sample consumption, and tolerance towards complex samples in the solid state. However, when analyses are focused on very complex samples, the lack of resolution, accuracy, and dynamic range of the classical TOF analyzer may be a major limitation. Recently, MALDI FTMS has been applied for lipid profiling of whole tissues and intact cells[42] and appears to be a very promising tool for peptidomics on intact cells also.

10.2.1.2   Mass Analyzer

The most common conventional analysers associated with ESI and MALDI sources are (see Table 10.1):

- the quadrupole (Q),
- the two-dimensional (2D) and three-dimensional (3D) ion traps,
- the time-of-flight (TOF) analyzer,
- Fourier transform (FT) detection either associated with ion cyclotron, resonance (ICR) traps, or the orbital electrostatic ion trap (Orbitrap),
- any combination of these analyzers.

Magnetic sector devices, which are still used in ion microscopy, are much less sensitive and are gradually being replaced by TOF or FT instruments. An analyzer is characterized by at least four key parameters:

- resolution,
- accuracy,
- accessible mass range,
- sensitivity.

The demand for powerful analytical capacities in the field of biochemical analysis of proteins is becoming more and more important, and instrumentation in mass spectrometry has evolved very quickly in the last few years. New types of analyzers and hybrid instruments have appeared, so supplying new opportunities in the study of highly complex proteomes.[43,44] In particular, exact mass measurements are more and more used in proteomics studies, not only to

Table 10.1   The most common conventional analyzers associated with ESI and MALDI sources.

Analyzer	Resolution ($M/\Delta M$)	Mass range (Da)	Accuracy (ppm)
3D ion trap	1 (full scan) to 1,000 (high res.)	4×10^3	100–300
2D ion trap	10 (full scan) to 10 000 (high res.)	4×10^3	100–300
Quadrapole	1	4×10^3	100–300
Time of flight	2×10^4	3×10^5	<30
Orbitrap	5×10^4	5×10^3	<5
FT-ICR	10^5	5×10^3	<3

FT-ICR, Fourier transform (FT) associated with ion cyclotron resonance (ICR).

identify proteins but also to characterize post-translational modifications, or interactions between proteins and other molecules in a less ambiguous way than before.

Some significant challenges in proteomics result from the great complexity of biological systems and from the range of concentrations of proteins in these systems. The example of the plasma or serum proteome is often quoted to illustrate this increasing demand in dynamic range. Indeed, in these systems, all the expressed proteins are potentially present and cover a range of concentration of at least 10 orders of magnitude which exceeds the dynamic range of any instrument used as a unique analytical method.[45] After enzymatic digestion, every protein is converted to some tens of peptides, which increases the mixture complexity. Taking into account the presence of multiple forms of the same protein (isoforms, PTMs, maturated and truncated forms), important analytical challenges are finally involved in the proteomic analysis. A fractionation strategy can be used, such as coupling MS with a multidimensional liquid chromatography separation.[46,47]

First of all, analysis of small samples requires high sensitivity. Analyzers such as ion traps or TOF can address this issue very well. In contrast, scanning analyzers, such as quadrupoles, could be very limited. However, multipole instruments suffer from a lack of resolution and accuracy in most of applications.

Mass accuracy and mass precision should be defined differently to avoid any confusion.[48] Accuracy is the degree of correspondence of the quantity measured with the true value. Precision is the degree with which the experimental measurements present reproducible results or at least quite similar results. Biological mass spectrometry aims to accurately determine the m/z ratio of the biomolecules of interest, to calculate their exact mass after isotopic deconvolution. Whatever the degree of purification and separation, robustness of peptide identification in MS or in MS/MS mode is highly dependent on the matching of experimental masses (peptide masses or peptide fragment masses) with theoretical masses deduced from sequence database (proteomic or genomic). It has been reported that the number of candidate amino acid sequences for one given mass decreases rapidly if the accuracy of mass measurement increases.[49–53] An accuracy of ± 1 ppm allows the exclusion of 99% of peptides with the same nominal mass but with different elemental compositions or with different amino acid composition.[53]

A high resolving power is required to reach high accuracy and exact mass measurement. Even if low-resolution mass analyzers can reach a high accuracy, their use is limited to specific simple mixtures. Compounds should be separated in the m/z domain (use of single quadrupole for selected reaction monitoring (SRM) experiments). Indeed, the need for high-resolution instruments is a prerequisite for some proteomics experiments; it is a real challenge for the exact and robust identification of mono-isotopic masses with a high dynamic concentration range and with overlapping m/z isotopic patterns. Among instruments able to reach simultaneously good sensitivity, accuracy below few ppm, and high resolving power for classical proteomics studies are FT-ICR instruments,[54] TOF instruments,[55] and tri-dimensional electrostatic ion traps (Orbitrap).[56] However, these analyzers are not sensitive enough to detect proteins that are expressed at low copy number per cell, because the most sensitive ones only reach attomole sensitivity.

10.2.2   Coupling Separation Techniques and Mass Spectrometry

10.2.2.1   LC-MS Coupling

The number and the robustness of peptide identifications can be drastically improved with the use of tandem mass spectrometry (MS^n),[57,58] with the combination of different fragmentation modes[59] or with high accuracy mass measurement.[49,50,53,60,61] Even though MS/MS analysis is of great interest for peptide and protein identification, the number of peptide species is very high.

This issue was immediately addressed by the use of liquid chromatography–tandem mass spectrometry coupling (LC-MS/MS). Even though LC-MS/MS is very efficient for the analysis of simple protein mixtures, in most cases the number of detected peptides eluted from a LC-MS/MS analysis widely exceeds the fragmentation capacity of a tandem mass spectrometer: there are too many peptides and not enough analysis time!

Moreover, the analysis of a proteome requires large-scale data collection to evaluate modification of proteins expression over the time. This involves high throughput MS capacity with high sensitivity for identification and quantification of proteins coming from very similar samples. These requirements can be fulfilled using approaches such as accurate mass and time (AMT) labels. If peptide mass and retention time can be accurately determined then the AMT label can be considered as unique and specific in the mass–time domain for each peptide candidate. If a database is developed according to a preliminary LC-MS/MS study, the subsequent analyses could be realized in LC-MS only.[62]

The ESI source remains the most frequently used ionization source associated with liquid chromatography. However, for complex mixtures the time scale for online coupling limits the amount of collected data. In fact MS acquisition and MS/MS acquisitions on the most abundant precursors alternate. The higher the resolution of the chromatographic separation is, the sharper peaks are. The available time for MS analysis decreases and if MS/MS acquisition lasts too much, other components will be totally eluted before they could be analyzed. On the other hand if LC is off-line coupled to a MALDI source, such limitations do not exist anymore. The whole set of collected MALDI spots can be analyzed several times independently of the retention time. The combination of the two sources has been proven to be complementary.[63]

10.2.2.2   Multidimensional Separations

Peak capacity gives a measure of the maximum number of peaks that can be theoretically separated in a given time window at a given resolution. Very often a monodimensional separation is not able to resolve a complex mixture of peptides generated by the hydrolysis of a complex protein mixture. A classical RPLC of 1 h gives an average peak capacity of a few hundreds,[64] even if developments have focused on the increase of the resolving power of mono-dimensional separations.[65] The peak capacity can be improved by the use of longer separation columns or smaller stationary phase particles size < 2 µm). Because these modifications induce an increase of the back pressure, specific instruments were also developed to work at thousands of bars in the so-called ultra-high-pressure liquid chromatography (UHPLC) approach.

Two-dimensional liquid chromatography (2D-LC) stands for an analytical strategy where two LC systems are combined for the analysis of a sample. According to Giddings[66] and Guiochon[67] such a coupling has a global peak capacity equal to the product of the independent peak capacity of each dimension, assuming that the two separations are based on independent physicochemical parameters. In this specific case, mechanisms of the two successive dimensions are considered as orthogonal.[66] If they are related the retention times of one given molecular species in each dimension will be correlated and the experimental 2D separation domain will not be optimal.

10.2.2.3 Coupling with Capillary Electrophoresis

Various electrophoretic techniques were coupled with ESI or MALDI mass spectrometry.[68] As quoted above one of the first single cell detections of protein by mass spectrometry involves capillary electrophoresis (CE)-ESI-MS coupling.[5] The CE-ESI interface is mainly based on LC-MS interfaces developed previously, with the supplementary constraint of checking the continuity of the electric circuit. Most of the time, an interface with a coaxial liquid sheath, using three concentric capillaries, is used.[69] The central capillary, used for CE separation, is inserted into the atmospheric pressure region of the ESI source with a metallic tube. This tube is concentric around the capillary, serves as the ESI needle and delivers the liquid sheath close to the electric circuit between the CE separation and the ESI needle. The third stainless steel concentric tube brings the nebulizing gas. The liquid sheath, having a flow of some microliters per minute, contains a volatile solvent (as methanol) to facilitate the nebulization. The choice of the electrolyte concentration has to be a compromise between the efficiency of the separation (requiring high ionic strengths) and the efficiency of the nebulization (requiring weak salt concentration). A variant consists in replacing the liquid sheath by a liquid complement *via* a tee. Due to the lack of sensitivity of classic ESI, nanospray interfaces were also developed and allowed to avoid the liquid sheath. However, an additional liquid may be added in parallel to stabilize the nebulization.[70] Complex proteomic samples have been analyzed by CE-MS. The first examples are urine and sera from patients with kidney disease,[71] diabetes or prostate hyperplasia.[72]

The same types of interface have been used for the coupling of capillary electrochromatography (CEC) to the MS and have been reviewed elsewhere.[73]

Capillary isoelectrofocalisation (CIEF) coupled to the MS has been considered has an alternative to 2D gel electrophoresis. This was first illustrated with the CIEF-ESI-FTICR-MS analysis of cell lysate from *Escherichia coli* and *Deinococcus radiodurans*.[74]

Finally, CE-MALDI coupling was developed with the automation of sample collection and spotting on the MALDI target after CE separation. It was validated on standard proteins and whole cell lysates before application to single cell analysis. Neuropeptides and hormones in a single neuron from *Aplysia californica* (40 μm) were detected using CE-MALDI MS.[75]

10.2.2.4   *Quantification*

Mass spectrometry is not an intrinsic quantitative analysis. Two major strategies have been developed to address this issue: stable isotope labeling for absolute and relative quantification and label-free analysis with comparison between samples for relative quantification only. In this last kind of approach, MS signals from samples to be compared are processed to evaluate the variation of their intensity, with the hypothesis that minor ion suppression effects occur. These suppression effects are minimized using a prior separation of the peptides of interest using, for instance, liquid chromatography. Some peptides have been described to be more relevant for quantification, the so-called proteotypic peptides.[76–78] They correspond to specific sequences of the protein of interest. They are more likely to be efficiently resolved in liquid chromatography, to be detected with a high signal-to-noise ratio, and to give rise to a clear sequencing pattern for unambiguous protein identification. They are the best candidates also for stable isotope labeling, a technique that takes advantage of the similar ionization efficiency of analogous molecular species differing only by their isotopic composition. If a known amount of a labeled peptide is co-ionized with the peptide of interest, the relative intensities of the associated signals are correlated with the relative abundances of the species. So the absolute amount of the peptide of interest can be calculated according to the amount of labeled reference species that was added to the sample. Labeled molecules can be added to the sample in different ways: metabolic or chemical labeling, enzymatic incorporation or addition of precursor of proteolytic peptides in the sample. The different strategies have been discussed in detail in a recent review.[79] Absolute quantification can be achieved only on known proteins according to the latest reports. Labeling procedures have also been reviewed recently by Pan *et al.*[80] The application of isotope labeling for single cell proteomics study has not been reported until now; however, it has been adapted for quantification of *Aplysia* insulin (AI) C_β-peptide and cerebrin prohormones on *Aplysia californica* single AI- and cerebrin-expressing neurons,[29] successfully using either spiking with the analyte of interest or a commercial iTRAQ labeling reagent kit. Once again the very low average amount of neuropeptides in single neurons (a few attomoles) is the major limitation of this challenging analytical technique.

10.3 Strategies for Studying Proteins in Low Amounts of Samples

The applications of proteomics are very various and give rise to new domains of study. Besides genomics and transcriptomics, the "omics" family has expanded with peptidomics for endogenous peptide characterization or degradomics for metabolite and protease activity studies. The very low amounts of available materials when dealing with human models have gradually required more and more sensitive methods for sample preparation and detection. The main idea to address sensitivity and recovery yield issues is miniaturization and integration to avoid non-specific loss by adsorption on the container's surface.

10.3.1 How to Enhance the Sensitivity: Miniaturization, Integration, and Automation

10.3.1.1 Micro-electro-mechanical Systems and ESI Ionization Interfaces

Optimal sensitivity in nanospray is obtained for flow rates of a few tens of nanoliters per minute.[81] At such flow rates, any dead volume should be eliminated in the interface between the analytical separation and mass spectrometry to prevent any loss of chromatographic resolution. On the other hand very small dimensions are very likely to present undetected leaks or to suffer from blockages. To improve the robustness of miniaturized systems, the first stage was to integrate the different steps until the ionization step into a microfluidic device. Micro-electro-mechanical system (MEMS) technology uses microfluidic channels with internal diameters ranging from 10 to 100 µm. They have been first used for lab-on-a-chip (LOC) electrophoresis systems. More complex LOC systems integrate the enzymatic proteolysis reactor after the electrophoretic separation.[82,83] For a recent review, see the report from Lee *et al.*[84]

More frequently, nanoESI interfaces have been coupled to microfluidics because they work with liquid samples which could be directly coupled. Different designs have been proposed since 1997[83,85] with the first studies from Ramsey's group[86] and Karger's group[87] on glass supports. However, the nanospray needle shape and the hydrophilic properties of silica–quartz or glass were too high. Another design used polycarbonate, which was better suited to wetting problems,[88] but which was less efficient for separation capacity. An intermediate design was proposed by Aebersold's group with the assembly of a microfluidic glass chip with a glued nanoESI capillary.[89] But one of the major drawbacks and limitations for robustness was the alignment of the capillary with microfluidic channels. Moreover, the risk of non-negligible dead volume is high. The next generation designs integrated progressively the nanoESI needle in the microfluidic device, on silica based or polymeric supports.

10.3.1.2 Lab-on-a-chip with Liquid Chromatography and NanoESI Interface

Current set-ups for shotgun proteomics studies associate nanoLC and MS with a preliminary step of enrichment and desalting. This is used for the concentration of the sample and the removal of non-volatile and/or biological salts. A microfluidic device combining these steps in one single chip has been commercialized.[90] However, this system was only interfaced with ion trap or TOF mass analyzers.[91] This technology is focused on the use of a polymeric microfluidic chip. Of the size of a credit card it integrates an enrichment cartridge, a chromatographic separation column in the nano-flow rate scale with multiple connections and a nanoESI needle. It avoids half of the classical connections and plumbing usually required by a classical nanoLC-MS system, thus decreasing risks of leaks and dead volumes. This chip is interfaced either with an ion trap or with a hybrid quadrupole–time of flight analyzer.[92] The chip is manufactured using polyimide films. Microchannels are etched with a laser and they are equipped with frits and solvent wastes. Enrichment and separation columns are packed with classical reversed phase medium. The chip itself is inserted in a solvent multichannel valve allowing loading and injection of the sample. On-line 2D liquid chromatography has been also integrated with strong cation exchange and reversed phase chromatography on a valve for more complex samples.[93]

10.3.1.3 Multinozzle Automated Nanospray

Another nanoESI infusion system was developed by Zhang and collaborators.[94] It was later commercialized by the company Advion.[95] The design is obtained by deep reactive ion etching (DRIE) on a silica plate with a single channel and connected on one side to a nozzle (10 μm ID×30 μm OD). A reservoir on the opposite face contains the liquid for direct infusion and the electric connection is obtained through a liquid junction on the back side. Liquid chromatography effluent can be coupled to this chip with zero dead volume. This interface has been adapted to the classical format of multi-well plates with 96, 384 or 1536 parallel nanoESI nozzles.[96] It is more specifically dedicated to high-throughput miniaturized screening. The whole interface is compatible with most commercial electrospray sources.

10.3.2 MALDI Interfaces

MEMS and MALDI interfaces with mass spectrometry have been reviewed in detail by Lee.[84,97] Integrated MALDI solutions with capillary electrophoresis[98] or off-line MALDI chip preparations[99,100] with capillary liquid chromatography have been reported. However, these interfaces are still at the prototype stage even if some of the work presented with this interface counted among the pioneering reports for single cell analysis of endogenous peptides by mass spectrometry. Indeed, the detection and identification (using mass and retention time of the peptides) on *Lymnaea stagnalis* neurons was reported in 1998.[99] Technical problems on the samples volumes and handling compatible with the laser focusing size are still being evaluated.

With the recent reports on the complementarities of ESI and MALDI in terms of peptide and protein coverage, these interfaces are likely to be of high interest in the future, and different groups are presently focusing their R&D efforts on MEMS/MALDI validation.

Conclusion

Despite numerous attempts to realize proteomics on single cell scale, these applications still remain a goal to be achieved. The lack of amplification techniques, the very high dynamic range of concentration of expressed proteins, the high molecular diversity subsequent to post-translational modifications and maturation processes, combined with the limited sensitivity of the analytical techniques, have hindered the success of such studies. Peptidomics on specific molecular targets or proteomics on small and fractionated samples have been reported. Another field of application not presented in the present text is molecular imaging or mass spectrometry imaging (MSI).[101] Briefly, as reviewed recently by several groups,[102–104] MALDI MSI allows direct detection and localization of proteins and peptides from whole tissue and cell. However, the identification of those species has to be done separately most of the time. Therefore, MALDI MSI should be considered a very powerful new technology for *in situ* biomarker detection and molecular histology but not yet as a new tool for *de novo* proteomics studies. The new era of sensitive, high resolution, and high accuracy analyzers combined with the MALDI mode might very soon open the way towards hybrid applications of AMT and MSI and towards (why not?) new isolated cell proteomics strategies.

References

1. N. L. Anderson and N. G. Anderson, *Electrophoresis*, 1998, **19**(11), 1853–1861.
2. D. N. Perkins, *et al., Electrophoresis*, 1999, **20**(18), 3551–3567.
3. E. Audinat, B. Lambolez and J. Rossier, *Neurochem. Int.*, 1996, **28**(2), 119–136.
4. P. A. Haynes, N. Fripp and R. Aebersold, *Electrophoresis*, 1998, **19**(6), 939–945.
5. S. A. Hofstadler, *et al., Anal. Chem.*, 1995, **67**(8), 1477–1480.
6. L. Li, *et al., Anal. Chem.*, 2000, **72**(16), 3867–3874.
7. V. Redeker, *et al., Anal. Chem.*, 1998, **70**(9), 1805–1811.
8. D. G. Grier, *Nature*, 2003, **424**(6950), 810–816.
9. M. Castelain, *et al., J. Chem. Phys.*, 2007, **127**(13), 135104.
10. G. D. Wright, *et al., Fungal Genet. Biol.*, 2007, **44**(1), 1–13.

11. I. Safarik and M. Safarikova, *J. Chromatogr. B: Biomed. Sci. Appl.*, 1999, **722**(1–2), 33–53.
12. J. An, *et al., Anal. Bioanal. Chem.*, 2009, **394**, 801–809.
13. X. Hu, *et al., Proc. Natl. Acad. Sci. U. S. A.*, 2005, **102**(44), 15757–15761.
14. M. B. Dainiak, *et al., Adv. Biochem. Eng. Biotechnol.*, 2007, **106**, 1–18.
15. H. B. Gutstein, *et al., Mass Spectrom. Rev.*, 2008, **27**(4), 316–330.
16. B. Herbert and E. Harry, *Methods Mol. Biol.*, 2009, **519**, 47–63.
17. B. A. Stanley, *et al., Dis Markers*, 2004, **20**(3), 167–178.
18. J. M. Robinson, D. D. Vandre and W. E. T. Ackerman, *Placenta*, 2009, **30**(Suppl A), S83–S89.
19. E. C. Schirmer and L. Gerace, *Trends Biochem. Sci.*, 2005, **30**(10), 551–558.
20. M. Bantscheff, *et al., Ernst Schering Found Symp. Proc.*, 2007, **3**, 1–28.
21. A. Amoresano, *et al., Methods Mol. Biol.*, 2009, **527**, 173–190.
22. S. Beranova-Giorgianni, D. M. Desiderio and F. Giorgianni, *Methods Mol. Biol.*, 2009, **519**, 383–396.
23. B. Domon, *et al., J. Proteome Res.*, 2009, **8**, 2633–2639.
24. M. L. Miller and N. Blom, *Methods Mol. Biol.*, 2009, **527**, 299–310.
25. A. Nita-Lazar, H. Saito-Benz and F. M. White, *Proteomics*, 2008, **8**(21), 4433–4443.
26. G. Pocsfalvi, *Methods Enzymol.*, 2009, **457**, 81–96.
27. C. R. Jimenez, *et al., J. Neurochem.*, 1994, **62**(1), 404–407.
28. R. W. Garden, *et al., J. Mass Spectrom.*, 1996, **31**(10), 1126–1130.
29. S. S. Rubakhin and J. V. Sweedler, *Anal. Chem.*, 2008, **80**(18), 7128–7136.
30. S. S. Rubakhin and J. V. Sweedler, *Nat. Protoc.*, 2007, **2**(8), 1987–1997.
31. S. Neupert, *et al., Anal. Chem.*, 2007, **79**(10), 3690–3694.
32. L. Li, R. W. Garden and J. V. Sweedler, *Trends Biotechnol.*, 2000, **18**(4), 151–160.
33. C. M. Whitehouse, *et al., Anal. Chem.*, 1985, **57**(3), 675–679.
34. J. B. Fenn, *et al., Science*, 1989, **246**(4926), 64–71.
35. M. Wilm and M. Mann, *Anal. Chem.*, 1996, **68**(1), 1–8.
36. Z. Takats, *et al., Science*, 2004, **306**(5695), 471–473.
37. M. Karas and F. Hillenkamp, *Anal. Chem.*, 1988, **60**(20), 2299–2301.
38. E. Nordhoff, H. Lehrach and J. Gobom, *Int. J. Mass Spect.*, 2007, **268**(2–3), 139–146.
39. J. C. Lee, *et al., Angew Chem. Int. Ed. Engl.*, 2006, **45**(17), 2753–2757.
40. N. Y. Hsu, *et al., Anal. Chem.*, 2008, **80**(13), 5203–5210.
41. M. L. Reyzer and R. M. Caprioli, *Curr. Opin. Chem. Biol.*, 2007, **11**(1), 29–35.
42. S. M. Batoy, *et al., Lipids*, 2009, **44**(4), 367–371.
43. M. A. Baldwin, *Methods Enzymol.*, 2005, **402**, 3–48.
44. B. Domon and R. Aebersold, *Science*, 2006, **312**(5771), 212–217.
45. N. L. Anderson and N. G. Anderson, *Mol. Cell. Proteomics*, 2002, **1**(11), 845–867.
46. M. P. Washburn, D. Wolters and J. R. Yates 3rd, *Nat. Biotechnol.*, 2001, **19**(3), 242–247.

47. D. A. Wolters, M. P. Washburn and J. R. Yates 3rd, *Anal. Chem.*, 2001, **73**(23), 5683–5690.
48. R. Zubarev and M. Mann, *Mol. Cell Proteomics*, 2007, **6**(3), 377–381.
49. K. R. Clauser, P. Baker and A. L. Burlingame, *Anal. Chem.*, 1999, **71**(14), 2871–2882.
50. T. P. Conrads, *et al.*, *Anal. Chem.*, 2000, **72**(14), 3349–3354.
51. L. Sleno, D. A. Volmer and A. G. Marshall, *J. Am. Soc. Mass Spectrom.*, 2005, **16**(2), 183–198.
52. E. J. Takach, *et al.*, *J. Protein Chem.*, 1997, **16**(5), 363–369.
53. R. A. Zubarev, P. Hakansson and B. Sundqvist, *Anal. Chem.*, 1996, **68**(22), 4060–4063.
54. M. B. Comisarow and A. G. Marshall, *J. Chem. Phys.*, 1976, **64**(1), 110–119.
55. J. Dawson and M. Guilhaus, *Rapid Commun. Mass Spectrom.*, 1989, **3**(5), 155–159.
56. A. Makarov, *Anal. Chem.*, 2000, **72**(6), 1156–1162.
57. S. A. Beausoleil, *et al.*, *Proc. Natl. Acad. Sci. U. S. A.*, 2004, **101**(33), 12130–12135.
58. J. V. Olsen and M. Mann, *Proc. Natl. Acad. Sci. U. S. A.*, 2004, **101**(37), 13417–13422.
59. M. L. Nielsen, M. M. Savitski and R. A. Zubarev, *Mol. Cell Proteomics*, 2005, **4**(6), 835–845.
60. W. Haas, *et al.*, *Mol. Cell Proteomics*, 2006, **5**(7), 1326–1337.
61. F. He, *et al.*, *J. Proteome Res.*, 2004, **3**(1), 61–67.
62. R. D. Smith, *et al.*, *Proteomics*, 2002, **2**(5), 513–523.
63. W. M. Bodnar, *et al.*, *J. Am. Soc. Mass Spectrom.*, 2003, **14**(9), 971–979.
64. M. Gilar, *et al.*, *J. Chromatogr. A*, 2004, **1061**(2), 183–192.
65. G. Guiochon, *J. Chromatogr. A*, 2006, **1126**(1–2), 6–49.
66. J. C. Giddings, *Anal. Chem.*, 1984, **56**(12), 1258A–1260A, 1262A, 1264A, passim.
67. G. Guiochon, *et al.*, *J. Chromatog.*, 1983, **255**(January), 415–437.
68. H. Stutz, *Electrophoresis*, 2005, **26**(7–8), 1254–1290.
69. J. Ding and P. Vouros, *Anal. Chem.*, 1999, **71**(11), 378A–385A.
70. F. Hsieh, *et al.*, *Rapid Commun. Mass Spectrom.*, 1999, **13**(1), 67–72.
71. S. Wittke, *et al.*, *J. Chromatogr. A*, 2003, **1013**(1–2), 173–181.
72. N. A. Guzman, *J. Chromatogr. B: Biomed. Sci. Appl.*, 2000, **749**(2), 197–213.
73. E. Barcelo-Barrachina, E. Moyano and M. T. Galceran, *Electrophoresis*, 2004, **25**(13), 1927–1948.
74. P. K. Jensen, *et al.*, *Electrophoresis*, 2000, **21**(7), 1372–1380.
75. J. S. Page, S. S. Rubakhin and J. V. Sweedler, *Anal. Chem.*, 2002, **74**(3), 497–503.
76. N. L. Anderson, *et al.*, *Mol. Cell Proteomics*, 2009, **8**(5), 883–886.
77. K. G. Kline, *et al.*, *J. Proteome Res.*, 2008, **7**(11), 5055–5061.
78. V. Lange, *et al.*, *Mol. Syst. Biol.*, 2008, **4**, 222.
79. A. Panchaud, *et al.*, *J. Proteomics*, 2008, **71**(1), 19–33.

80. S. Pan, *et al., J. Proteome Res.*, 2009, **8**(2), 787–797.
81. A. Schmidt, M. Karas and T. Dulcks, *J. Am. Soc. Mass Spectrom.*, 2003, **14**(5), 492–500.
82. A. Dodge, *et al., Analyst*, 2006, **131**(10), 1122–1128.
83. A. G. Hadd, *et al., Anal. Chem.*, 1997, **69**(17), 3407–3412.
84. J. Lee, S. A. Soper and K. K. Murray, *J. Mass Spectrom.*, 2009, **44**(5), 579–593.
85. W. C. Sung, H. Makamba and S. H. Chen, *Electrophoresis*, 2005, **26**(9), 1783–1791.
86. R. S. Ramsey and J. M. Ramsey, *Anal. Chem.*, 1997, **69**(6), 1174–1178.
87. Q. Xue, *et al., Anal. Chem*, 1997, **69**(3), 426–430.
88. J. Wen, *et al., Electrophoresis*, 2000, **21**(1), 191–197.
89. D. Figeys, Y. Ning and R. Aebersold, *Anal. Chem.*, 1997, **69**(16), 3153–3160.
90. H. Yin, *et al., Anal. Chem.*, 2005, **77**(2), 527–533.
91. H. Yin and K. Killeen, *J. Sep. Sci.*, 2007, **30**(10), 1427–1434.
92. M. H. Fortier, *et al., Anal. Chem.*, 2005, **77**(6), 1631–1640.
93. M. Vollmer, *et al., J. Sep. Sci.*, 2006, **29**(4), 499–509.
94. G. A. Schultz, *et al., Anal. Chem.*, 2000, **72**(17), 4058–4063.
95. S. Zhang, C. K. Van Pelt and J. D. Henion, *Electrophoresis*, 2003, **24**(21), 3620–3632.
96. J. M. Dethy, *et al., Anal. Chem.*, 2003, **75**(4), 805–811.
97. D. L. DeVoe and C. S. Lee, *Electrophoresis*, 2006, **27**(18), 3559–3568.
98. J. Liu, *et al., Anal. Chem.*, 2001, **73**(9), 2147–2151.
99. S. Hsieh, *et al., Anal. Chem.*, 1998, **70**(9), 1847–1852.
100. T. Miliotis, *et al., J. Mass Spectrom.*, 2000, **35**(3), 369–377.
101. A. Walch, *et al., Histochem. Cell Biol.*, 2008, **130**(3), 421–434.
102. I. Fournier, M. Wisztorski and M. Salzet, *Expert Rev. Proteomics*, 2008, **5**(3), 413–424.
103. M. Wisztorski, *et al., Dev. Neurobiol.*, 2008, **68**(6), 845–858.
104. W. M. Hardesty and R. M. Caprioli, *Anal. Bioanal. Chem.*, 2008, **391**(3), 899–903.

CHAPTER 11

Microfluidics: Basic Concepts and Microchip Fabrication

CONNI VOLLRATH AND PETRA S. DITTRICH

Department of Chemistry and Applied Biosciences, Wolfgang-Pauli-Str. 10, CH-8093 Zurich, Switzerland

Abstract

Microfluidic devices are nowadays versatile platforms for analytical applications and chemical syntheses, and are frequently used for cell culturing and cell studies. What are the unique features of a microfluidic device and how do these properties support the development of novel analytical methods for investigations on the single cell level? This chapter introduces to the basic concepts and aspects of microfluidics. It gives newcomers an understanding of the special characteristics such as the laminar flow and segmented flow. Furthermore, a short overview of fabrication methods is given highlighting the most popular approaches, particularly soft lithography. Finally, various materials useful for microchip fabrication are described and discussed with a special emphasis on their applicability for cell analytical devices.

11.1 Size Matters: An Introduction

In our daily life, we regularly use and apply small-scale devices. Of course, it is most evident for electronic gadgets such as mobile phones and laptops, or highly sophisticated electronic toys for children. In these items, we directly enjoy the most obvious benefit of miniaturization: the small size, and therewith the small weight, having an amazingly efficient machine in our hands at the

RSC Nanoscience & Nanotechnology No. 15
Unravelling Single Cell Genomics: Micro and Nanotools
Edited by Nathalie Bontoux, Luce Dauphinot and Marie-Claude Potier
© Royal Society of Chemistry 2010
Published by the Royal Society of Chemistry, www.rsc.org

same time. Moreover, miniaturized components are incorporated frequently in large instruments and machines, *e.g.* in domestic appliances or cars. Often, these components are sensors, which are by definition small instruments capable of measuring certain parameters such as temperature, pressure, pH, and humidity.

Test strips for medical diagnostics such as pregnancy tests or glucose monitoring instruments are other examples where size matters. Here, the portability of the device helps to perform the diagnosis wherever it is necessary and to continuously monitor the health state. At the same time, the volume of sample (*e.g.* blood) can be kept as small as possible as well (until detection of small amounts becomes a major issue).

These examples demonstrate that miniaturization is indeed very beneficial, supporting our daily tasks, and wherever small size is more comfortable for us, the miniaturization process is pushed quickly.

In chemical and biological research, the trend towards miniaturization is obvious as well, and nowadays it is impossible to imagine chemistry or biology without micro- and nanotechnology. An increasing demand for automated and high-throughput methods leads necessarily to small-scale instruments, providing many convincing benefits, particularly for analytical and diagnostic applications, *e.g.* smaller analyte consumption, and faster heat and mass transport to increase controllability as well as efficiency of reactions and separations.

It is intriguing to replace current standard instruments (many of them are benchtop devices) and integrate entire process sequences from synthesis to purification and analysis on a single microchip (Figure 11.1). In many studies, highly integrated systems have already been presented in which automated multi-step processes could be performed. Indeed, a trend to complete and sophisticated devices can be noticed having outstanding performance with respect to quality, sensitivity, and speed. Furthermore, the dimensions of structures fabricated by means of microsystem technology are matching biological relevant length scales; hence, valuable platforms for cell analysis and cultivation can be designed, and completely new applications without macroscopic equivalent can be developed. Examples are miniaturized sensors, microelectro-mechanical systems (MEMS) and biochips for DNA analysis; in the following, we focus on miniaturized fluidic systems, and use the terms "microfluidics" or "lab-on-chip technology".

Microfluidics is a very interdisciplinary research field that combines various disciplines such as engineering sciences, natural sciences, material sciences, and life sciences. Of course, it is neither possible to cover all aspects nor to go into details without going far beyond the scope of this chapter. The intention is rather to give newcomers and interested researchers an introduction into microfluidics and a glance at standard fabrication methods. For further reading several review articles and specialized books are mentioned throughout the chapter, and it should be pointed out that several scientific journals such as *Lab on a Chip* contain current scientific work and many internet platforms offer specific information.[1–4]

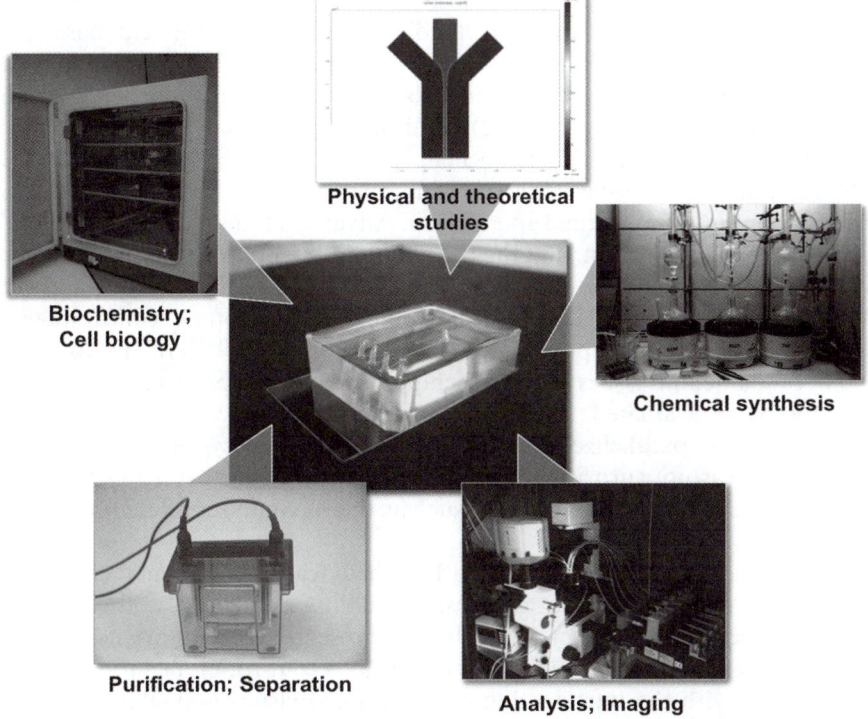

Physical and theoretical studies

Biochemistry; Cell biology

Chemical synthesis

Purification; Separation

Analysis; Imaging

Figure 11.1 Can microfluidics technology improve and replace conventional chemical and biological methods in the laboratory?

11.2 A Short Chronology of Microfluidics Research

Microfluidics can extensively be found in natural systems, *e.g.* phleom and xylem in plants or blood vessels in animals, and the desire to understand the principles of such systems has evolved over a long time. Systematic studies of fluid behavior on small scales (*e.g.* investigation of capillarity), the use of capillaries in instruments as well as the development of small test strips were the first approaches to microfluidics. However, the history of microfluidics as an enabling engineering science dates back only 20 years (except for the gas chromatograph developed in the 1970s by Terry[5]). Researchers at the beginning of the 1990s became aware of the potential of fluidic miniaturized systems for analytical and synthetic chemistry, and started to design and fabricate devices with planar fluidic networks. This was only possible as appropriate fabrication processes and materials were then available. In particular, in the 1980s the fabrication technology had been developed to build micro-electro-mechanical systems (MEMSs). MEMSs combine electronic and mechanic components and typically require multi-step fabrication processes based on lithography and various thin-film technologies. These developments paved the way for micro-fluidic systems, which in terms of fabrication are usually much simpler than

MEMSs. The first examples of microfluidic systems were a combination of sensors and a component for fluid handling used to perform acid–base titration,[6] and to measure pH in blood.[7] The field emerged when microfluidic systems were developed for separation processes, particularly chromatography and electrophoresis[8] (Figure 11.2), and high-speed and high-performance separations were achieved. The benefit of microchip systems compared to column- and capillary-based analytical separation devices is not only an improvement in terms of time and quality. Advanced fluid handling for sample pretreatment and injection, and the possibility to integrate several process steps and detectors on the same microchip can also dramatically improve and simplify analytical methods. This idea of a "miniaturized total chemical analysis system" (μTAS) was proposed by Manz *et al.* in 1990.[10,11] In the following years, fundamental issues such as chip materials, chip designs, and fluid transport were evaluated. Various chip designs and injection methods were introduced, first parallelized and multi-step automated systems were demonstrated.[12,13] It is not surprising that miniaturized analytical systems were soon available on the commercial market, such as the Bioanalyzer offered by Agilent Technologies.

Besides glass and silicon chips with etched features, polymer chips were produced by molding and hot embossing enabling rapid, fast, and cheap prototyping. The accessibility of microchips for researchers who are not familiar with micro-engineering or without a nearby clean-room facility increased dramatically with the introduction of poly(dimethylsiloxane) (PDMS), which can be processed by a simple molding technique.[14,15] Although PDMS is certainly not a perfect material[16] (see section 11.4.3.3.1), it exhibits a number of favorable properties and its elastomeric nature enables the formation of

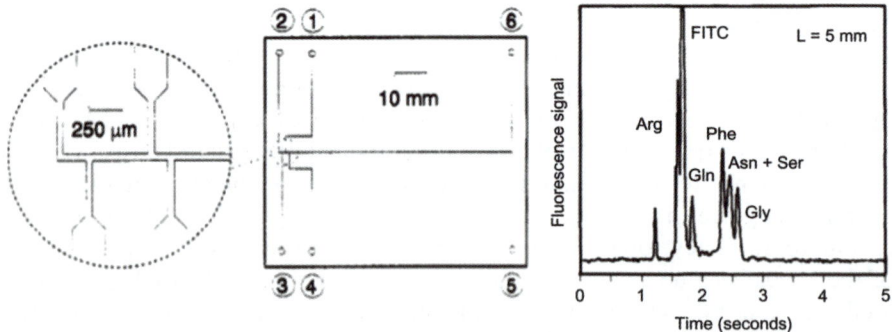

Figure 11.2   Scheme of a glass microchip, designed in 1993, for electrophoretic separation. Fast and efficient separation of mixtures of amino acids tagged with a fluorescent dye was achieved within seconds as shown in the electropherogram. Furthermore, the microchip design facilitates automated repetitive sample injection on a time scale of seconds. (Reprinted, with permission, from Effenhauser *et al.*[9] © 1993 American Chemical Society.)

membranes and pneumatic/hydraulic valves.[17] The simple integration of such valves in a microchip (by a process termed multilayer soft lithography, see section 11.4.2) opened the way to fabricate dense fluidic circuits that individually address hundreds of reaction chambers with volumes of picoliters to nanoliters. These NanoFlex™ valves are the heart of the microfluidic systems for genetic analysis and protein crystallization now sold by Fluidigm Corporation.

The advantages of microfluidic channel networks for fluid handling are most impressively demonstrated in the formation of microsized droplets. Small volumes of fluids can reproducibly be produced, transported, manipulated, and analyzed. This field of so-called digital microfluidics includes water droplet formation in an immiscible hydrophobic carrier solution,[18] but also manipulation of water droplets on electrode arrays (electrowetting[19]). The picoliter to nanoliter volumes can serve as small reactors for chemical and biological reactions, particle and crystal formation or single cell analysis.

In parallel with the promising developments in analytical chemistry, efforts to miniaturize chemical reaction systems were made in the middle of the 1990s. Here, great success can be found for fast and highly exothermic or endothermic reactions.[20] Since fast heat transfer can be achieved, temperature-dependent side reactions can be reduced. Benefits are reduced reaction times, lower energy consumption, high product yields, and improved selectivity.[21] Moreover, safe handling of toxic and radioactive reagents is possible.[22]

A further milestone in this respect was in 1998 the realization of the biological reaction for DNA amplification (polymerase chain reaction, PCR) on a microchip, illustrating the feasibility of precise heating and cooling on a microchip.[23] Further advances for DNA amplification,[24] sequencing,[25] and analysis[26] followed in the subsequent years. Combinations of DNA stretching in confined sections of a micro- or nanochannel and sensitive detection on single molecule level enabled determination of DNA–protein interactions[27] and DNA mapping.[28]

The biggest impact of microfluidics in academic research can be found in cell biology, systems biology, and pharmaceutical and medical applications, where size matters, *e.g.* to manipulate the cell environment, or to select, lyse, and analyze a single cell.[29–31] Microfluidic chips are well-suited to generate controllable and reproducible microenvironments.[32] The first cell analytical devices were developed for analysis of suspended cells, *i.e.* the miniaturization of cell counters and cell sorting devices as well as the caging of suspended cells in dielectrophoretic traps.[33,34] During the last decade, microchips have been presented for studies on adherent cells and cell cultures providing new insights into cellular processes and cell screening.[35–37] In other exciting studies, organ- or cell-like systems are created to mimic natural conditions,[38] or cells are exploited to act as a pump or to deliver other kinds of energy[39–41] (see section 11.4.2).

Nowadays, microfluidic-related research activities are performed in many fields and include research on fabrication issues, theoretical studies, experimental work on fundamental micro- and nanofluidic phenomena as well as a huge number of applied studies in the fields of chemistry, biology, and medicine. Furthermore, microfluidic technology can be considered as a tool to bridge micro- and nanotechnology, and thereby enables the utilization and integration of nanoscopic objects into microchip formats and the construction of molecular machines.

11.3   Microfluidics: Some Basics

Microfluidics refers to research on liquids or gases confined in channels and reservoirs with length scales of a few up to a few hundred micrometers, corresponding to volumes of a few femtoliters up to a few nanoliters (Figure 11.3). It covers theoretical aspects such as simulation and modeling of fluid streams, distribution of compound concentration and temperature, and experimental research and applications. Microfluidic systems provide a number of advantages compared to large-scale devices, and compared to non-fluidic systems:

- Low sample/analyte consumption (and as a consequence, safe handling of toxic or radioactive substances),

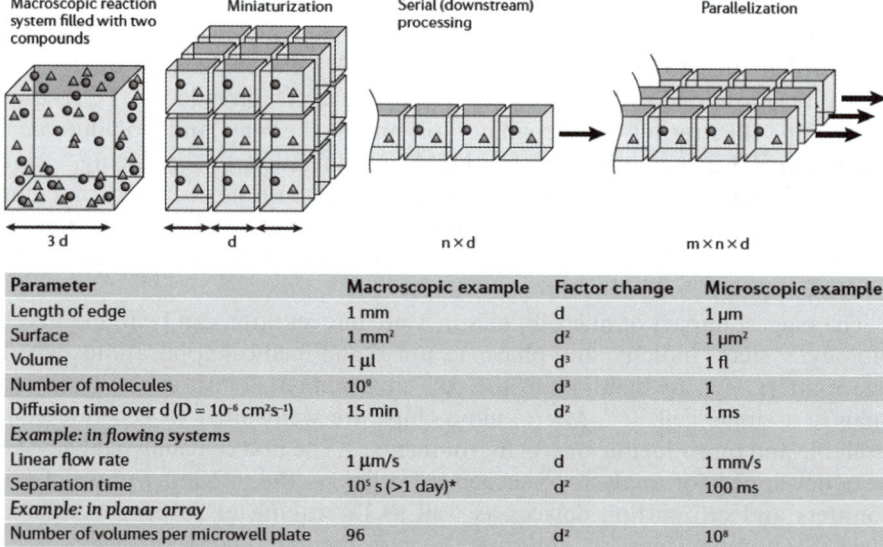

Parameter	Macroscopic example	Factor change	Microscopic example
Length of edge	1 mm	d	1 µm
Surface	1 mm²	d²	1 µm²
Volume	1 µl	d³	1 fl
Number of molecules	10⁹	d³	1
Diffusion time over d (D = 10⁻⁶ cm²s⁻¹)	15 min	d²	1 ms
Example: in flowing systems			
Linear flow rate	1 µm/s	d	1 mm/s
Separation time	10⁵ s (>1 day)*	d²	100 ms
Example: in planar array			
Number of volumes per microwell plate	96	d²	10⁸

*Typically, for example, high-performance liquid chromatography in packed column.

Figure 11.3   Simple scaling considerations demonstrate the benefits in micro-sized reaction systems. The short distances enable fast heat and mass transport; high surface-to-volume ratios support effective heat exchange to the environment; in a flowing system, serial processing is possible; due to small scales, massive parallelization is feasible. (d: length of edge, n and m: numbers of reaction systems serial and parallel, respectively). (Reprinted, with permission, from Dittrich and Manz.[42] © 2006 Macmillan Publishers Ltd.)

- Portability, *e.g.* useful for point-of-care diagnostics,
- Fast mass and heat transfer; small (defined) residence time distribution,
- Integration of various components is feasible, as well parallelization of various processes,
- Control over small fluid volumes,
- Similar dimensions to cells and cell organelles: excellent tool to handle/manipulate microscopic objects.

Although, at a first glance, the nature of a chemical reaction or a biochemical process will not change in a miniaturized system as long as we are far away from molecular dimensions (*i.e.* a few tens of nanometer or larger), simple scaling considerations predict an increase of efficiency, performance and high through-put for reactions and analyses. Furthermore, decreasing the dimension of a reaction system gives rise to phenomena that we usually do not observe on the macroscopic scale. For example, in small dimensions, surface-to-volume ratios increase with the consequence that the interfaces between channel surface and liquid become larger. Any interaction between molecules in the liquid and the surface can occur more efficiently than in a macroscopic system. Hence, the understanding of surface effects is crucial in many microchip applications. Phenomena such as adsorption, wetting, and capillary forces have to be considered.

In the following, some general aspects of fluid movement in microchannels are addressed, *i.e.* the generation of flow, the consequences of the laminar flow regime present in a microfluidic device, and the formation of segmented flow. For further reading we refer to specialized review articles and books.[43–45]

11.3.1 Flow Generation

Different methods to explore fluid movement through micro- and submicrometer-sized channels have been explored such as hydrodynamic and electro-osmotic flow, flow induced by centrifugation and shear-driven flow between two plates being moved against each other. More easily, capillary forces can drive the fluids in small microchannels, this is often exploited to fill a microfluidic network.

Pressure-driven (hydrodynamic) flow is most frequently used. It results in a parabolic flow profile, *i.e.* the flow speed is maximum in the channel centre, while decreasing to zero at the channel walls (Figure 11.4A). The difference in flow velocities has an important consequence: compounds are transported with different speed depending on their position resulting in a characteristic (so-called) dispersion.

Experimentally, the pressure difference of input and output channels can conveniently be achieved by connecting tubes from the microchannels to syringe pumps that provide a precise volume flow rate. Typically, flow rates from nanoliters per minute to milliliters per minute can be achieved. At low flow rates, the stepwise movement of the pump motor is reflected in a pulsed flow.

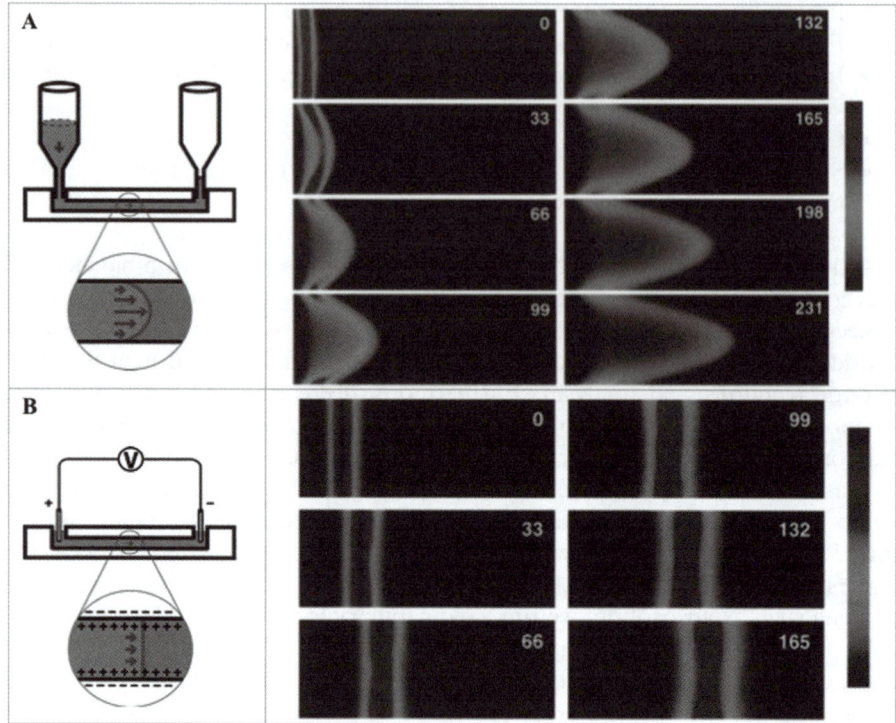

Figure 11.4   Generation of flow in a microchannel. A parabolic flow velocity profile
emerges from a pressure-driven, hydrodynamic flow (A), while electro-
osmotic flow results in a constant flow velocity across the channel (B). In
the figures at the right side, the flow is visualized by activation of a caged
fluorescent dye at a certain position, and taking fluorescent images at
selected time delays. (Reprinted, with permission, from Paul *et al*.[47] ©
1998 American Chemical Society.)

Hence, flow rates of few nanoliters per minute or below are best realized with
systems, in which the gas pressure onto the fluid reservoir is controling the flow
rate. Furthermore, pressure differences between input and output can be
achieved by a number of other methods including bubble formation or
acoustically driven pressure.[46]

Fluids can also be moved by means of an electric field. The electro-osmotic
flow (EOF) relies on the movement of mobile ions (or in general, charged
particles) relative to an immobile (stationary) charged surface upon application
of a potential across the microchannel. The requirement for EOF is the presence of an electrical double layer at the interface between the channel wall and
solution. For most materials, *e.g.* glass or PDMS, the surface of the microchannel exhibits negative charges over a broad pH range. Hence, cations
concentrate near the channel wall. Upon application of an electric potential
the cations migrate towards the cathode dragging along the rest of the solution,
i.e. neutral molecules as well as negatively charged ions. (In EOF, the fluid

moves as a whole. The application of strong electric fields induces another electrokinetic phenomenon: electrophoresis. It refers to the movement of suspended charged particles depending on their size and valence. This movement can be opposed to the EOF). The advantage of EOF is the constant flow velocity across the microchannel (Figure 11.4B). However, besides the intensity of the applied voltage, several further parameters influence the EOF, including surface charge, pH and the ion concentration of the buffer. As a consequence, it is crucial to carefully control these conditions.

An alternative fluid transport method that should be briefly mentioned is the use of centrifugal forces. It is experimentally achieved by rotating the microfluidic chip so that the fluid moves from the central position to the outside of the microchip, depending on the rotation speed. Since the centrifugal forces must be higher than the capillary forces in order to move the liquid, valves can be created simply by changing the channel diameter, or the surface properties, while the on/off switch is achieved by changing the rotation speed. This method easily enables parallelization of process steps.[48] Centrifugal microfluidics can be performed on a standard compact disk (CD) with imprinted channels, and thus readily adapted to standard interfaces. (Such CD microlaboratories are commercially available from Gyros.)

11.3.2 Laminar Flow

One of the most often mentioned peculiarities of flow movement in microchannels is laminar flow, *i.e.* the absence of turbulence (Figure 11.5). In general, the flow regime is characterized by the dimensionless Reynolds number, which represents the ratio of inertial to viscous forces (Box 11.1). In a microfluidic system with typical dimensions between 1 and 100 μm and flow velocities in the range of micrometers per second to millimeters per second, the Reynolds numbers range between 0.001 and 1, *i.e.* the inertial forces are low compared to viscous forces. The lack of unpredictable turbulences has several implications, positive or negative, depending on which application is requested. An insight into how laminar flow has been used for chemical and biological reactions and analysis systems is given below.

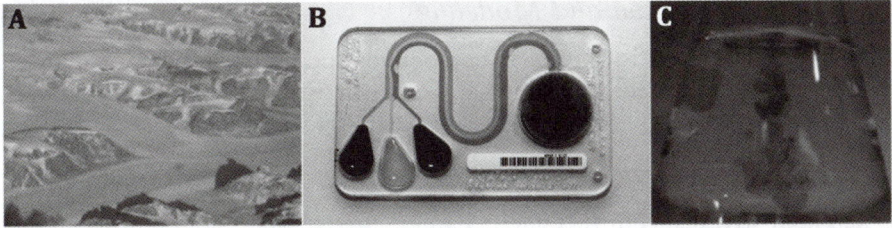

Figure 11.5 When viscous forces become important over inertial forces, we can observe laminar flow. Glacier (A), and co-flowing solutions in a microfluidic chip (B, microchip from Micronics, Inc.). In contrast, the mixing of an analyte in a glass flask is usually turbulent (C).

Box 11.1 Dimensionless numbers

Dimensionless numbers help to describe the properties of a given system. More specifically, they can be used to predict the changes due to miniaturization of a system, or due to changes in the flow velocity. Three examples mentioned in the text are given below.

The **Reynolds number (*Re*)** describes the fluid flow regime. It predicts whether turbulences can be expected, or laminar flow is present. (Turbulent flow is present for $Re > 2300$)

Viscous forces/inertial forces	$Re = \frac{\rho \cdot l \cdot U}{\eta}$	ρ: density of the solvent; l: typical length scale, *e.g.* radius of the channel; U: flow velocity; η: viscosity of the solvent

The **Péclet number (*Pe*)** predicts the channel length for a given flow velocity, required for diffusive mixing; or in other words: The number indicates whether diffusive mixing is sufficient in a given channel and at a given flow velocity.

Convection/diffusion (here of the particle)	$Pe = \frac{l \cdot U}{D}$	l: typical length scale, *e.g.* radius of the channel; U: flow velocity; D: Diffusion coefficient

The **Capillary number (*Ca*)** describes the relation of viscous forces and surface (interfacial) tension present at the interface between gas/fluid or two immiscible fluids in a system.

Viscous forces/surface tension	$Ca = \frac{\eta \cdot U}{\gamma}$	U: flow velocity; η: viscosity of the solvent; γ: surface tension

11.3.2.1 Simulation and Modeling

Fluid movement, *e.g.* distribution of compounds and transport times of analytes, in microchannels is predictable and can be calculated (by the Stokes equations).

11.3.2.2 Mixing and Diffusion

When fluid streams from two microchannels are merged, the solutions are not actively mixed. Mixing will occur, but based on molecular diffusion only. Molecular diffusion in a time τ_D can be estimated according the Einstein–Smulochowski relation: $\tau_D = l^2/2D$, where l is the distance the molecule has traversed; and D is the diffusion coefficient.

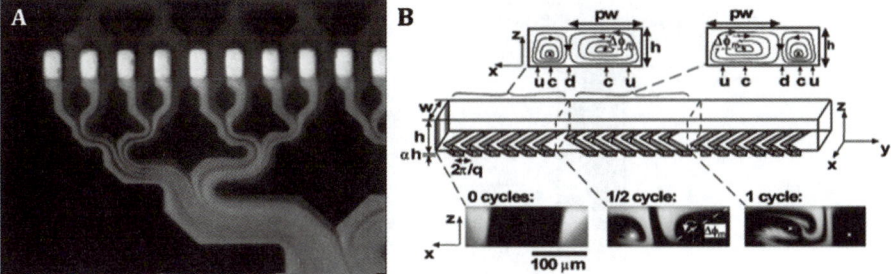

Figure 11.6 Examples of mixing in a microfluidic device. (A) Laminating mixer (Reprinted, with permission, from Stroock *et al.*[52] © 1999 Royal Society of Chemistry.) (B) "Staggered herringbone mixer", in which ridges placed on the channel floor induce chaotic mixing of water and dye solution as demonstrated in the confocal images. (Reprinted, with permission, from Oddy *et al.*[53] © 2001 American Chemical Society.)

For species with diffusion coefficients in the order of 10^{-9} (solute ions) to 10^{-11} m^2 s^{-1} (*e.g.* proteins), it will take about 10 s (ions) up to 16 min (proteins) to cross a distance of 100 µm. Hence, for a given flow velocity, it is possible to calculate the length of the channel required for complete mixing. In general, this relation of convection to diffusion is expressed by the Peclet number (Box 11.1). A closer look at the concentration distribution in a microchannel reveals that reality is frequently more complicated; in hydrodynamic flows, the parabolic flow profile has to be taken into account.[49] In case a chemical reaction is induced by co-flowing solutions, the local concentration of analyte depends additionally on the reaction rate.[50]

To enhance mixing, narrow microchannels are favorable, though not always practicable, because of their high fluidic resistance. Various kinds of mixers have been developed which exploit specific channel designs or use actively controlled mixers with external energy sources to induce turbulences. Examples of such mixers are the so-called laminating mixers, in which several streams are separated and united in turn to decrease the diffusion distance,[51] the implementation of grooved surfaces[52] or the use of electrokinetically induced instabilities introduced by Oddy *et al.*[53](Figure 11.6).

11.3.2.3 Reactions at the Interface of Two Laminar Streams

Sometimes, mixing is not required: stable interfaces between co-flowing solutions provide unique properties for fast reactions (Figure 11.7A). For example, the formation of a silver wire was achieved at the interface of a stream containing a silver salt, and a stream containing a reductant.[54]

11.3.2.4 Separation and Filtration

The laminar flow regime enables separation of molecules with different diffusion coefficients in a microfluidic device. Simple planar devices, in which two or

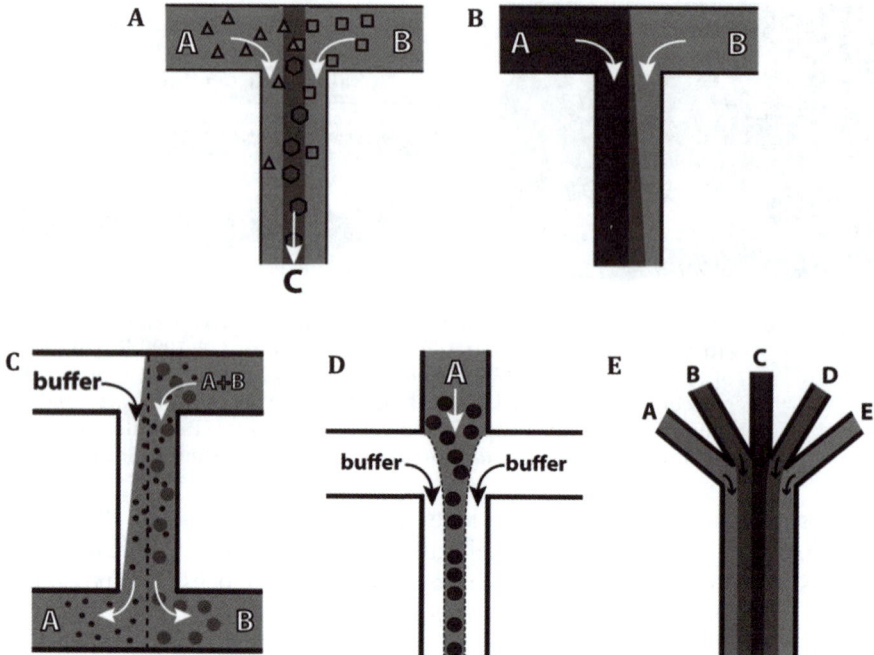

Figure 11.7 Schematic drawings to illustrate the consequences of the laminar flow present in microfluidic devices. (A) The analyte is formed at the interface between the laminar flow of reagents. (B) Analyte concentration and diffusion coefficients can be measured in this geometry called T-sensor. (C) Molecules are filtered using this channel network, called an "H-design". (D) The analyte stream is focused by two side streams. (E) Multilaminar stream can be used to generate stepwise gradients of analytes.

three channels merge into a main channel, allow small particles/molecules to diffuse across the channel, while larger particles remain at the original position. For example, a so-called T-sensor (Figure 11.7B) has been employed to measure concentrations and diffusion coefficients,[55] as well as to characterize antigen–antibody binding at the interdiffusion zone.[56] A channel network with an "H-design" could be used to filtrate molecules (Figure 11.7C).[57]

11.3.2.5 *Focusing of an Analyte Stream*

Knight *et al.* first characterized the process of hydrodynamic focusing of a stream by two side streams (Figure 11.7D).[58] Several modifications were introduced, such as elektrokinetic focusing[59] and two-dimensional focusing.[60] Focusing of the analyte stream can be of high importance to increase the detection efficiency. For example, it is frequently used in (miniaturized) cytometers to facilitate excitation of individual cells by laser light.[61] The extent of focusing, *i.e.* the width of the focused stream, depends on the applied flow rates. By varying the flow rates of the side streams, the position of the focused stream can be controlled. This can be exploited for directing the focused stream into a certain outlet channel.

11.3.2.6 Multilaminar Streams and the Generation of Stepwise and Continuous Gradients

Several adjacent laminar streams can be used to create defined gradients of analytes (Figure 11.7E); stepwise changes of analytes or continuous concentration gradients can be realized.[62] Among the exciting applications of multilaminar streams is the analysis of (single) cells or cell populations that are placed within the microchannels. Depending on the position in the microchannels, the cells are exposed to different streams, *i.e.* treated in a different way. This method has been employed for probing cellular response on different stimulating reagents, *e.g.* chemotaxis,[63] local activation of gene expression[64] or even subcellular changes upon local treatment.[65] An interesting variation is the study on *Drosophila* embryos, exposed to two adjacent streams of different temperature.[66]

The formation of laminar streams can be achieved for aqueous, or generally for miscible, solutions. In the case of non-miscible fluidic streams the interfacial tension between the two solutions should be small, and/or microchannel surfaces have to be modified properly. For example, Kitamori and co-workers[67] established, on a microchip, two-phase flows between liquid and gas, which may be useful for multiphase reactions[68] or extraction of molecules.

However, when two immiscible fluids are brought together in a microchip, we usually make another observation, *i.e.* the formation of droplets, which will be described in the next section.

11.3.3 Digital Microfluidics: Segmented Flow

Monodisperse microdroplets can be formed on a microchip by introducing an aqueous flow into a hydrophobic oil phase. In these processes, capillary forces play a crucial role as they are competing (and exceeding) the viscous forces (see Box 11.1 for an explanation of the capillary number). In other words, due to the surface tension, the interfacial area is forced to minimize, while viscous stress causes the surfaces to drag into the downstream flow. As a result, a water stream breaks into droplets when it comes in contact with another immiscible fluid stream. A microchip design for droplet formation is shown in Figure 11.8.

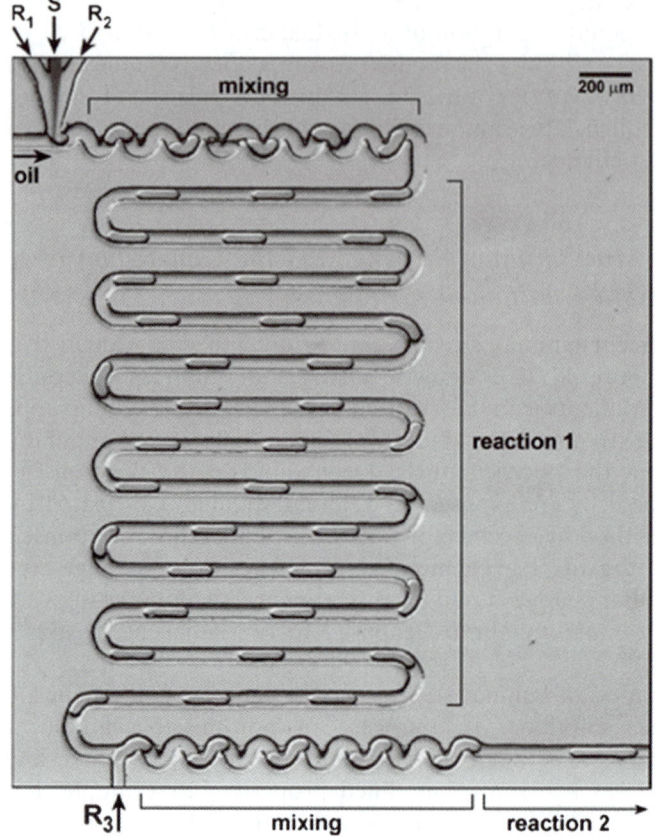

Figure 11.8 Microchip design for droplet formation and subsequent reaction in the channel. (Reprinted, with permission, from Shestopalov *et al.*[69] © 2004 Royal Society of Chemistry.)

Since the first demonstration in 2002 by Quake and co-workers,[18] the field of so-called digital microfluidics, has attracted great interest. Digital microfluidics facilitates the fast generation of femtoliter to nanoliter volumes, in which compounds for chemical reactions, or analytes or cells can be enclosed. The droplets can be filled at the very moment of formation, and the tiny, finite volume enables rapid mixing of reagents and transport along channel networks, without dispersion of reagents or contamination between individual droplets. Hence, microdroplets offer extremely valuable conditions for many screening applications. In addition, the ease of formation, *i.e.* the use of a simple planar design, certainly supports the overwhelming interest in microdroplet research. Various chip designs for formation of emulsions and double/triple emulsions have been presented, and methods for fusion, fission and directed movement of droplets are available.[70]

Droplet-based microfludics has been utilized for conduction of chemical reactions,[71] kinetic investigations,[72] particle formation,[73] and studies on crystallization of water.[74] Biological applications include PCR,[75] protein crystallization,[76] cell-free protein expression,[30] and single cell analysis.[77] A comprehensive discussion of droplet microfluidic for single cell analysis is given in chapter 17. In order to fully exploit the unique properties of droplet-based microfluidics, the analysis of individual droplets by comprehensive detection methods is key. While the majority of detection techniques are based on fluorescence assays or optical microscopy, promising approaches to extract droplets from the hydrophobic carrier and to interface the microchip with electrospray ionisation mass spectrometry have recently been developed (ESI-MS).[78,79] This is an important step towards the use of this method for proteomics research.

11.4 Fabrication Techniques and Materials

The development of fabrication processes has certainly been pushed by miniaturization demands of the electronics industry and many protocols have directly been overtaken for the production of MEMSs and microfluidic systems. However, the fabrication of useable small-scale sensors and particularly microfluidic channels has peculiarities and challenges that need to be addressed individually. Optimization and often simplification of fabrication processes, as well as the search for new suitable materials (particularly, polymers and hydrogels) and coatings are the focus of current research. Another issue is the development of interfaces between the microchip and the macroscopic peripheral devices. Finally, integration of different techniques, materials, and components is required.

In the following, a short overview of photolithography is given, and the most common materials for production of microfluidic devices are listed and evaluated. Among the alternatives for materials, the polymer PDMS plays a distinguished role, as it is by far the most common material for rapid prototyping of devices and used in many applications, especially in analytical ones. Therefore, the techniques described in this chapter have a special emphasis on the protocols for making planar PDMS devices. We furthermore give some remarks on interfaces and coatings. Detailed descriptions of microfabrication can be found in specialized textbooks, such as *Fundamentals of Microfabrication – The Science of Miniaturization* by M. Madou[80] or *Handbook of Microlithography, Micromachining and Microfabrication* by P. Rai-Choudhury.[81]

11.4.1 Photolithography

Whenever defined features with dimensions below $\approx 300\,\mu m$ are required, standard milling machines are not suitable anymore, and the use of microfabrication technologies becomes necessary. The standard processes include optical photolithography in combination with etching and deposition techniques. In bulk micromachining, three-dimensional patterns are etched into a substrate, typically a silicon wafer. In contrast, surface micromachining relies on the sequential deposition of layers of different materials such as metals, silicon dioxide or silicon nitride, into which three-dimensional structures are created.

Photolithography refers to the transfer of a pattern into a substrate, which is accomplished by exposure of a photosensitive resist deposited on the substrate to light. A photolithography mask placed between the light source and the substrate determines the pattern of light exposure. For microsized features >1 μm, exposure to UV light (from a mercury lamp, *e.g.* 365 nm) through a photolithography mask is most suitable. Another option is the direct writing by focused visible light, X-ray or an electron beam. E-beam lithography enables the creation of high-resolution patterns in the nanometer range, but the patterning of large areas is time-consuming and expensive.

The mask defines the pattern and, hence, it should be of high resolution. Photolithography masks are typically chromium-coated glass wafers with the pattern, made by e-beam lithography or direct laser writing, etched into the chromium. A low-cost alternative for feature sizes >10 μm are film masks obtained by high-quality printouts.

For photolithography, the substrate, *e.g.* a silicon wafer, is prepared with a thin layer of photoresist. This process is done on a spin coater, a device that rotates the substrate while holding it on a plate by suction. The resist dissolved in a volatile solvent is deposited onto the substrate and homogeneously distributed at rotation speeds of several thousands rpm. The substrate is then placed on a heating plate to remove the solvent, and to partly harden the thin film.

The mask is then placed on top of the photoresist layer and both are subjected to light for typically a few seconds. Photosensitive resists absorb light efficiently, but become transparent after light-induced reactions. During illumination, the resist undergoes chemical and physical rearrangements that alter its solubility in a particular solvent. Once the wafer is placed into the developer solution the photoresist layer partly dissolves. Whether the illuminated or the non-illuminated areas are eliminated by the developer solution depends on the type of photoresist. So-called positive resists become soluble in developer solvents after light exposure, and negative resists become insoluble after light exposure. Hence, a positive resist transfers the transparent areas of the mask into holes, grooves, and chambers after light exposure and development, whereas a negative resists would produce columns, ridges, and plateaus (Figure 11.9). Positive photoresists are often chosen when the pattern is transferred into the substrate, *e.g.* glass or silicon. Negative photoresists are frequently taken when a mold is needed, *e.g.* for PDMS molding. Most important to mention is SU-8, a negative photoresist that can form thick photoresist layers on the substrate (several hundred micrometers are reported),

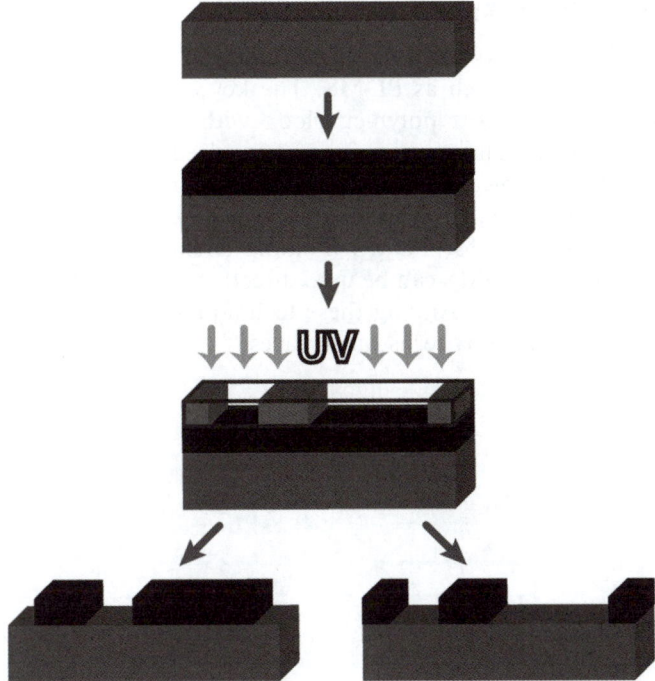

Figure 11.9 Photolithography. The substrate for the wafer (*e.g.* silicium) is coated with a photoresist. Afterwards a mask is placed onto this photoresist layer and exposure to UV light follows. During incubation in a developer solution non-illuminated structures are eliminated using a negative photoresist (left) or illuminated structures are eliminated using a positive resist (right).

and features with high aspect ratios can be processed. After development and post-baking the wafer can directly be used as mold for soft lithography (see next section).

Regardless of the type of photoresist, the preciseness of pattern transfer is limited even if the light source produces parallel beams of same intensity over the entire substrate area. Diffraction patterns on the exposed areas of the resist and at the side walls result in deviation from a perfect image of the mask in the photoresist. Typically, the precision of the features is better for thin (layers of a few micrometers thickness) than for thick resist layers, and for positive better than for negative photoresists.

Photolithography is a planar fabrication technique, *i.e.* patterns are transferred onto the microchip layer by layer. Usually, the final step is a bonding process, during which the open structures (open channels) are sealed, *e.g.* with a glass slide or with a polymer foil.

11.4.2   Soft Lithography

Soft lithography refers to molding and patterning techniques related to the use of a soft elastomer such as PDMS. The key element of soft lithography is the fabrication of a soft polymer block with patterned relief structures on its surface.[15] Although a master form is required initially for this, the following processes are typically cheap, simple, convenient, and can be done in a standard laboratory instead of a cleanroom, which explains the wide use of soft lithography. The scheme for PDMS molding is shown in Figure 11.10 and described in Box 11.2. PDMS can be used directly as a microfluidic chip, or act as a mold, stamp or mask. Among these techniques, microcontact printing is commonly used to create patterns on surfaces.[15]

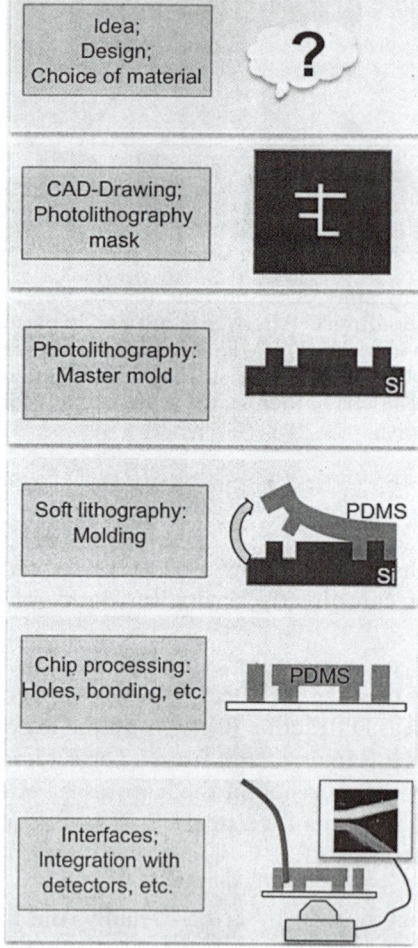

Figure 11.10   Sequence for fabrication of a polymer (PDMS) microchip by soft lithography – from design to utilization.

Box 11.2 Procedure of PDMS microchip molding

PDMS is commercially available as a two-component system, which consists of the oligomer rubber (vinyl-terminated PDMS) and a curing agent, which is a mixture of a platinum catalyst and a copolymer of methylhydrosiloxane and dimethylsiloxane (see figure below). The mixing ratio of both solutions (typically, 10:1 by weight) determines the rigidity of the hard polymer. To prepare PDMS chips, the viscous and degassed mixtures of the oligomer and curing agent is poured on top of a master form and hardened in an oven or on a hot plate. PDMS spreads out on the wafer, *i.e.* for thick microchips of a few millimeters, a frame is required that can be placed around the whole master form or the area of interest. Temperature and time of hardening can be varied over a broad range. Typically, temperatures between 70 and 150 °C are chosen and times between 10 min and 3 h. PDMS shrinks during the process, depending on the curing conditions, up to a few percent of the original size, which has to be considered if several layers are produced and overlaid.[82] Afterwards, the microchip is peeled off the wafer, cut into size if required, and holes are punched for the connection of tubing and syringes. Freshly made microchips are sticky to glass and other materials and hence, can directly be attached onto a glass slide. Stronger bonding is required for high flow rates. This is accomplished by activation of the PDMS microchip and the glass cover by means of a plasma.

Formation of reticulated PDMS (right) from oligomer and copolymer (left).

Due to the flexible nature of PDMS, it is possible to deform the microchannels. This is a disadvantage on the one hand, *e.g.* it is difficult to obtain features with high aspect ratios; on the other hand, it facilitates the controlled deformation of microchannels and thin PDMS layers to create valves. A simple version of a microvalve is a small machine screw embedded in a PDMS layer, positioned above a microchannel. Turning the screw results in deformation and finally in blocking the microchannel.[83] A similar very elegant valve and pump system has been presented by Gu *et al.*[84] They interfaced a Braille display (the system that enables blind people to read) with a microfluidic chip made of PDMS. The moving pins of the Braille display closing locally the microchannels are computer controlled, thus allowing fast and programmable processes to be conducted.

Another very viable solution to deform a PDMS microchannel is the implementation of hydraulic valves. This principle has been developed by Quake and co-workers,[17] and the fabrication process is termed multilayer soft-lithography. For construction of the valves, two PDMS layers are required (see Figure 11.11). The first (bottom) layer comprises the microfluidic channel

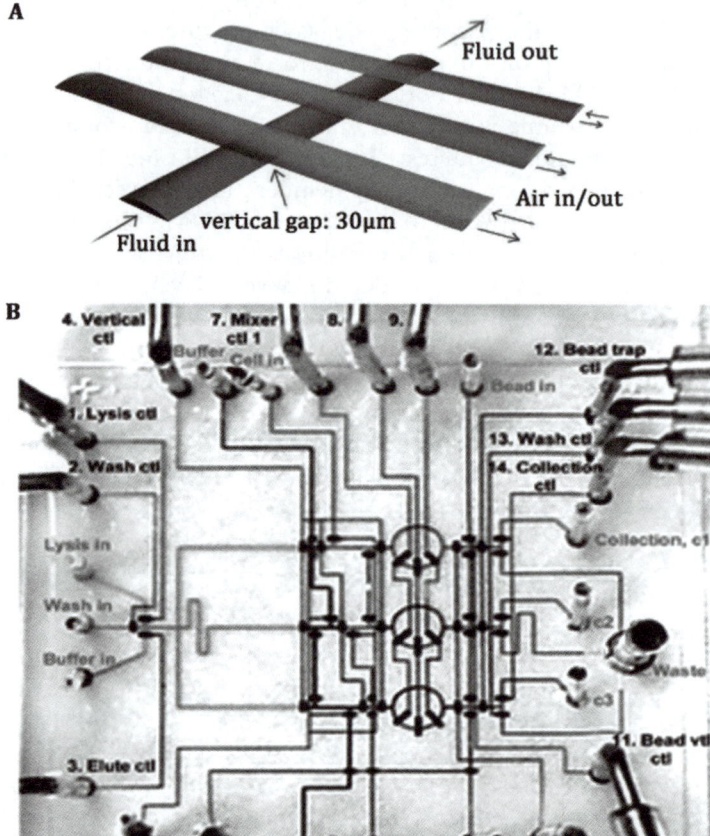

Figure 11.11   Multilayer soft lithography is used to form valves. Two layers of PDMS are bonded on top of each other. The bottom layer contains the microfluidic network, while the top layer contains dead end channels that can be filled by gas. High gas pressure seals the microchannels underneath at the cross-section of the top and bottom channel. In this way, multiple, individually addressable valves can be easily incorporated in a microdevice. A detailed sketch (A) and the whole device (B) for automated nucleic acid purification are shown. ((A) Reprinted, with permission, from Unger et al.[17] © 2000 American Association for the Advancement of Science. (B) Reprinted, with permission, from Hong et al.[85] © 2004 Macmillan Publishers Ltd.)

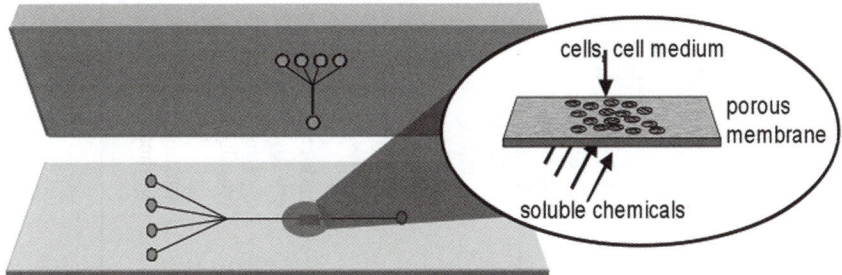

Figure 11.12 Scheme of a bilayer chip with an integrated porous membrane.

network for the transport of the sample solution. It is a thin layer; above the microchannels is a PDMS membrane of only few micrometers thickness. The second layer, situated on the top, represents the control layer. By applying pressurized gas in the top channels, the bottom layers are closed at the intersections of top and bottom layer. As pointed out by Unger *et al.*,[17] the shape of the bottom channels is important for optimal closing. Complete sealing of the channels is observed with rounded channels as it gives a continuous contact edge joining the left and right edges of the flow channel. The advantage of this method is the construction of highly complex valve patterns, which enable the formation of small liquid volumes employed for fluid metering in protein crystallization studies[86] or single-cell assays.[18]

Packaging of multiple layers to form a single microchannel network is an approach for other applications as well. Bilayer microchips, in which a porous membrane made of polycarbonate or polyester separates top and bottom microchannel networks (see Figure 11.12), have been used in cell culture studies. With this kind of device Takayama and co-workers[87] cultured pulmonary epithelial cells on the porous membrane and could mimic the flow conditions present in many pulmonary diseases ("lung-on-a-chip"). The researchers could even reproduce the typical sound created in the lung of patients suffering from lung injuries, *e.g.* asthma. In another study, a similar device was used to active the protein (GFP) expression in yeast cells; multiple laminar streams in the bottom channel were generated to define local microenvironments. The pattern of activated (fluorescent) yeast reflected the conditions created by the laminar streams.[64] Bilayer microchips could also be adapted as (micro)bioreactors for tissue growth.[88] Instead of the porous membrane, a silicon slide with integrated pores of well-defined sized and at well-defined positions has been sandwiched between two PDMS layers for temporally and spatially resolved activation of PC12-cells ("artificial synapse")[89] and formation of liposomes and lipid membrane tubules.[30]

11.4.3 Microchip Materials

Although the list of possible materials is long and we seem to be spoiled for choice, usually the requirements given by the application will lead straight to

Table 11.1 Comparison of various materials used for microchip fabrication.

	Silicon	*Glass*	*PDMS*	*Hydrogels*
Fabrication processes	Lithography; wet and dry etching	Lithography; wet etching	Soft lithography (molding); master form required	Typically no direct fabrication; combination with other materials
Bonding/sealing	Anodic bonding (and others)	Thermal bonding (and others)	Adhesion to glass, or plasma-activated bonding	By chemical modification
Price	Medium	Medium	Low	Depending on hydrogel: low to high
Optical properties	Not transparent for visible light, transparent for infrared light	Transparent for visible light	Transparent for visible light	Transparent for visible light
Electrical properties	Semiconductor	Insulator	Insulator	Conductor
Mechanical stability	High	High	Low	Very low
Thermal stability	High	High	Up to ~200 °C	Low (loss of water)
Chemical resistance	Acids, bases, organic solvents	Acids, bases, organic solvents	Acids and bases; swelling in organic solvents	Typically low
Cell adhesion on unmodified surface (typical)	Yes	Yes	No	Natural gels: yes; synthetic gels: no

one specific material. The criteria listed below have to be considered when choosing the chip material:

- Fabrication protocols: availability and price,
- Bonding and patterning,
- Optical properties: transparency (for UV, VIS, IR or all), refractive index,
- Electrical properties (*e.g.* insulating),
- Mechanical properties (*e.g.* elastomeric),
- Thermal conductivity and stability,
- Surface (energy, stability, adsorbance of material),
- Chemical properties: chemical patterning and surface modifications, resistance to chemicals,
- Toxicity, biocompatibility.

So far, there is no material that perfectly fulfils all the desired properties. In the following, the most common materials are briefly described and their strengths and weaknesses are discussed (see also Table 11.1). Alternative materials for microfluidic devices include paper,[90] metals[91] or ceramics.[92]

11.4.3.1 Silicon

Before polymers took over, most microfluidic systems were made in silicon or glass. Silicon is extensively used as a substrate for MEMS. The large number of well-known processing techniques for silicon enables the formation of complex structures, and the integration of many functional modules. Silicon is a good choice for more complex microdevices, in which sensors or detectors are integrated.[80]

Properties:

- Fabrication: many well-established fabrication processes are available. Silicon substrates can be etched isotropically or anisotropically (Figure 11.13). Integration of electronic circuits is possible.
- Silicon is not transparent for visible light, but transparent for infrared light,
- Mechanically stable.
- Resistant against most chemicals and organic solvents.
- Bonding techniques available to cover silicon microchips, *e.g.* with a glass plate.

Silicon wafers are often employed as master for planar polymer chip fabrication (Figure 11.14), or as a component of hybrid microchips made of several materials (Figure 11.15).

11.4.3.2 Glass

Glass is a perfect material for many chemical and biological applications. It is optically transparent and chemically resistant to many chemicals including

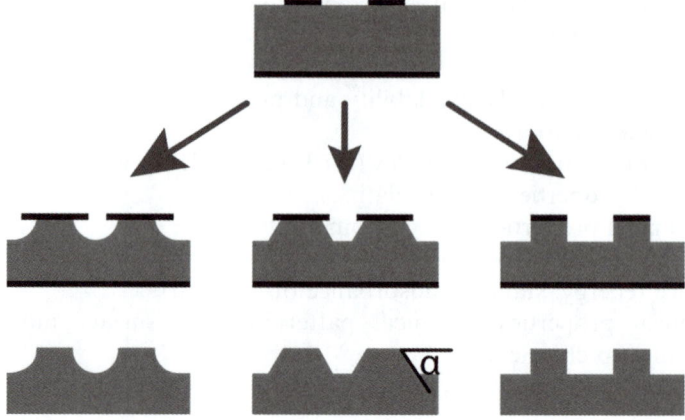

Figure 11.13   Different etching methods for silicon. Isotropic wet etching (left), *i.e.*
the etching rate is independent of the direction, is achieved using a
mixture of acids ($HF/HNO_3/CH_3COOH$). In contrast, in anisotropic
wet etching (middle), the etch rate depends on the plane of the crystalline silicone, *i.e.* in KOH the (111) plane is etched much slower than
the other planes resulting in the typical angle of $\alpha = 54.7°$ of the side
walls. Vertical side walls and high aspect ratios can be generated by
anisotropic dry etching techniques such as reactive ion etching or deep
reactive ion etching (right).

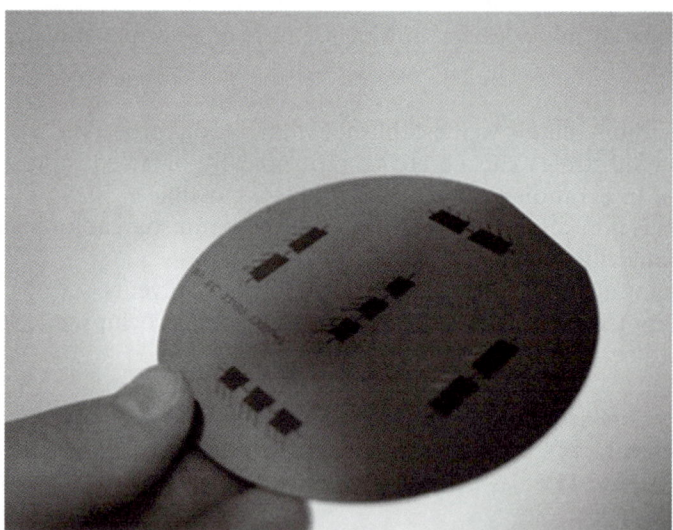

Figure 11.14   Photography of a mold used for soft lithography. The negative photoresist SU-8 on the silicon wafer contains the reliefs of five different
microchannel designs.

Figure 11.15 Photograph of a microchip made of a PDMS layer, covered with a silicon slide. Pipette tips are used for fluid control.

organic solvents. It is a dielectric material, and widely employed for microchip electrophoresis and, in general, for experiments where electro-osmotic flow is applied. Furthermore, glass is the most common material for microreactors employed in chemical synthesis.[21,93]

Glass consists mainly of non-crystalline silicon dioxide, and is available with different additives such as Na_2O or K_2O that modify its properties, *e.g.* transition temperature or dielectric constant. Borofloat (Pyrex®) glass is used very often for microstructuring, and quartz/fused silica is chosen for applications where optical transparency of ultraviolet light is required.

In spite of the convincing advantages of material properties, glass microchips are less common than polymers. The reason lies in the demanding and time-consuming manufacturing, while fewer options and fewer experiences exist for glass than for silicon micromachining. In consequence, glass chips are typically cleaned and re-used.

Structuring of glass wafers is performed most often by lithography and isotropic wet etching with hydrofluoric acid. Depending on the concentration of the acid, etching rates of a few micrometers per minute are possible.

Glass chips can be bonded to another glass plate by thermal fusion bonding. In this process, the glass substrates are thoroughly cleaned and placed in a furnace on top of each other, weighted with a load. The glass is heated to nearly transition temperature to obtain chemical bonding. Naturally, the glass substrates need to be absolutely plain and smooth, and special care is needed for cleaning. Nevertheless, although protocols for the process are available, bonding of glass chips requires special skills and expertise – and some luck. Thus, alternative bonding techniques for glass-to-glass sealing are still development.[94,95] Packaging of several glass wafers to form multilayer chips is possible as well,[96] but most glass chips have a planar design.

11.4.3.3   Polymers

Polymers are widely used for microchip fabrication. Among various techniques available for the fabrication of polymer microchips the most common are casting, injection molding, and hot embossing, which all require the fabrication of a master mold (Figure 11.14). For most polymers, techniques for electrode patterning and surface modifications are developed, and bonding is possible by thermal pressure, use of solvents or glue. Polymeric materials with various properties are available, with different mechanical stability and deformability, thermal characteristics, and surface chemistry.[97] Most of them have a good stability against aqueous acids and bases, and do not absorb water, but dissolve or swell in organic solvents.

The thermoplastics poly(methyl methacrylate) (PMMA), polycarbonate (PC) and cyclo-olefin polymers are commonly used for microfluidic chip fabrication. By far the most often utilized polymer, however, is PDMS and will be discussed separately in the following section.

11.4.3.3.1   Poly(dimethylsiloxane).   Among the polymers, PDMS plays a special role, since it is very valuable in research applications, in which the fluid is water or an aqueous buffer. Molding of PDMS microchips by soft lithography (section 11.4.2) is particularly simple, easy to learn and suitable for rapid prototyping. Feature sizes lie typically between $\sim 0.5\,\mu m$ up to a few hundred micrometers. High aspect ratios of the features are critical, *e.g.* a design with low height but large width will cause sagging of the channel top. Integration into larger systems and creation of composite devices is possible.

PDMS has some valuable properties, including:

- Fabrication of PDMS chips by molding, suitable for rapid prototyping, most processes can be done in standard laboratories.
- Low price per copy (initial costs can be high), often single-use microchips.
- Optical characteristics: transparent for visible light down to ~ 300 nm, *i.e.* straightforward adaptation to optical microscopy.
- Flexibility: the elasticity of PDMS still allows tight bonding if tubes and other components are incorporated into the microchip, as well as simple formation of multilayer chips and valves, and thin membranes.
- Insulating material.
- Thermal stability up to $200\,°C$.
- Good stability when aqueous liquids are used, but swells in most organic solvents.

- It is generally regarded as a non-toxic material, *e.g.* useful for cell culture chips as long as only channel walls are made of PDMS and the ground plate of the microchip is glass or another cell adhesion supporting material. Depending on the cell type, direct growth of (viable) cells on PDMS is weak or not possible.
- Hydrophobic surface: can be transferred to a transient hydrophilic surface by plasma exposure, which facilitates bonding or surface coating.

However, PDMS also has some weaknesses, including:

- Aging effects: a plasma-treated, hydrophilic PDMS surface converts back into a hydrophobic surface over time (Figure 11.16). In general, the surface hydrophobicity is difficult to precisely predict because any contact to air affects the surface.
- Gas permeable: a positive characteristic for many applications. During long-term use of PDMS microchips, annoying formation of air bubbles can be observed. The consequences of evaporation through PDMS for long-term cell studies on microchips, *e.g.* osmolarity shifts, have been described by Heo *et al.*[98]
- Likewise, PDMS absorbs small hydrophobic molecules, such as lipids or dye molecules, that diffuse afterwards into the bulk polymer. This dynamic process changes the concentration of molecules in solution, which finally can impact the result of the experiment and need to be taken into consideration.[99]

Several studies report on derivatives of PDMS, additives to PDMS, or surface modifications of PDMS[100] to change material properties. For example, the stiffness/elasticity of PDMS is of importance for precise molding of nanometer-sized features.[101] Some recent developments on modified PDMS that can be directly patterned by photolithography should be mentioned as well. It combines the advantage of a photoresist that allows pattern transfer from a mask by exposure to UV light, while exhibiting most of the advantages of PDMS. Detailed properties of commercially available photopatternable PDMS, *i.e.* the silicon resin dissolved in a plasticizing matrix, are described by J. Voldman and co-workers.[102] In another study by Choi and Rogers,[103] a prepolymer consisting of long linear PDMS chains was prepared, and urethane methacrylate groups served as rigid and photocurable cross-linkers.

11.4.3.4 Hydrogels

Hydrogels are biocompatible, porous, cross-linked networks of natural or synthetic polymers. While the polymers are insoluble in water, they can take up a high amount of water (90–99% w/w) due to hydrophilic groups. Hydrogels are usually mechanically unstable, *i.e.* it is difficult to imprint small structures in hydrogels and hence to use them directly as chip substrate. Few examples exist

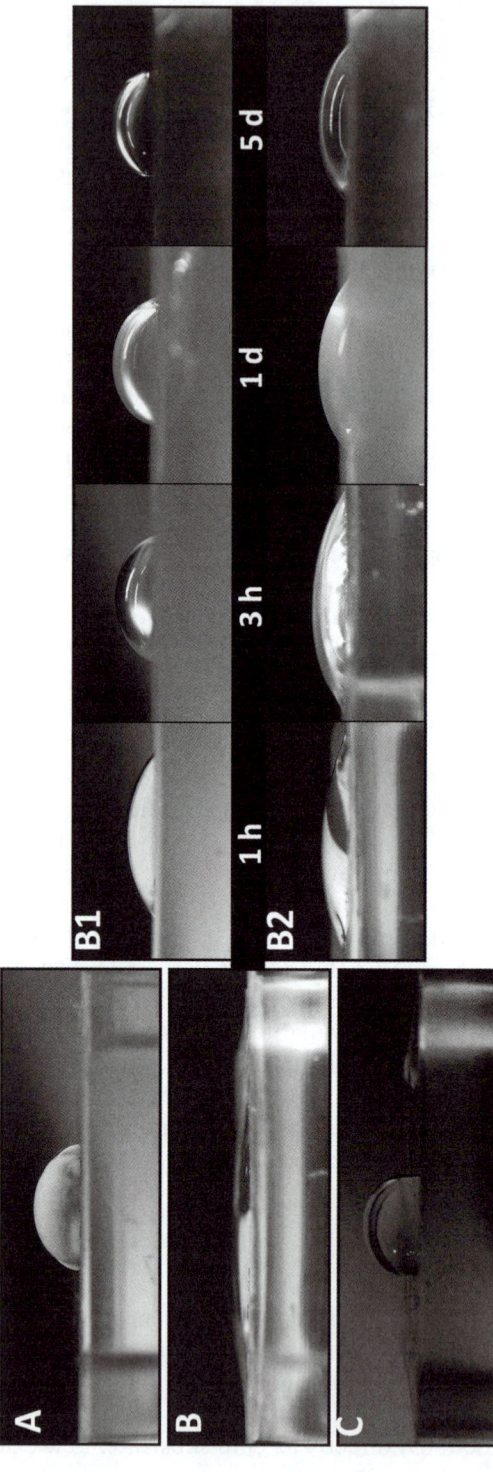

Figure 11.16 Hydrophobicity of the PDMS surface, illustrated by placing a water droplet on top of the surface. Freshly prepared PDMS is hydrophobic (A), but can be transferred to hydrophilic surface by plasma treatment (B). However, the surface converts back over time (C). Within a few hours up to a few days, the surface properties alter significantly. After a few days storage in ambient air (B1), the hydrophobicity is even stronger than for freshly prepared PDMS. The plasma-activated, hydrophilic surface can be partly retained when the microchip is stored in water (B2).

where the full microchip was prepared from a hydrogel, *e.g.* from alginate.[104] Another example is the alginate stamp to produce patterns of bacterial cells on a surface.[105] Typically, hydrogels are used in hybrid devices that are made of a mechanically stable material, while the hydrogel is introduced into the micro-chambers or microchannels, where, for example, it forms the functional matrix for tissue formation.[106] In Table 11.2 a selection of materials used as hydrogels in microfluidics is listed. Their physical properties facilitate reversible volume shrinking or expansion in response to applied signals such as electric field, ionic strength, light, pH, small (bio)molecules, solvent or temperature. Such responsive hydrogels are employed as functional elements like valves or sensors. A comprehensive review on this issue is given by Eddington and Beebe.[107]

Hydrogels are divided into two classes, depending on their natural or synthetic origin.

Properties of natural hydrogels:

- Polymer chains are cross-linked *via* ionic or physical bonds.
- Reversible cross-linkage.
- Assure an environment of endogenous signals for cellular interactions to form tissue.
- Variability in composition, complexity, and properties make it difficult to analyze cellular behavior of embedded cells.

Properties of synthetic hydrogels:

- Polymer chains are cross-linked via covalent bonds.
- Most common gelation mechanisms are radical chain polymerization and chemical cross-linking.
- Consistent composition.
- Predictable and controllable manipulation of properties is possible.
- Permanent cross-linkage.
- Lack of functional sites. Modifications with natural chemistries can avoid this disadvantage in many biological applications.

The huge variability of precise controllable physical and chemical properties of hydrogels allow applications in many biological research areas such as biological sensing, cell encapsulation, (co)culturing of mammalian cells, drug delivery, construction of an extracellular matrix or tissue engineering.[108–111] Each application requires special and high demands of the used polymeric network. In tissue engineering (Figure 11.17), for instance, mechanical properties are very important, as the gelation process must not affect the cell vitality. The hydrogel has to generate and preserve enough space for tissue development, and its degradation by hydrolysis, enzyme activity or dissolution must directly follow tissue growth. With this aspect, low toxicity of degradation products as well as cross-linking agents is necessary. With respect to biological applications, it is noteworthy that the permeability for oxygen, nutrients, and waste is generally high, and inert surfaces in modified hydrogels can prevent

Table 11.2 Selection of hydrogels for cell culture and tissue engineering used in microfluidic devices.

Material	Used for	Embedded cell type
Natural polymers		
Agarose	Engineered 3D tissue model[113]	Murine embryonic fibroblast cells (NIH-3T3)
Alginate	Co-culturing cells[109]	Human fetal lung fibroblast cells (HFL1) and human hepatocellular liver cells (HepG2)
	Microfluidic patterning[112]	Yeast cells
	Tissue engineering[106]	Hepatoblastoma cells (HepG2/C3A)
	Co-culturing cells[114]	Breast cancer cells (MDA-MB-231) and macro-phages tumour cells (RAW-264.1)
	Modeling cellular environment[115]	Murine embryonic fibroblast cells (NIH-3T3)
Collagen and gelatin	Formation of hydrogels containing microfluidic networks[116]	Human dermal fibroblasts
	Vitrified collagen as functional and sacrificial material for 3D cell culture[117]	Epithelial cells, keratocytes
Synthetic polymers		
Poly(acrylic acid) and derivatives	3D channel topography for cell capture[118]	K562 and L929 cells
Poly(ethylene oxide) and copolymers	Polymerization of hydrogel microstructures[119]	Cluster of differentiation 4 T-helper cells (CD4 + T)

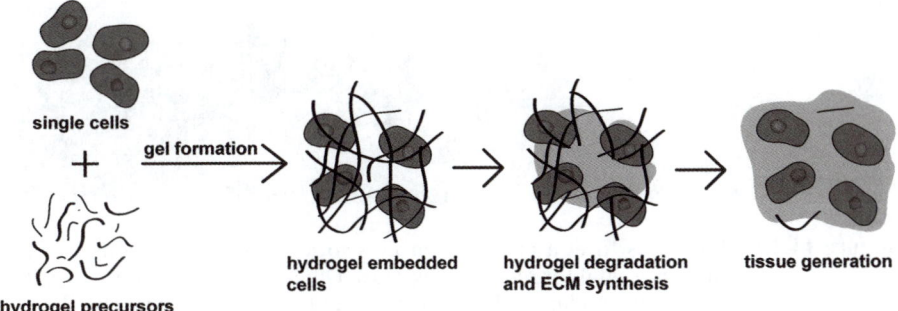

Figure 11.17 Cell encapsulation process. After the mixing of single cells with hydrogel precursors in a liquid solution, the gel is formed. Encapsulated cells synthesize the extracellular matrix (ECM). For tissue engineering the dence of the cross-linked scaffold and its degradation is important to enable proper ECM synthesis.

unspecific adsorption of proteins. Furthermore, it is possible to incorporate peptides like Arg-Gly-Asp or Ile-Lys-Val-Ala-Val to form bioactive molecules, or receptors to obtain bioresponsive/biointeractive materials.[108]

For applications, especially in microfluidic devices, the weak adhesion of hydrogels to surfaces must be considered. However, coating the channel surfaces strongly improves the adhesion, *e.g.* a poly(ethylenimine) (PEI)-coated alginate gel adheres tightly to glass.[112] The strong correlation between cell vitality and nutrient profiles across cell-laden microfluidic 3D hydrogels can also not be disregarded. Due to nutrient consumption, diffusion of nutrients is only possible near the media source and hence, cell death increases with greater distance between cells and media.[113]

11.4.4 From Fabrication to Application

11.4.4.1 Interfacing

The replication of microchip devices is straightforward, once the fabrication protocols are established. The connection to peripheral devices, however, often requires additional care. Bridging the microscopic and the macroscopic world, as well as connecting one microchip module to another one, is not always straightforward.[120] Methods for liquid supply, interfaces for flow control, and imaging/detection systems are required. Figure 11.18 shows some examples of how microchips can be connected to tubings. However, the solutions for these challenges have to be solved separately for each kind of material and application.

Another issue is the adaptation to conventional instruments. Solutions to simply mount a microfluidic chip onto an existing device would certainly help to further distribute the use of microchips. For example, microsystems with dimensions according to standard multi-well plates could be directly integrated

Figure 11.18   Interfaces form a micro- to macro-world. Leakage-free connection to
tubings is vital for reliable flow rates. Different solutions are shown: (A)
PDMS microchip with metal plugs and silicone tubings; (B) PMMA
microchip with NanoPorts (Upchurch Scientific); (C) Casing made of
polycarbonate (PC) with fluidic and electronic connections. A PDMS
microfluidic chip is embedded.

on standard microscope stages and instruments.[121] With respect to analytical
applications, the development of interfaces to mass spectrometers, NMR
devices or X-ray instruments is a focus of research. With this respect, the
miniaturization of the entire analytical techniques is intriguing (*e.g.* micro-
scopes-on-chip,[122] MS-on-chip[123] or NMR-on-chip[124]).

A further approach to simplify the use of microchips for non-experts and to
increase the flexibility of microchips is the construction of a complete system
from individual assembly blocks made of PDMS[125,126] or for chemical syntheses blocks made of metals.[127]

11.4.4.2   *Coatings and Surface Patterning*

As pointed out before, physical and chemical characteristics of the surfaces can
become important for particular applications. Hence, it is often required to
modify the surface, either during microchip fabrication (*e.g.* before sealing) by
dipping the entire microchip into the solution with modifying reagent, or by
treatment with a vaporized compound. Flushing the channel with the dissolved
coating material and allowing some time for adsorption to and reaction on the
channel walls enable a covalent or transient coating of microchannels. Sometimes, the surface has to be activated. This can be achieved for PDMS and glass
by means of oxygen plasma. The process increases the wettability of PDMS[128]
and allows direct binding of ligands. For example, Delamarche and co-workers[129]
derivatized the plasma-activated surface of PDMS with poly(ethyleneglycol)
(PEG)-disilane to prevent adsorption of proteins. Other surface coatings are
employed to render the surface passive, to obtain a stable hydrophobic surface
for immobilizing proteins and allowing cell adhesion.[100,130] Some examples of
surface modifications are given in Table 11.3.

Heterogeneous surface modifications, *i.e.* patterned surfaces, are valuable to
locally vary the reaction or binding conditions. For the creation of pattern,
several versatile techniques are available, such as dip-pen lithography, micro-
contact printing or selective chemical treatment. For example, Yousaf and

Table 11.3 Selection of recent surface modifications to change the cell adhesion of materials used in microfluidics.

Material	Cell adhesion	Shown for
3-aminopropytriethoxysilane (APTES)	Enhance	Glass[133]
Collagen	Enhance	Glass[133]
Fibronectin	Enhance	PDMS[134]
Gelatin	Enhance	PDMS[135]
Glutaraldehyde	Enhance	Glass[133]
Poly(ethylenimine) (PEI)	Enhance	PDMS[136]
Laminin	Enhance	PDMS[136]
Magnetron-sputter deposition of aluminium	Enhance	PDMS[137]
Oligopeptides containing cell adhesion sequence	Enhance	Gold-coated glass[138]
Poly-L-lysine	Enhance	PDMS[134]
Alkanethiol	Prevent	Gold/silver coated glass[139]
Pluronic (Tween 20)	Prevent	PDMS[134]
Poly(acrylamide) (PAAm)	Prevent	Silicon[140]
Poly(ethylene glycol) (PEG)	Prevent	PDMS[135], Silicon[141]
Poly(ethylene oxide) (PEO)	Prevent	PDMS[134]
Poly(dimethylsiloxane) (PDMS)	Prevent	Glass[133]
Poly(tetrafluoroethylene) (PTFE)	Prevent	Glass[133]

co-workers[131] altered a hydroxy-terminated surface by oxidization at two different conditions. Flushing a mild oxidant through a microchannel that was positioned on top of the initial modified surface induced the local oxidation. A short oxidation time transferred the hydroxy groups into an aldehyde, longer times resulted in formation of carboxylic acid tail groups. The resulting surfaces are used afterwards for selective immobilization of primary amines or amino acids, to form either an oxime or an amide. In another valuable method, surfaces are first patterned with a metal layer such as gold by means of standard photolithographic processes and sputtering. Afterwards, self-assembled monolayers are formed, cross-connected by linkers that have a thiol group as alkanethiols (see, for example, Roberts *et al.*[132]).

11.5 Concluding Remarks

The development and applications of microfluidic devices have increased dramatically in the last decade, particularly for use in research fields related to life sciences. The benefits that microfluidic devices provide for analytical and diagnostic applications are convincing, *e.g.* smaller analyte consumption, faster heat and mass transport increase the controllability as well as the efficiency of reactions and separations. However, throughput in terms of volume per time is usually much smaller than in macrofluidic devices, and further downscaling requires a trade-off between internal volumes and detection limits. In other words, the number of analytes that can be detected is dramatically decreased in a small volume, which sets a limit to device miniaturization for many analytical applications. Nevertheless, much effort focuses on further miniaturization of fluidic systems to explore the unique properties and phenomena in the nanometer regime, where device dimensions are close to molecular length scales.

On one hand, there are many activities on the commercial market and numerous companies that offer services on microchip fabrication, microchip modules such as valves or pumps, or complete microfluidics systems. On the other hand, the implementation of microfluidic components into commercial instruments is certainly not fully established. However, research in microfluidics frequently bears novel and exciting ideas, applications, and usage, which surely are not fully explored. Besides the key applications including drug discovery, systems biology, and regenerative medicine research, we anticipate further impact for synthetic biology and the creation of biomimicking systems as well as for the generation of soft matter devices, where microfluidics helps to engineer, handle, and manipulate small-scaled objects.

References

1. www.biomems.net.
2. www.lab-on-chip.com.
3. www.memsnet.org.
4. www.rsc.org/Publishing/Journals/lc/Chips_and_Tips/index.asp.
5. S. C. Terry, J. H. Jerman and J. B. Angell, *IEEE Trans. Electron Devices*, 1979, **26**, 1880.
6. W. Olthuis, B. H. van der Schoot, F. Schavez and and P. Bergfeld, *Sens. Actuators*, 1989, **17**, 279.
7. S. Shoji, M. Esashi and T. Matsuo, *Sens. Actuators*, 1988, **14**, 101.
8. D. J. Harrison, K. Fluri, K. Seiler, Z. Fan, C. S. Effenhauser and A. Manz, *Science*, 1993, **261**, 895.
9. C. S. Effenhauser, A. Manz and H. M. Widmer, *Anal. Chem.*, 1993, **65**, 2637.
10. A. Manz, N. Graber and H. M. Widmer, *Sens. Actuators B1*, 1990, 244.
11. A. Manz, Y. Miyahara, J. Miura, Y. Watanabe, H. Miyagi and K. Sato, *Sens. Actuators B1*, 1990, 249.
12. D. R. Reyes, D. Iossifidis, P.-A. Aurox and A. Manz, *Anal. Chem.*, 2002, **74**, 2623.
13. P.-A. Aurox, D. Iossifidis, D. R. Reyes and A. Manz, *Anal. Chem.*, 2002, **74**, 2637.
14. D. C. Duffy, J. C. McDonald, O. J. A. Schueller and G. M. Whitesides, *Anal. Chem.*, 1998, **70**, 4974.
15. Y. Xia and G. M. Whitesides, *Annu. Rev. Mater. Sci.*, 1998, **28**, 153.
16. R. Mukhopadhyay, *Anal. Chem.*, 2007, **79**, 3248.
17. M. A. Unger, H.-P. Chou, T. Thorsen, A. Scherer and S. R. Quake, *Science*, 2000, **288**, 113.

18. T. Thorsen, S. Maerkl and S. R. Quake, *Science*, 2002, **298**, 580.
19. A. Wheeler, *Science*, 2008, **322**, 539.
20. K. Jähnisch, V. Hessel, H. Löwe and M. Baerns, *Angew. Chem. Int. Ed.*, 2004, **43**, 406.
21. H. Pennemann, P. Watt, S. J. Haswell, V. Hessel and H. Löwe, *Org. Process Res. Dev.*, 2004, **8**, 422.
22. A. M. Elizarov, *Lab Chip*, 2009, **9**, 1326.
23. M. U. Kopp, A. J. de Mello and A. Manz, *Science*, 1998, **280**, 1046.
24. E. A. Ottesen, J. W. Hong, S. R. Quake and J. R. Leadbetter, *Science*, 2006, **314**, 1464.
25. C. P. Fredlake, D. G. Hert, C.-W. Kan, T. N. Chiesl, B. E. Root, R. E. Forster and A. E. Barron, *Proc. Natl. Acad. Sci. U. S. A.*, 2008, **105**, 476.
26. J. O. Tegenfeldt, C. Prinz, H. Cao, R. L. Huang, R. H. Austin, S. Y. Chou, E. C. Cox and J. C. Sturm, *Anal. Bioanal. Chem.*, 2004, **378**, 1678.
27. Y. M. Wang, J. O. Tegenfeldt, W. Reisner, R. Riehn, X.-J. Guan, L. Guo, I. Golding, E. C. Cox, J. Sturm and R. H. Austin, *Proc. Natl. Acad. Sci. U.S.A.*, 2005, **102**, 9796.
28. E. Y. Chan, N. M. Goncalves, R. A. Haeusler, A. L. Hatch, J. W. Larson, A. M. Maletta, G. R. Yantz, E. D. Carstea, M. Fuchs, G. G. Wong, S. R. Gullans and R. Gilmanshin, *Genome Res.*, 2004, **14**, 1137.
29. J. El-Ali, P. K. Sorger and K. F. Jensen, *Nature*, 2006, **442**, 403.
30. P. S. Dittrich, M. Heule, P. Renaud and A. Manz, *Lab Chip*, 2006, **6**, 488.
31. P. Yager, T. Edwards, E. Fu, K. Helton, K. Nelson, M. R. Tam and B. H. Weigel, *Nature*, 2006, **442**, 412.
32. S. Takayama, E. Ostuni, P. LeDuc, K. Naruse, D. E. Ingber and G. M. Whitesides, *Chem. Biol.*, 2003, **10**, 123.
33. S. Fiedler, S. G. Shirley, T. Schnelle and G. Fuhr, *Anal. Chem.*, 1998, **70**, 1909.
34. A. Y. Fu, C. Spence, A. Scherer, F. H. Arnold and S. R. Quake, *Nature*, 1999, **17**, 1109.
35. F. K. Balagaddé, L. You, C. L. Hansen, F. H. Arnold and S. R. Quake, *Science*, 2005, **309**, 137.
36. A. Groisman, C. Lobo, H. Cho, J. K. Campbell, Y. S. Dufour, A. M. Stevens and A. Levchenko, *Nat. Methods*, 2005, **2**, 685.
37. C. J. Ingham, A. Sprenkels, J. Bomer, D. Molenaar, A. van den Berg, J. E. T. van Hylckama Vlieg and W. M. de Vos, *Proc. Natl. Acad. Sci. U.S.A.*, 2007, **104**, 18217.
38. T. H. Park and M. L. Shuler, *Biotechnol. Prog.*, 2003, **19**, 243.
39. Y. Tanaka, K. Morishima, T. Shimizu, A. Kikuchi, M. Yamato, T. Okanobe and T. Kitamori, *Lab Chip*, 2006, **6**, 362.
40. M. J. Kim and K. S. Breuer, *Small*, 2008, **4**, 111.
41. F. Qian, M. Baum, Q. Gu and D. E. Morse, Lab Chip, 2009, 9, doi: 10.1039/b910586g.
42. P. S. Dittrich and A. Manz, *Nat. Rev. Drug Discovery*, 2006, **5**, 210.
43. T. M. Squires and S. R. Quake, *Rev. Mod. Phys.*, 2005, **77**, 977.

44. P. Tabeling, *Introduction to Microfluidics,* Oxford University Press, 2005.
45. O. Geschke, H. Klank and P. Telleman, *Microsystem Engineering of Lab-on-a-Chip Devices,* Wiley-VCH, Weinheim, 2008.
46. S. M. Langelier, D. S. Chang, R. I. Zeitoun and M. A. Burns, *Proc. Natl. Acad. Sci. U. S. A.*, 2009, **106**, 12617.
47. P. H. Paul, M. G. Garguilo and D. J. Rakestraw, *Anal. Chem.*, 1998, **70**, 2459.
48. D. C. Duffy, H. L. Gillis, J. Lin, N. F. Sheppard and G. J. Kellog, *Anal. Chem.*, 1999, **71**, 4669.
49. R. F. Ismagilov, A. D. Stroock, P. J. A. Kenis, G. Whitesides and H. A. Stone, *Appl. Phys. Lett.*, 2000, **76**, 2376.
50. J.-B. Salmon, C. Dubrocq, P. Tabeling, S. Charier, D. Alcor and L. Jullien, *Anal. Chem.*, 2005, **77**, 3417.
51. F. G. Bessoth, A. J. de Mello and A. Manz, *Anal. Commun.*, 1999, **36**, 213.
52. A. D. Stroock, S. K. W. Dertinger, A. Ajdari, I. Mezic, H. A. Stone and G. M. Whitesides, *Science*, 2002, **295**, 647.
53. M. H. Oddy, J. G. Santiago and J. C. Mikkelsen, *Anal. Chem.*, 2001, **73**, 5822.
54. P. J. A. Kenis, R. F. Ismagilov and G. M. Whitesides, *Science*, 1999, **285**, 83.
55. B. H. Weigl and P. Yager, *Science*, 1999, **283**, 346.
56. A. Hatch, A. E. Kamholz, K. R. Hawkins, M. S. Munson, E. A. Schilling, B. H. Weigl and P. Yager, *Nat. Biotechnol.*, 2001, **19**, 461.
57. J. P. Brody and P. Yager, *Sens. Actuators A: Phys.*, 1997, **58**, 13.
58. J. B. Knight, A. Vishwaath, J. P. Brody and R. H. Austin, *Phys. Rev. Lett.*, 1998, **80**, 3863.
59. S. C. Jacobson and J. M. Ramsey, *Anal. Chem.*, 1997, **69**, 3212.
60. C. Simonnet and A. Groisman, *Appl. Phys. Lett.*, 2005, **87**, 114104.
61. P. S. Dittrich, M. Jahnz and P. Schwille, *ChemBioChem*, 2005, **6**, 811.
62. J. Pihl, J. Sinclair, E. Sahlin, M. Karlsson, F. Petterson, J. Olofsson and O. Orwar, *Anal. Chem.*, 2005, **77**, 3897.
63. W. Saadi, S.-J. Wang, F. Lin and N. L. Jeon, *Biomed. Microdev.*, 2006, **8**, 109.
64. F. Kurth, C. A. Schumann, L. M. Blank, A. Schmid, A. Manz and P. S. Dittrich, *J. Chromatogr. A*, 2008, **1206**, 77.
65. S. Takayama, E. Ostuni, P. LeDuc, K. Naruse, D. E. Ingber and G. M. Whitesides, *Nature*, 2001, **411**, 1016.
66. E. M. Lucchetta, J. H. Lee, L. A. Fu, N. H. Patel and R. F. Ismagilov, *Nature*, 2005, **434**, 1134.
67. A. Hibara, S. Iwayama, S. Matsuoka, M. Ueno, Y. Kikutani, M. Tokeshi and T. Kitamori, *Anal. Chem.*, 2005, **77**, 943.
68. J. Kobayashi, Y. Mori, K. Okamoto, R. Akiyama, M. Ueno, T. Kitamori and S. Kobayashi, *Science*, 2004, **304**, 1305.
69. I. Shestopalov, J. D. Tice and R. F. Ismagilov, *Lab Chip*, 2004, **4**, 316.

70. D. R. Link, E. Grasland-Mongrain, A. Duri, F. Sarrazin, Z. Cheng, G. Cristobal, M. Marquez and D. A. Weitz, *Angew. Chem. Int. Ed.*, 2006, **45**, 2556.
71. H. Song, J. D. Tice and R. F. Ismagilov, *Angew. Chem. Int. Ed.*, 2003, **115**, 791.
72. H. Song and R. F. Ismagilov, *J. Am. Chem. Soc.*, 2003, **125**, 14613.
73. A. Günther, S. A. Khan, M. Thalmann, F. Trachsel and K. F. Jensen, *Lab Chip*, 2004, **4**, 278.
74. C. A. Stan, G. F. Schneider, S. S. Shevkoplyas, M. Hashimoto, M. Ibanescu, B. J. Wiley and G. M. Whitesides, *Lab Chip*, 2009, **9**, 2293.
75. M. Curcio and J. Roeraade, *Anal. Chem.*, 2003, **155**, 791.
76. B. Zheng, L. S. Roach and R. F. Ismagilov, *J. Am. Chem. Soc.*, 2003, **125**, 11170.
77. M. He, J. S. Edgar, G. D. M. Jeffries, R. M. Lorenz, J. P. Shelby and D. T. Chiu, *Anal. Chem.*, 2005, **77**, 1539.
78. L. M. Fidalgo, G. Whyte, B. T. Ruotolo, J. L. P. Benesch, F. Stengel, C. Abell, C. V. Robinson and W. T. S. Huck, *Angew. Chem. Int. Ed.*, 2009, **48**, 3665.
79. R. T. Kelly, J. S. Page, I. Marginean, K. Tang and R. D. Smith, *Angew. Chem. Int. Ed.*, 2009, **48**, 6832.
80. M. J. Madou, *Fundamentals of Microfabrication: The Science of Miniaturization*, CRC Press LLC, Boca Raton, FL, 2002.
81. P. Rai-Choudhury, *Handbook of Microlitography, Micromachining and Microfabrication*, SPIE Optical Engineering Press, Bellingham (WA, USA) and London (UK), 1997.
82. S. W. Lee and S. S. Lee, *Microsyt. Technol.*, 2008, **14**, 205.
83. D. B. Weibel, M. Kruithof, S. Potenta, S. K. Sia, A. Lee and G. M. Whitesides, *Anal. Chem.*, 2005, **77**, 4726.
84. W. Gu, X. Zhu, N. Futai, B. S. Cho and S. Takayama, *Proc. Natl. Acad. Sci. U. S. A.*, 2004, **101**, 15861.
85. J. W. Hong, V. Studer, G. Hang, W. French Anderson and S. R. Quake, *Nat. Biotechnol.*, 2004, **22**, 435.
86. C. L. Hansen, E. Skordalakes, J. M. Berger and S. R. Quake, *Proc. Natl. Acad. Sci. U. S. A.*, 2002, **99**, 16531.
87. D. Huh, H. Fujioka, Y.-C. Tung, N. Futai, R. Paine, J. B. Grotberg and S. Takayama, *Proc. Natl. Acad. Sci. U. S. A.*, 2007, **104**, 18886.
88. G. Mehta, M. J. Kiel, J. W. Lee, N. Kotov, J. J. Linderman and S. Takayama, *Adv. Funct. Mater.*, 2007, **17**, 2701.
89. M. C. Peterman, J. Noolandi, M. S. Blumenkranz and H. A. Fishman, *Proc. Natl. Acad. Sci. U. S. A.*, 2004, **101**, 9951.
90. X. Li, J. Tian, T. Nguyen and W. Shen, *Anal. Chem.*, 2008, **80**, 9131.
91. L. Martínez-Latorre, P. Ruiz-Cebollada, A. Monzón and E. García-Bordejé, Catal. Today, 2009, doi: 10.1016/j.cattod.2009.07.008.
92. C. M. Mitchell and P. J. A. Kenis, *Lab Chip*, 2006, **6**, 1328.
93. K. Sato, A. Hibara, M. Tokeshi, H. Hisamoto and T. Kitamori, *Anal. Sci.*, 2003, **19**, 15.

94. L. Chen, G. Luo, K. Liu, J. Ma, B. Yoa, Y. Yan and Y. Wang, *Sens. Actuators, B*, 2006, **119**, 335.
95. P. B. Allen and D. T. Chiu, *Anal. Chem.*, 2008, **80**, 7153.
96. A. Daridon, V. Fascio, J. Lichtenberg, R. Wütrich, H. Langen, E. Verpoorte and N. F. de Rooij, *J. Anal. Chem.*, 2001, **371**, 261.
97. H. Becker and C. Gärtner, *Anal. Bioanal. Chem.*, 2008, **390**, 89.
98. Y. S. Heo, L. M. Cabrera, J. W. Song, N. Futai, Y.-C. Tung, G. D. Smith and S. Takayama, *Anal. Chem.*, 2007, **79**, 1126.
99. M. W. Toepke and D. J. Beebe, *Lab Chip*, 2006, **6**, 1484.
100. H. Makamba, J. H. Kim, K. Lim, N. Park and J. H. Hahn, *Electrophoresis*, 2003, **24**, 3608.
101. H. Schmid and B. Michel, *Macromolecules*, 2000, **33**, 3042.
102. S. P. Desai, B. M. Taff and J. Voldman, *Langmuir*, 2008, **24**, 575.
103. K. M. Choi and J. A. Rogers, *J. Am. Chem. Soc.*, 2003, **125**, 4060.
104. M. Cabodi, N. W. Choi, J. P. Gieghorn, C. S. D. Lee, L. J. Bonassar and A. D. Stroock, *J. Am. Chem. Soc.*, 2005, **127**, 13788.
105. D. B. Weibel, A. Lee, M. Mayer, S. F. Brady, D. Bruzewicz, J. Yang, W. R. DiLuzio, J. Clardy and G. M. Whitesides, *Langmuir*, 2005, **21**, 6436.
106. N. W. Choi, M. Cabodi, B. Held, J. P. Glenghorn, L. J. Bonassar and A. D. Stroock, *Nat. Mater.*, 2007, **6**, 908.
107. D. T. Eddington and D. J. Beebe, *Adv. Drug Deliv. Rev.*, 2004, **56**, 199.
108. R. V. Ulijn, N. Bibi, V. Jayawarna, P. D. Thornton, S. J. Todd, R. J. Mart, A. M. Smith and J. E. Gough, *Mater. Today*, 2007, **10**, 40.
109. M. C. W. Chen, M. Gupta and K. C. Cheung, *30th Int. IEEE EMBS*, 2008, 4848.
110. G. D. Nicodemus and S. J. Bryant, *Tissue Eng. Part B*, 2008, **14**, 149.
111. M. W. Tibbitt and K. S. Anseth, *Biotechnol. Bioeng.*, 2009, **103**, 655.
112. R. M. Johann and P. Renaud, *Biointerphases*, 2007, **2**, 73.
113. Y. S. Song, R. L. Lin, G. Montesano, N. G. Durmus, G. Lee, S.-S. Yoo, E. Kayaalp, E. Hæggström, A. Khademhosseini and U. Deirci, *Anal. Bioanal. Chem.*, 2009, **395**, 185.
114. C. P. Huang, J. Lu, H. Seon, A. P. Lee, L. A. Flanagan, H.-Y. Kim, A. J. Putnam and N. L. Jeon, *Lab Chip*, 2009, **9**, 1740.
115. A. P. Wong, R. Perez-Castillejos, J. C. Love and G. M. Whitesides, *Biomaterials*, 2008, **29**, 1853.
116. A. P. Golden and J. Tien, *Lab Chip*, 2007, **7**, 720.
117. C. M. Puleo, W. McIntosh Ambrose, T. Takezawa and J. Elisseeff, T.-H. Wang, Lab Chip, 2009, 9, doi: 10.1039/b908332d.
118. X. Liu, Q. Wang, J. Qin and B. Lin, *Lab Chip*, 2009, **9**, 1200.
119. J. Liu, D. Gao, H.-F. Li and J.-M. Lin, *Lab Chip*, 2009, **9**, 1301.
120. C. K. Fredrickson and Z. H. Fan, *Lab Chip*, 2004, **4**, 526.
121. S. C. Oppegard, K.-H. Nam, J. R. Carr, S. C. Skaalure and D. T. Eddington, *PLoS ONE*, 2009, **4**, e6891.
122. X. Cui, L. M. Lee, X. Heng, W. Zhong, P. W. Sternberg, D. Psaltis and C. Yang, *Proc. Natl. Acad. Sci. U.S.A.*, 2008, **105**, 10670.

123. J.-P. Hauschild, E. Wapelhorst and J. Müller, *Int. J. Mass Spectrom.*, 2007, **264**, 53.
124. H. Lee, E. Sun, D. Ham and R. Weissleder, *Nature Med.*, 2008, **14**, 869.
125. M. Rhee and M. A. Burns, *Lab Chip*, 2008, **8**, 1365.
126. P. K. Yuen, *Lab Chip*, 2008, **8**, 1374.
127. www.ehrfeld.com.
128. S. Bhattacharya, A. Datta, J. M. Berg and S. Gangopadhyay, *J. Microelectromech. Syst.*, 2005, **14**, 590.
129. A. Papra, A. Bernard, D. Juncker, N. B. Larsen, B. Michel and E. Delamarche, *Langmuir*, 2001, **17**, 4090.
130. A. J. Garcia, *Biomaterials*, 2005, **26**, 7525.
131. A. Pulsipher, N. P. Westcott, W. Luo and M. N. Yousaf, *J. Am. Chem. Soc.*, 2009, **131**, 7626.
132. C. Roberts, C. S. Chen, M. Mrksich, V. Martichonok, D. E. Ingber and G. M. Whitesides, *J. Am. Chem. Soc.*, 1998, **120**, 6548.
133. X. Zhang, P. Jones and S. J. Haswell, *Chem. Eng. J.*, 2008, **135**, 82.
134. M.-H. Wu, *Surf. Interface Anal.*, 2009, **41**, 11.
135. S. Sugiura, J.-I. Edahiro, K. Sumaru and T. Kanamori, *Colloids Surf. B*, 2008, **63**, 301.
136. M. N. De Silva, J. Paulsen, M. J. Renn and D. J. Odde, *Biotechnol. Bioeng.*, 2006, **93**, 919.
137. N. Patrito, C. McCague, P. R. Norton and N. O. Petersen, *Langmuir*, 2007, **23**, 715.
138. S. Zhang, L. Yan, M. Altman, M. Lässle, H. Nugent, F. Frankel, D. A. Lauffenburger, G. M. Whitesides and A. Rich, *Biomaterials*, 1999, **20**, 1213.
139. M. Mrksich, L. E. Dike, J. Tien, D. E. Ingber and G. M. Whitesides, *Exp. Cell Res.*, 1997, **235**, 305.
140. B. J. Kirby, A. R. Wheeler, R. N. Zare, J. A. Fruetel and T. J. Shepodd, *Lab Chip*, 2003, **3**, 5.
141. S. Mandal, J. M. Rouillard, O. Srivannavit and E. Gulari, *Biotechnol. Prog.*, 2007, **23**, 972.

CHAPTER 12

Cell Capture and Lysis on a Chip

SÉVERINE LE GAC AND ALBERT VAN DEN BERG

BIOs The Lab-on-a-Chip Group, University of Twente, P.O. Box 217, 7500 AE Enschede, The Netherlands

Abstract

Single cell analysis in a microfluidic device proceeds through a certain number of steps. The two first steps of the protocol are the manipulation of individual cells in the microfluidic device and their positioning in precise locations and subsequently the process to rupture the cell membrane to retrieve the cellular content. In this chapter, we will focus on these two first steps of cell trapping and single cell lysis, and review different approaches and techniques found in the literature for these two steps to achieve these in a microfluidic format. We will also discuss the advantages and limitations of the different techniques of manipulation, trapping and lysis.

12.1 Introduction

The two first steps of the whole protocol for analyzing a single cell in a microfluidic device are firstly the manipulation of individual cells and their isolation in chosen locations in the microdevice, and secondly, the lysis of the individual cells. Following this, the resulting lysate or intracellular content is used for on-line analysis or for specific isolation of certain biomolecules, as described in the next chapters of this book.

RSC Nanoscience & Nanotechnology No. 15
Unravelling Single Cell Genomics: Micro and Nanotools
Edited by Nathalie Bontoux, Luce Dauphinot and Marie-Claude Potier
© Royal Society of Chemistry 2010
Published by the Royal Society of Chemistry, www.rsc.org

Analyzing a single cell in a microfluidic device firstly implies the capability of both handling individual cells in the device and subsequently precisely positioning them in a defined place where they will be analyzed. A great variety of principles have been employed to displace individual cells in a microfluidic device and to immobilize them in precise locations. Some techniques have already been demonstrated in a conventional environment (*e.g.* dielectrophoretic trapping, chemical patterning, optical trapping) while a number of techniques are specific to a microfluidic environment (droplet-based isolation, microstructure-based trapping). Still, it is worth mentioning that, in general, a microfluidic format is particularly adequate for handling small objects such as cells; the confinement found in microsystems assists in the reliable and precise manipulation of cells, and dedicated microstructures can be added for cell isolation and trapping.

A second key element for the analysis of a single cell is the possibility to separately access the content of individual cells. This requires the rupture of the cell membrane, also known as cell lysis. Different methods are again possible for that purpose, and every technique presents specific characteristics. As pointed out later, of importance is the time scale of the lysis process with respect to the time scale of two other phenomena: (1) biological processes and reactions occurring inside the cells; and (2) molecular diffusion in the microfluidic device. This has two main consequences. Firstly, if biological processes are affected by the lysis or are quicker than the time required for complete lysis of the cell, the analysis will not be reliable. However, this limitation is not specific to microfluidic systems and single cell analysis. Secondly, if the lysis process is too slow, biological material will be dispersed in the solution and lost, due to diffusion phenomena. This last aspect becomes critical when using reduced amounts of material and analyzing single cells.

The two sections of this chapter will describe these two steps which are essential for single cell analysis: cell trapping and cell lysis in a microfluidic format.

12.2 Cell Capture on a Chip

A first crucial step to make single cell investigation possible is the manipulation of individual cells. This includes not only cell handling in a microfluidic platform but also their positioning in precise locations for experimentation, characterization or analysis. The manipulation of cells has been demonstrated using various principles, and we will present here in detail (1) hydrodynamic and mechanical trapping, (2) electrical trapping (electrophoretic and dielectrophoretic), (3) fluidic trapping *e.g.* in droplets, (4) chemical trapping, (5) optical trapping, and (6) magnetic trapping of cells. However, as some of these techniques apply at the level of a cell population and are mostly used for sorting purposes, they will only be briefly mentioned here. For the different trapping principles, we will also discuss the easiness of implementation in a microfluidic format, the possible scaling-up of the trapping to a large amount of cells, the reversibility of the trapping, and the suitability of the technique for the purpose of single cell analysis.

12.2.1 Mechanical Trapping

A first approach for cell manipulation in a microfluidic format relies on the use of microfabricated structures added in the device for mechanical trapping of cells. Under the umbrella of mechanical trapping, several strategies are found. A first strategy combines the use of hydrodynamics and microfabricated structures for passive cell manipulation and positioning in a microfluidic device. While this strategy is easy to implement in a microfluidic format, scalable and possibly automated for the positioning of a large number of cells, trapping mainly remains static and is not compatible with dynamic manipulation of cells. It should also be noted that the cells are not always easy to release from the traps. Another option relies on the use of active structures (*e.g.* valves or a microgripper) to trap cells in a given location in a microchannel. While this strategy requires the addition of more sophisticated structures in the system design, it is potentially amenable to the trapping of a large number of cells and more interestingly, trapping becomes dynamic.

We can distinguish three levels of mechanical trapping depending on the trapping density which is conceivable, and we classify these here as zero-dimensional, one-dimensional, and two-dimensional trapping of cells. Three-dimensional trapping would be more tissue-like applications. In this case, cells are immobilized in a matrix, a hydrogel for instance, but this goes beyond the scope of this chapter. We will in the following present the three levels of (1) zero-dimensional, (2) one-dimensional, and (3) two-dimensional trapping.

12.2.1.1 Zero-dimensional Trapping using Active Structures

The zero-dimensional strategy makes use of active structures such as either a series of independent valves or a microgripper to dynamically trap a single cell.

In the former case, the valves enable the creation of a small chamber in a microfluidic channel where a cell can be isolated. The trapped cell can thereafter be studied, subjected to various chemical treatments,[1] and eventually lysed for the analysis of its intracellular content. The fabrication of valves is now well-established in microfluidics, and their fabrication relies on the use of a soft material and a three-layer microfluidic structure. The bottom layer corresponds to the microfluidic channel, the intermediate layer to a thin membrane made from a soft and easily deformable material and the top layer is used to control the deflection of the membrane using a liquid or a gas flow, and to thereby close the microfluidic channel.[2] This valve-based zero-dimensional trapping appears as an elegant and reversible manner to create higher confinement in a microfluidic system and the volume of the chambers defined by the valves can be decreased down to some tenths of picoliters, or several times the volume of a single mammalian cell. These small chambers are popular to confine chemical reactions, but they are also highly appropriate for single cell isolation and investigation.

For instance, Wu *et al.* make use of this strategy not only for the isolation of a single cell, but also for its treatment using precisely defined volumes of liquids.[1] For that purpose, a picopipette structure is added to the set of valves to precisely meter the amount of chemical solution delivered to the trapping chamber after isolation of a cell therein. This platform has not only been used for highly controlled stimulation of cells with the injection of chemical stimulants, but also for the cell lysis in a confined area after addition of a lysis buffer. In this configuration, the use of detergents for the lysis of the cell is also applicable as the confinement brought by the chamber limits the loss of biological material by diffusion (see section 12.3). This approach can notably be compared to earlier single cell experimentation performed in the same group.[3] There, a single cell was isolated between microstructures placed along a channel wall, in an open space and sequentially subjected to different well-defined chemical plugs. Cell lysis would in that case be less straightforward to implement as the open configuration of the trapping site would lead to the loss of biological material *via* diffusion phenomena.

A similar approach has been adopted by Marcus *et al.*[4] or Bontoux *et al.*[5] to isolate a single cell before its analysis; in that scenario, the cell is again trapped in a small volume defined by a series of valves. Opening of the valves enables to release the isolated single cell and to inject it in a nanoliter rotary mixer for its lysis, retrieval of its mRNA content and the RT reaction to convert it into cDNA (see Chapter 16).

Alternatively, another more elegant and more sophisticated strategy has been described for active single cell manipulation in a lab-on-a-chip (LOC) device. It relies on the integration of a dedicated microgripper on which cells can be hooked to be translated on long distances in a microchip in a harmless manner.[6] Sadani *et al.* particularly describe two different translators, a vertical capture translator and a lateral capture translator, which are aimed at oocyte manipulation in a microfluidic platform and their transportation to specific sensors. Both translators include an inverted pyramidal structure etched in a silicon substrate. In the first case the pyramidal structure is located on the top of the translator, which implies lowering the microgripper for loading and unloading of cells. Consequently, in a second approach, the pyramidal structure is placed on the edge of the translator (see Figure 12.1A) so that its aperture is connected to the side, and this was found to facilitate loading and release of the cells.

12.2.1.2 One-dimensional Trapping Using Lateral Microstructures

The second strategy which is found for trapping a larger number of cells is "one-dimensional". It relies on the use of lateral trapping sites where cells are immobilized with the help of side (nano)-channels[7] or dam structures.[8] With this strategy, cells can be studied and processed in parallel. However, they are mostly not fully and separately isolated as they are all trapped in a common channel (see Figure 12.1B) and depending on the device layout, trapping is also

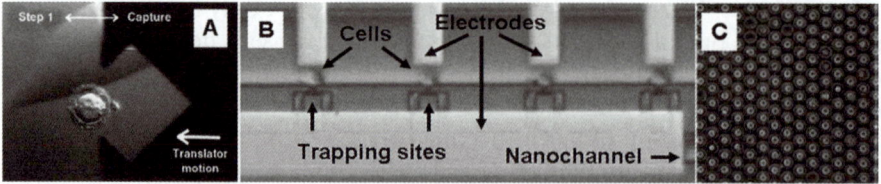

Figure 12.1 Mechanical trapping of cells using a zero-dimensional (A), a one-dimensional (B), or a two-dimensional approach (C). (A) Trapping and active manipulation of an oocyte using a silicon-based microgripper integrated in a microfluidic platform. Shown here is the capture of the occyte on a lateral translator in which the occyte is loaded from the side into a pyramidal structure etched in the silicon. Reprinted, with permission, from Sadani *et al.*[6] (B) One-dimensional trapping of cells in lateral dedicated microstructures in a single cell electroporation platform. Individual cells sit in trapping sites which are etched in a silicon substrate; the trapping site consists of narrow channel connected to a nanochannel from which a negative pressure is applied to monitor cell trapping, and each site is electrically addressed by two electrodes as indicated on the figure. Here four C2C12 cells are shown immobilized in four independent trapping sites. Reprinted, with permission, from Valero *et al.*[7] (C) Two-dimensional trapping of cells in a microwell array molded in PDMS. RBL-1 cells are individually isolated in independent microwells by simply seeding them on the microwell array for about 25 min. Shown here are the arraying of cells in microwells having optimal dimensions for RBL-1 single cell trapping (20 µm diam.; 27 µm depth). Reprinted, with permission, from Rettig and Folch.[13]

monitored from a common channel.[7] This trapping strategy has consequently mostly been applied for other purposes such as cellular patch-clamp[9] and single cell electroporation,[7,8] and not for single cell analysis. For the purpose of single cell analysis, the design of the microsystem must be adapted with for instance the addition of individual analysis channels behind each trapping site in which the content of the individual cells can be separately retrieved and analyzed.[10]

12.2.1.3 Two-dimensional Arraying of Cells Using Lateral Structures or Microfabricated Wells

The last mechanical technique for trapping individual cells with the help of microstructures can be categorized as two-dimensional trapping. This strategy makes use of two types of structures arranged as an array in a micro-chamber: either lateral structures as found in the one-dimensional approach, or bottom structures or microwells fabricated in the substrate located under the cells.

The first approach of lateral trapping has been proposed by Di Carlo and co-workers for creating a 2D array of cells in a microfluidic chamber.[11] Cells are individually trapped in suspended cup-shaped obstacles regularly placed in the microchamber, and they are subsequently imaged and their reaction to specific chemical stimuli studied. The same array has been recently slightly adapted by Skelley *et al.* for controlled cell pairing using two different cell types prior to their fusion.[12]

The second approach consists of planar trapping. It makes use of structures machined in the bottom substrates on which cells are seeded to give a large scale array of trapped cells. In this category are found microwells and microwhole-based structures.

For instance, Rettig *et al.* describe a device for passive seeding and arraying of up to 18 000 cells in individual poly(dimethylsiloxane)-based microwells.[13] Crucial parameters to the successful large-scale arraying of individual cells are the dimensions of the wells (depth and diameter of the wells) as well as the seeding time given to cells to sediment in the microwells while the seeding density is of minor importance. Interestingly, these parameters vary as a function of the cell line, as demonstrated for two cell types, fibroblasts (3T3) and rat basophilic leukemia cells (RBL-1). Using optimized parameters, 90% trapping efficiency is easily reached without any sharp expertise as illustrated in Figure 12.1C. This platform lends itself well to cell imaging for short-term cellular assays but for longer term experimentation, the microwells should be functionalized to create a better environment for the cells. The same strategy has been employed since then by two other groups for cell imaging followed by active sorting of cells (see also non-conventional optical),[14] or for the characterization of individual cells using sensors placed in the microwells.[15]

As a conclusion, these different mechanical protocols for trapping and eventual isolation of single cells are, in general, simple and easily implementable in a microfluidic device. However, it should be noted that they are either dynamic (zero-dimensional trapping) when they employ active structures such as valves or a microgripper but limited in throughput, or static (one- and two-dimensional trapping) but amenable to large-scale manipulation. To the best of our knowledge, none of the mechanical techniques described in the literature till now combined both advantages of being both dynamic – unless another technique is used for cell release from the trapping structures – and scalable.

12.2.2 Electrical Trapping

Another integrated strategy for cell manipulation in a LOC device relies on electrical forces. This includes electrophoretic or dielectrophoretic techniques, exploiting, respectively, the negative charges located at the surface of the cell and its dielectric properties. In both cases, a series of electrodes must be added in the microdevices for localized trapping, and the higher the amount of trapping sites, the higher the number of electrical connections. To these two limitations, the fabrication of electrodes in the device and the limited upscaling of the trapping, adds the potential damage an electric field can cause on cells yielding to low survival rate. Still, these electrical techniques are highly popular in the LOC field for a number of reasons. Firstly, cell trapping is reversible and cells can easily be released by simply switching off the trapping voltage. Secondly, shaping the electric field is straightforward in a microformat *e.g.*, by simply adding microstructures in the device. Finally, the same electrodes can be used for cell trapping and their subsequent lysis to retrieve the cellular content for its on-line analysis.

12.2.2.1 Electrophoretic Trapping

A first example of active manipulation of cells using electrophoretic trapping and optical release out of the traps is found in the work of Ozkan *et al.*[16] In their system an array of electrodes is fabricated on the bottom substrate of a microdevice, and cells are placed on the individual electrophoretic traps upon application of a DC voltage (<20 V, 1 min). The traps are made from a transparent semi-conductor material, Indium Tin Oxide (ITO), to allow for cell imaging. Besides, agarose microwells (20–100 µm diameter) are patterned on the ITO substrates for cellular confinement down to the single cell level. Once the cell array is formed, individual cells can be released from the traps using optical techniques and miniaturized sources of infrared (IR) light (vertical cavity surface emitting lasers (VCSELs)) integrated in the individual microwells and acting as optical tweezers. In a first stage, the whole system and manipulation protocol has been validated using 20 µm negatively charged polystyrene beads, and subsequently applied to neural stem cells.

Toriello *et al.* report another methodology for arraying cells on gold electrodes using electrophoresis.[17] The gold electrodes are patterned using silicon dioxide to give rows of $16 \, \mu m^2$ spots where cells are attracted and immobilized. Cellular immobilization on the surface is irreversible and is controlled by previous functionalization of the cell membranes by thiolated Arg-Gly-Asp (RGD) peptides, the RGD peptides interacting with the cell membrane while the thiol moieties enabling cell anchoring on the gold surface. The technique is suitable for patterning single cells by controlling the application time of the electric field. For instance, immobilized cells were studied for their response to chemical stimuli. On other aspects by controlling the application of the voltage on certain electrodes only, several cell populations can be successively patterned in the same device.

12.2.2.2 Dielectrophoretic Trapping

In this category, of particular interest is the work of Taff *et al.* who have developed an easily scalable positive-dielectrophoretic cell sorting array.[18] This two-dimensional array is built using a novel strategy where individual trapping structures are addressed as rows and columns. The first obvious advantage of this strategy is the lower number of electrical connections required for addressing all trapping sites: it varies as the square root of the number of trapping sites and not anymore as the number of trapping sites as would happen using a conventional electrode design. It makes this strategy easily scalable compared to other electrode designs. Furthermore, the different traps can still be addressed individually by switching off or on the two electrodes (column and row) crossing on the trapping site. On other aspects the ring-dot trapping structures generate a higher trapping force than other trap designs, and the trapping remains efficient also upon application of a flow of 10–50 $\mu L\,min^{-1}$.

The whole platform has been applied not only for arraying cells but also for cell sorting. After trapping, cells are separately imaged and release of a single cell (depending on its pattern of fluorescence) is triggered by grounding the two electrodes (column and row) crossing on that particular trapping site.

12.2.3 Fluidic Trapping

12.2.3.1 Hydrodynamic Trapping

In LOC devices, an obvious and simple manner to manipulate cells involves playing with the flow. Cell focusing is notably widely used in the technique of flow cytometry to create a file of single cells[19] that can separately and successively be studied *e.g.* for their fluorescence pattern. This technique of hydrodynamic manipulation and cell focusing does not really enable the immobilization and creation of a pattern of cells but is still of great interest for sorting/separating cells as a function of their size,[20,21] and directing them in a microchannel. Eventually, microstructures can be added to help direct the flow or for sorting the cells.[22] It is still worth mentioning that such an approach has been applied for isolating a single cell in a channel before its lysis and on-line separation using capillary electrophoresis (CE).[23,24]

12.2.3.2 Droplet-based Trapping

Droplet-based platforms are rapidly emerging as platforms in the field of microfluidics.[25] For biological applications, for instance, aqueous droplets are generated in an oil phase and stabilized with a surfactant. The droplet size is highly controlled and the generation frequency now reaches the kHz range. These platforms enable the creation of well-defined compartments with a higher level of confinement than conventionally found in a LOC device. This leads to much lower consumption of chemicals, and on other aspects, mixing in droplets is more efficient due to the smaller distances molecules need to travel, and can further be promoted by chaotic advection.[26] Droplets consist of picoliter to nanoliter size reactors which are attractive not only for conducting chemical reactions but also for the isolation of (single) cells in a reduced volume (Figure 12.2). In that respect, the droplet appears as a closed space where substances produced or secreted by the cell(s) are retained, and can be analyzed

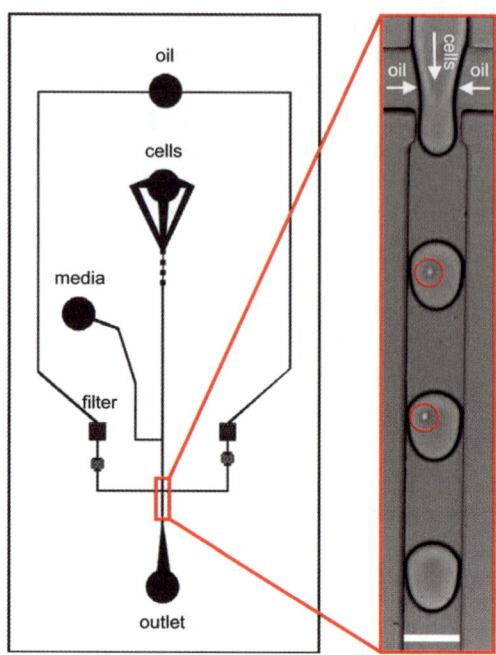

Figure 12.2   Droplet-based encapsulation and manipulation of individual cells in a microfluidic platform. *Left*: General design of the microfluidic platform to generate aqueous droplets in a continuous oil phase, and for the encapsulation of single cells in individual droplets. A cell solution is first diluted in medium and injected in an oil phase to form droplets in which single cells are encapsulated. *Right*: Enlarged view on the encapsulation zone in the device showing the merging of the two aqueous and oil flow and the generation of the aqueous droplets in the continuous oil flow. Jurkat cells (highlighted by red circles) are encapsulated in individual 660 pL droplets at a frequency of 800 Hz. Scale bar 100 nm. Reprinted, with permission, from Clausell-Tormos *et al.*[28]

in situ without any further dilution, as would still happen in a microchamber. Moreover, the droplet reactors can be stored in conventional glassware and be re-injected in a microfluidic platform for readout of the studied process.[27] Droplet-based platforms have already been widely applied for experimentation on cells, single cell treatment, drug screening (on bacteria), single cell analysis, or even single cell electroporation.[28–31]

12.2.4   Alternative Trapping Techniques

12.2.4.1   Chemical Trapping

Chemistry provides an attractive approach for immobilizing cells along given patterns in a microfluidic environment. Proteins of the extra-cellular matrix

(ECM) such as fibronectin are first patterned using microfabrication techniques (mostly micro-contact printing), often in combination with a cell-repellent coating such as PEG (polyethylene glycol) outside the defined protein patterns. This technique lends itself well to the immobilization of a large number of (individual) cells, as one step patterning is sufficient to generate the chemical functionalization, and there is virtually no limit to the surface area which is patterned. Still, cell immobilization is in general irreversible unless the surface is fully exposed to trypsin to release the cells or a "smart" immobilization strategy is employed for reversible anchoring of cells. This technique has notably been applied to investigate the behaviour and the differentiation of cells with respect to the geometry of the chemical patterns (Figure 12.3) or to the nature of the chemicals spotted on the surface.[32–36]

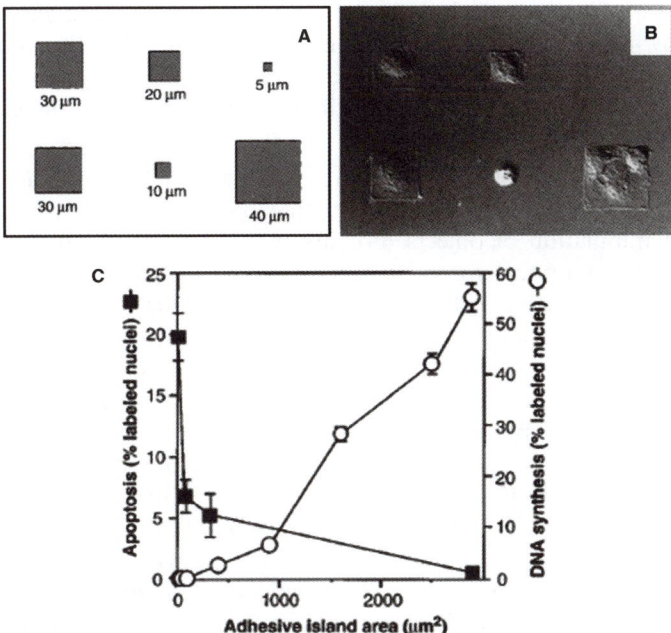

Figure 12.3 Chemical trapping of cells on fibronectin patterns of various sizes and prepared by micro-contact printing: effect of spreading on cell growth and apoptosis. (A) Illustration of the protein pattern distribution and sizes (5–40 μm side squares) on the surface and corresponding pictures (B) of the bovine adrenal capillary endothelial cells adherent to the patterned substrate (Nomarski views). (C) Apoptosis level measured in the cell population (TUNEL assay) and DNA synthesis level (BrDU assay; percentage of stained nuclei) as a function of the size of the protein pattern. Data obtained using islands containing single adherent cells (human or bovine endothelial cells). Reprinted, with permission, from Chen *et al.*[32]

12.2.4.2   Optical Trapping

12.2.4.2.1   Optical Tweezers.   Optical forces consist of an alternative technique for cell handling in a microfluidic platform. The most popular technique found in this category is optical trapping whereby a focused laser beam generates forces in the piconewton range and these forces are exploited to precisely monitor the displacement of small particles in three dimensions, and trapping in that case is reversible.[37] Optical tweezers are becoming more popular among biologists not only for single molecule studies but also for manipulation and sorting of single cells, as well as in a microfluidic environment. This is not only due to the non-invasive character of the technique but also to the higher number of commercial platforms with a user-friendly interface, and the more and more enhanced resolution gives a hope for manipulation of sub-cellular entities. On other aspects, while the technique is often restricted to a single cell, as most optical trapping systems present only one laser beam, the combination of the technique to a holographic plate allows for the generation of a large number of individual traps from a single laser beam.[38] Still, optical trapping remains a cost-effective, low-throughput technique (unless it is combined with a holographic plate), and the force generated by the beam must be high enough to trap cells or particles which are also submitted to a drag force in a microfluidic system.

12.2.4.2.2   Non-conventional   Optical   Methods.   Besides   non-standard approaches adding to optical trapping, they indirectly benefit from a laser beam for manipulation of objects. For instance, Kovac and Voldman demonstrated the use of optical forces to remove cells from microwells for further analysis.[14] In a first stage, cells were seeded in individual microwells as described previously (see section 12.2.1.3) and continuously imaged. Subsequently, depending on the fluorescence pattern of the cells, they can be isolated and retrieved individually for further experimentation and analysis. In this set-up, the scattering force from a focused IR laser is employed to raise single cells and eventually push the cell out of the well into the flow for further investigation, as illustrated in Figure 12.4. A similar approach was reported to expel individual cells from microwells using an IR laser, as briefly mentioned earlier.[16]

On other aspects, Allbritton and co-workers report an innovative methodology where cells are grown on SU-8 pallets fabricated on a surface and indirectly manipulated *via* a laser-assisted expulsion of the individual pallets.[39] Briefly, the energy deposited under the pallet by the focused laser pulse results in the sudden formation of a bubble, which ejects the pallet in the solution without damaging the cells thereon. The technique was first applied at the single cell level, but has recently been successfully demonstrated for the screening and manipulation of colonies of mouse embryonic stem cells.[40]

12.2.4.3   Magnetic Trapping

A technique that is becoming increasingly popular for cell manipulation is magnetic trapping, where cells are specifically retained on magnetic beads *via*

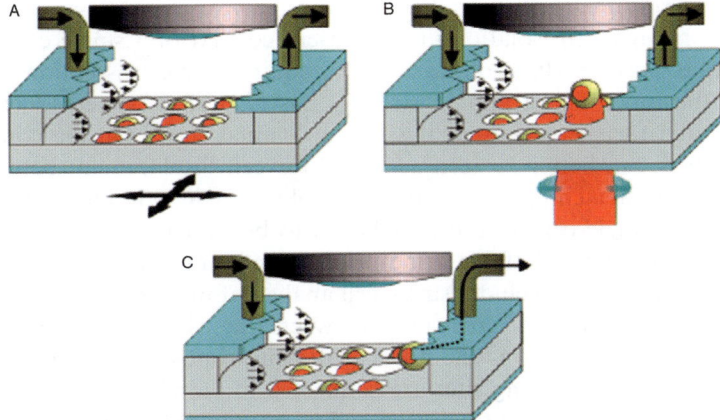

Figure 12.4 Optical manipulation of single cells using non-conventional optical techniques. Cells are first seeded individually in an array of micro-fabricated wells (A) and continuously imaged. When a cell shows a distinct fluorescent pattern for instance, it is individually released from its microwell (B and C). For this, an infrared (IR) laser beam is focused on the cell of interest; this enables cells to rise into the flow field with the help of the optical scattering force (B). Once the cell is in the flow field, the fluid drag overcomes lateral optical forces, releasing the cell and washing it downstream for later further analysis (C). Reprinted, with permission, from Kovac and Voldman.[14]

appropriate chemical functionalization of the beads. However, this technique is more amenable for flow-cytometry applications: large-scale manipulation of cells such as the separation of various cell types,[41] the enrichment[42] or isolation of sub-population of cells, depending on their surface chemistry or their size,[43] and it has not been downscaled yet for real single cell manipulation in a microfluidic network.

12.2.5 Conclusion on Cell Trapping

We have reviewed here different techniques and principles that have been applied for manipulating and trapping cells in a microfluidic format. As mentioned in the introduction, some techniques are generic and can also be found for lab-scale cellular experimentation while a number of techniques have only been demonstrated in combination with microfluidics. Beside, not all the aforementioned techniques enable single cell trapping and some lend themselves more to the manipulation or isolation of cell populations.

Still, while the scalability of cell trapping and its reversibility are key-aspects for a number of applications (*e.g.* cell imaging, cell electroporation), these may not be the most important characteristics for the foreseen context of single cell analysis. In addition to this, not every trapping technique presented in this section is applicable for single cell analysis in a microfluidic format. For instance, zero-dimensional trapping using valves appears as a suitable strategy for single cell isolation in the present context of a single cell analysis platform, as only one type of structure is required for cell isolation and lysis most of the time. This trapping strategy has also already been successfully demonstrated for the genetic analysis of individual cells.[5] On the contrary, a microgripper[6] can better be seen as a tool for cell displacement between different locations in a chip, and it could be employed in a preliminary step for the isolation of a single cell from a population to bring it to a place for analysis inside a microfluidic platform. The "open configuration" found with one-dimensional trapping precludes single cell analysis if no particular care is taken to alleviate the existing cross-talk between the different trapping sites. Of interest in this context is the novel platform proposed by Le Gac *et al.*,[10] where the different trapping sites are connected to individual channels for the analysis of isolated cells, but work on cell lysis and analysis is still under way and has not been demonstrated yet. Finally, cell arraying, whatever technique is employed for this purpose,[11,13,14,16,18,32,39,44] is obviously ideal for large-scale cell immobilization but may not be really adequate for single cell analysis: it does not provide either the required confinement for single cell lysis or a simple manner to retrieve the content of individual cells after lysis. As highlighted in the examples described previously, the implementation of a single cell analysis protocol together with two-dimensional trapping of cells requires the capability to release the cells individually, and subsequently implies dynamic trapping of cells. This dynamic trapping is made possible by the use of an IR laser[14,16,39] or individual electrical addressing of dielectrophoretic traps.[18] On the contrary, it is precluded if the cells are chemically anchored on the surface,[17,32] or if the electrodes are not individually addressed for single cell release. In this scenario, once cells have been retrieved from the traps they must be redirected to another section of the microfluidic platform where the actual analysis takes place, and this last step has not been demonstrated yet in combination with cell arraying. A noticeable advantage of this is the possible combination of single cell analysis with cell imaging and cell selection before their analysis. Alternatively, the lysis can be done in the wells where cells are immobilized, and the cell content simultaneously sucked in capillary tubing for its analysis.[44]

Let us now consider hydrodynamic trapping and its suitability in a single cell analysis protocol. Hydrodynamic trapping has already been employed for the isolation of a single cell before its analysis. On other aspects, the use of a droplet platform is an elegant approach for the isolation of individual cells. While it is comparable with a valve-based zero-dimensional trapping of cells, this strategy lends itself better to scaling-up and the simultaneous analysis of a large number of individual cells suing a simple microfluidic designs, *i.e.* without the need for an additional microfluidic network for the actuation of the valves.

To the best of our knowledge, optical trapping has now not yet been used for microfluidics-based single cell analysis although it would be a possible approach for single cell isolation in this context. Non-conventional optical manipulation, as described above, is better applicable for the controlled release of individual cells and is ideal in combination with cell arraying, for instance.

12.3   Cell Lysis in a Chip

Cell lysis is defined as the irreversible disruption of a cell membrane. This mostly aims to access and retrieve the intracellular content, and is followed by the analysis of targeted molecules contained in the lysate. A wide range of applications is found, among which expression analysis (mRNA and protein levels), protein kinase activity and pathogen identification (genetic analysis). Different techniques can be employed to achieve electrical lysis, chemical lysis, alkaline lysis, thermal lysis, mechanical lysis, acoustic lysis or optical lysis. The choice of the lysis protocol is often dictated by the molecules to be analyzed or the envisioned analysis technique, as discussed in detail later.

We focus here on single cell analysis; this translates into the lysis of a single cell followed by the analysis of the resulting lysate. Using conventional labware, a 1 pL cell is lysed in an amount of liquid of circa 1 μL. Subsequently, any molecule present in the cell is diluted with six orders of magnitude and this dilution issue strongly hinders molecular analysis. When using microfluidics, this dilution issue can be limited as cells are handled on low nanoliter volumes, and the same lysis techniques can be used *a priori* as in a lab-scale experiment. Other advantages of the microfluidic format are found in the simplicity of the protocol; while off-chip protocols are often time-consuming and involve extensive and multiple washing steps, purification, and extraction processes, a single cell lysis is mostly sufficient on a chip, and molecules of interest are directly isolated for further analysis or processing. For the rest of the chapter we will focus on single cell lysis in a microfluidic setting.

The same issues and requirements are found for the lysis step in this more focused context than in a conventional set-up. If this single cell approach in a microfluidic format presents numerous advantages, some issues found with lysis become relatively more important when working at the single cell level. In addition to this, the lysis in a microfluidic format is mostly coupled to an on-line analysis step, and this adds other challenges regarding the loss of biological material contained in the lysate.

A key parameter is the temporal resolution of the lysis; it can be defined as the time required for disrupting the cell membrane and strongly varies with the technique used for lysis. This first parameter goes together with a second issue, the absence of induced stress response of the cell or disturbance of any biological processes within the cell as this would bias the analysis of the cellular content. This criterion is indeed notably met when lysis is very fast, and at least faster than typical biological processes in the cell. In that case, biomolecules are not "changed" during the lysis, biological processes are frozen and this is particularly crucial when investigating enzymatic processes in a cell (protein kinase analysis, for instance) whose timescale lies in the low ms range. In addition to this, the technique used for lysis should not deteriorate the molecules to be analyzed. Harsh conditions are used for some lysis techniques, and this may harm some classes of molecules. Contamination of the lysate should also be avoided; this is detrimental to most of the separation protocols. As a consequence, care should be taken to choose a lysis technique which is appropriate for the envisioned later analysis and processing of the lysate.

To these general issues there are additional microfluidic-specific challenges. Firstly, the lysis technique must be readily transposable and applicable to a microfluidic format, and should not result in complex microfluidic designs or fabrication processes. Interestingly, as seen later in this chapter new lysis techniques and protocols have been created for a microfluidic setting to alleviate complex system design and fabrication. Secondly, the lysis step is mostly directly followed by processing or analysis of the lysate. As all steps are integrated in a single platform, the lysis technique should be easily coupled to the other analytical steps. For instance, if the cell content is to be separated on-line by capillary electrophoresis, care must be taken to locally keep the lysate concentrated and prevent any loss by diffusion in the surrounding medium. Thereby, not only the injection prior to the analysis is optimized but band broadening during the separation is also limited. In that context again, a fast lysis technique is preferred.

12.3.1 Thermal Lysis

Thermal lysis is the most popular and most often employed technique in a conventional set-up. It relies on the exposure of the cells to a high temperature for a short period of time; thereby pores spontaneously form in the cell membrane which is irreversibly damaged. The popularity of this technique is understandable as this high-temperature protocol can easily be followed by on-line PCR for the analysis of nucleic acids.

This technique is notably reported by Waters *et al.* in a microfluidic format for DNA analysis.[45] The microsystem integrates the different steps of cell lysis, multiplexed amplification using PCR, and the on-line electrophoretic sizing of the amplified genetic material. The lysis step is "integrated" in the PCR protocol and performed for 2 min at 94 °C, *i.e.* the temperature used for the first step of the polymerase chain reaction (PCR) thermal cycle. The chip is placed in a commercial thermal cycler and undergoes a 94 °C–50 °C–72 °C conventional thermal cycle for the PCR amplification, as well as two additional minutes at 94 °C for the lysis of the cell before the PCR cycling. This straightforward and integrated protocol is demonstrated here on *Escherichia coli*. Cells are introduced in the chip in suspension in the PCR mixture, lysed at 94 °C, and the DNA resulting from the PCR amplification is analyzed on-line. The easiness of the protocol is emphasized in the article. However, the authors do not discuss in detail the advantages of this lysis technique in comparison with alternative methodologies.

While this lysis technique is popular in a conventional set-up, it presents a number of issues at the microscale. Here, Waters *et al.*[45] use an external bath to regulate the temperature. Subsequently, the whole chip is exposed to the same temperature and the temperature cannot be monitored locally in the chip. This precludes, for instance, the preparation of more sophisticated devices including more analytical steps taking place at various temperatures. In the latter case, electrodes or a cooling/heating system must be added in the device to locally lyse the cells (high temperature) and subsequently cool the lysate. Nonetheless, this integrated approach adds in complexity to the fabrication of the micro-fluidic device. Next to this, thermal lysis has a limited applicability: while this system offers a promising approach for fast and integrated analysis of nucleic acids, the same lysis technique cannot be applied for the analysis of more fragile molecules such as proteins which are damaged and denatured upon exposure to a very high temperature.

As a conclusion, the use of a high temperature for a short period of time is an interesting and straightforward route for on-chip cell lysis but the suitability of the method highly depends on the analytes of interest. Thermal lysis is notably precluded for protein investigation as the use of a high temperature will cause denaturation.

12.3.2 Chemical Lysis

The second most popular technique for lysing cells in a conventional set-up consists of using detergents that incorporate in the cell membrane and subse-quently solubilize the phospholipid molecules and the proteins composing the cell membrane, and thereby the cell membrane is destructed. Different types of detergents can be used – ionic, non-ionic, and zwitterionic – and the choice of detergent affects the lysis rate and the protein extraction efficiency out of the cells. For instance, the strong and ionic detergent sodium dodecylsulfate (SDS) gives very fast lysis processes in the second time range, but it simultaneously denaturates the proteins. Triton X is a milder non-ionic detergent; lysis is slightly slower, protein denaturation occurs to a lesser extent, and protein complexes remain intact. Zwitterionic detergents do not change the charge of proteins but may affect the electro-osmotic flow (EOF) if pumping in the device is controlled by electro-osmosis. The main issue found with chemical lysis in a microfluidic platform is the time required for the lysis, *i.e.* the rate at which the surfactant is added to the cell suspension and, as a consequence, the mixing efficiency of the two cell and detergent solutions, and the time needed for the lysis reaction. This notably implies a good mixing strategy in the microfluidic devices, and three of the numerous examples found in the literature and pre-sented in detail below illustrate three strategies to address this issue of mixing.

Chen *et al.* compare two mixing procedures, a T-type model and sandwich-type one, for microfluidic chemical lysis of cells using blood samples, and use two system designs with straight or coiled channels.[46] In both cases, mixing occurs only by diffusion (laminar flow configuration) either upon merging of two solutions in a T-shape microdevice or in a three-phase flow configuration, respectively. However, it should be noted that when using coiled channels the

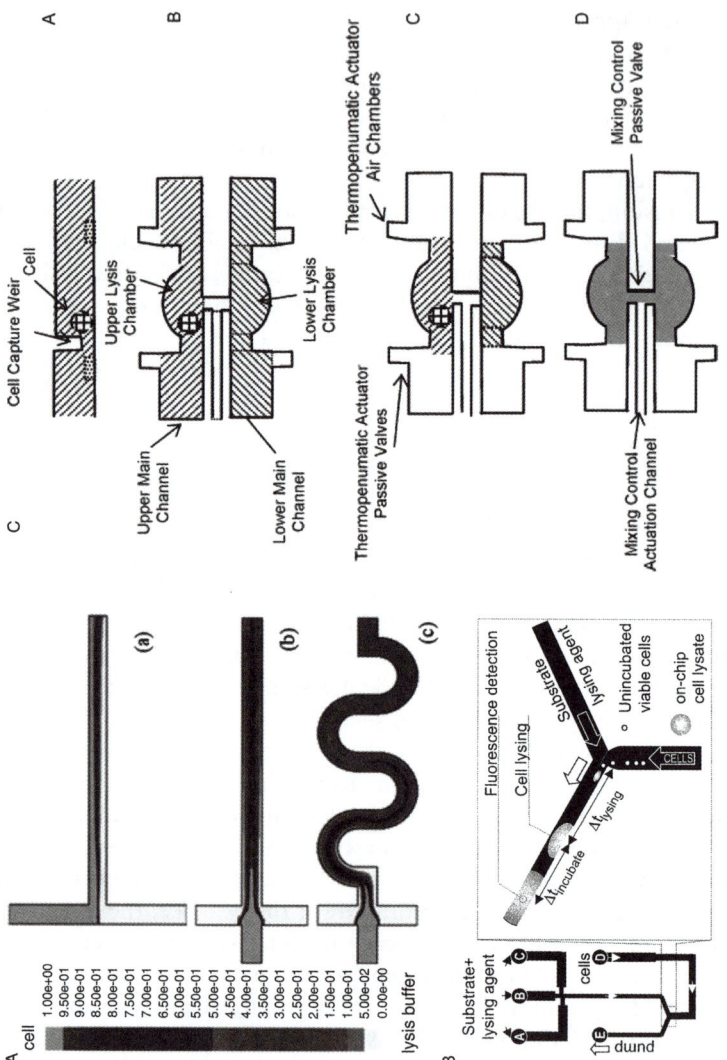

overall mixing efficiency is enhanced as chaotic mixing in the coiled structures adds to diffusion phenomena, as illustrated in Figure 12.5A In both cases, the mixing efficiency depends on the channel dimensions but also on the velocities of the solutions to be mixed; the lower the cell:detergent velocity ratio, the shorter the diffusion length and subsequently, the more efficient the mixing. Best mixing is found for a sandwich-type system and coiled channels, with a cell:detergent solution velocity ratio of 1:5, and this configuration gives an 80% mixing efficiency between blood samples and a detergent buffer (4 M GuSCN, 1% Triton-X 100, pH 6.7). After lysis, DNA is extracted on-chip with the help of a porous silicon solid support placed in the coiled channel. DNA purification in 1 µL blood sample is demonstrated within less than 20 min, and quantified using a fluorescent probe to give a 39.7 ng (or 80%) extraction yield per microliter of whole blood (against 20–30 ng using conventional commercial kits).

Ocvirk *et al.* also employ a Y-shape configuration for on-stream mixing of the cell solution with the lysis agent.[47] The cells are introduced as a focused stream with a single cell file towards the Y intersection where the lysis buffer is introduced by suction from the other channel (Figure 12.5B). They demonstrate cell lysis using two types of detergent solutions, Triton-X 100 and SDS, with lysis times of 30 and 2 s, respectively. For instance, when using Triton and a flow-rate of 40 µm s^{-1}, the lysis is completed within 1.2–2.4 mm downstream to the Y junction. The lysis step is followed by the on-chip indirect analysis of a protein enzyme, β-galactosidase, by measuring its enzymatic activity in the lysate. For that purpose, the lysis buffer (Triton 0.1%) is supplemented with a

---

Figure 12.5   Three strategies for enhanced mixing of the detergent solution with the cell suspension for chemical lysis of cells. (A) Three microfluidic configurations for diffusion-based mixing in a laminar flow profile. (a) T-type mixing model, (b) sandwich type mixing model with lined channel, and (c) sandwich-type mixing model with a coiled channel. Numerical simulation results for the species concentration distributions when $v_{cell} = 0.005$ m s^{-1}, $v_{buffer} = 0.0025$ m s^{-1}. The color in this figure refers to the concentration intensity of the pure cell sample and the pure lysis buffer, respectively. Reprinted, with permission, from Chen *et al.*[46] (B) On-stream diffusion-based mixing in a Y-shape configuration. The substrate and lysing agents are brought from ports A–C, and cells from port D. Suction is applied at port E, causing mixing of cells and a lytic solution containing enzyme substrate at the intersection. A fluorescence detector was located downstream of the mixing point, and an observation microscope sat over the intersection. Reprinted, with permission, from Ocvirk *et al.*[47] (C) Mixing by merging two small-sized volumes (25 pL) containing one cell and the lyzing agent, respectively. A. One cell is introduced with the fluid in the upper main channel and captured in the cell lysis chamber by a dam-like structure. B. Lysing solution is introduced into the lower main channel. C. Closed volume fluid compartments are formed by the coordinated action of the four thermopneumatic actuators. D. Cell lysis is achieved by removing air from the mixing channel, allowing the contact and mixing of the fluids. Reprinted, with permission, from Irimia *et al.*[48]

fluorogenic substrate specific to this enzyme (fluorescein-di-β-D-galactopyr-anoside), whose degradation gives rise to the production of a fluorescent product (fluorescein mono-β-D-galactopyranoside). The whole protocol takes place in less than a few minutes, and appears as a straightforward strategy for enzymatic analysis at the single cell level. In that precise example, a chemical protocol for cell lysis is attractive as there is no real issue of loss of biological material and no need for fast lysis; the enzymatic reaction takes place continuously as soon as the lysis is initiated and as soon as the substrate can react with the released enzyme. Moreover, it should be noted that such an enzymatic protocol at the single cell level is only possible in a microfluidic format where the enzymatic reaction occurs faster as a result of the shorter distance the molecules must travel and as the much lower dilution rate of the cell content.

Finally, Irimia *et al.* use another strategy for enhanced chemical lysis of cells;[48] lysis takes place in a 50 pL chamber and the use of such a small reaction volume brings about enhanced mixing performance of the cell solution with the lysis buffer. First, they actually create two independent liquid plugs of 25 pL which are defined by virtual air walls, one containing a single cell and the other the lysis buffer. These two plugs are subsequently merged to form a 50 pL lysis reactor. This mixing strategy for cell lysis is illustrated in Figure 12.5C. The precise manipulation of liquids and the creation of well-defined liquid plugs are achieved with a combination of both passive and active control structures, *i.e.* hydrophobic walls and thermopneumatic actuators. In a first step, a cell is trapped, the lysis buffer is introduced in the second half of the system, and two 25-pL plugs are created by coordinate activation of four thermopneumatic actuators. Subsequently, the air plug present between the two liquid plugs is removed by aspiration in an actuation channel, and mixing is initiated. Mixing is governed by diffusion of the detergent molecules in the 50 pL volume, and cell lysis occurs within 0.5–1 s for guanidine thiocyanate (GTC), 3 s for 0.2% SDS and 10 s for 10% SDS. This lysis strategy is illustrated for quantitative analysis of fluorescent markers either preliminarily introduced in the cell (cell tracker) or added to the lysis solution and specifically binding to molecules of interest (actin). It is worth mentioning that with this approach, other issues often found with chemical lysis, *i.e.* the dilution lysate and the dispersion and subsequent loss of biological material, are also alleviated. Here, the lysis reaction takes place in a confined environment and biological components are retained in that small reaction volume. The 1 pL cell content is retrieved in a 50 pL volume, giving a dilution factor of approximately 50 to be compared to the six orders of magnitude found with a conventional approach (1 μL).

As a conclusion, while chemical lysis would appear as an easy and straightforward approach for on-chip cell lysis, a number of issues arise from its alliance with microfluidics, such as the mixing time and efficiency, the diffusion of analytes out of the lysate and their subsequent loss, and the dilution of the lysate although this last point is improved in a microfluidic setting compared to a lab-scale approach. These first series of issues can be addressed by a smart design of the microfluidic system to enhance mixing,[46–48] to limit the dilution factor[48] and to prevent molecular loss.[48] Still, a compromise must be found in the choice of the detergent to optimize the lysis time while minimizing molecular damage (protein denaturation). Consequently, chemical lysis is an attractive method but its application is limited and depends on the targeted molecules to be analyzed.

12.3.3 "Alkaline" or Electrochemical Lysis

An alternative chemical route relies on the use of alkaline conditions that result in a progressive destruction of the cell membrane. The dissolution of the membrane is triggered here not with the help of detergent molecules that break phospholipid molecules, but through the hydrolysis of the individual phospholipid molecules (cleavage of the fatty acid–glycerol bond) that compose the membrane once they are placed in highly alkaline conditions. This lysis strategy is used extensively for recovery of plasmid DNA[49] and intracellular recombinant products[50,51] from large bacterial cultures.

Di Carlo *et al.* propose one microfluidic platform for alkaline lysis of cells that include two electrodes for water electrolysis (Figure 12.6).[52] On the first electrode (cathode), OH^- ions are generated to give a local and high increase in the pH of the medium, while H^+ ions are concomitantly produced on the second electrode (anode) so that the cell lysate solution remains neutral:

$$\text{Cathode} \quad 4H_2O + 4e^- \leftrightarrow 4OH^- + 2H_2$$
$$\text{Anode} \quad 2H_2O \leftrightarrow 4H^+ + 4e^- + O_2$$

Consequently, no extra washing step is required to produce a neutral lysate and molecules of interest are not endangered by a basic medium. In addition to this, the microdevice format enables real-time the lysis conditions to be tuned in by changing the concentration of hydroxide ions. According to off-chip preliminary experiments, a concentration of 20 mM hydroxide ions has been assessed to be optimal for fast lysis of cells in the second time range, and the lysis efficiency on-chip depends on the concentration of hydroxide ions at 15–$30\,\mu L\,min^{-1}$. Experiments conducted on-chip using various cell lines (CHO, HeLa, macrophages, RBC) show a cell line dependent lysis behavior: HeLa cells are lysed quickly at a pH higher than 11.7 while macrophages are lysed within less than 60 s at pH 13 against more than 2 min at pH 12.

The technique appears to be attractive as it brings a level of control on the lysis process depending on the cell line. A lower voltage is required here compared to conventional electrical lysis, resulting in lower power consumption, and the authors confirm that no electrical lysis occurs as the lysis takes place in a zero-electric-field zone of the device. Although the exposure of proteins to basic conditions leads to their denaturation, the authors claim that a short exposure is not detrimental and that the proteins return to their native configuration once placed in a neutral environment. This is notably detailed in a second article from the same group with the observation of green fluorescent protein (GFP) release in the lysate of GFP-expressing cells.[53] Last, the lysate can readily be used for further on-line investigation, and the protocol is

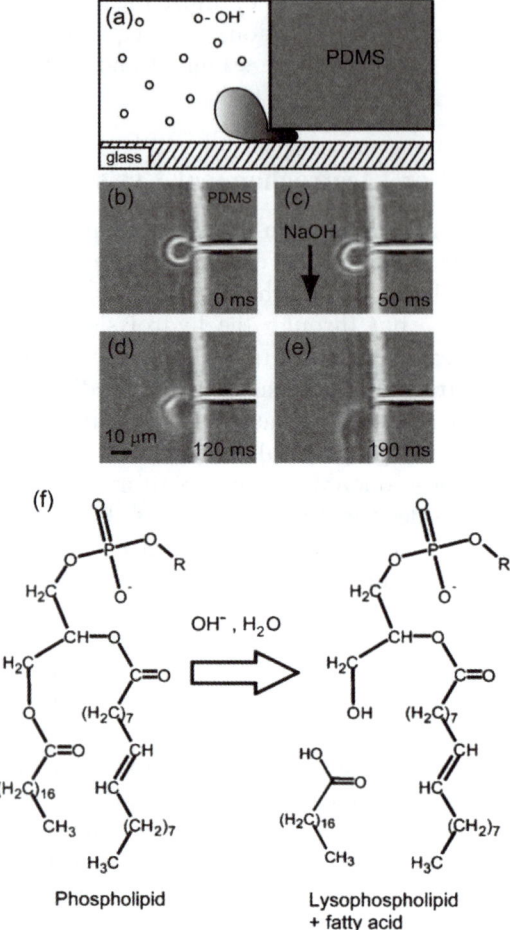

Figure 12.6 Alkaline lysis of a single cell. (a) Schematic representation of a cell
trapped in a PDMS dam structure and exposed to hydroxide ions. The
cell is trapped by applying suction to a small channel on the side of a
larger chamber. (b–e) Frames from a movie of alkaline lysis for a single
cell trapped in a microfluidic device. An isotonic alkaline solution
(50 mM NaOH in PBS) is introduced into the main chamber from the top
(arrow, c). Colloid osmotic lysis happens over a time scale of 190 ms. (f)
Scheme of the chemical reaction responsible for cleaving the fatty acid
groups of membrane phospholipids, leading to the creation of lysopho-
spholipids which have been shown to induce membrane permeabilization. Reprinted, with permission, from Di Carlo *et al.*[52]

applicable to any kind of biomolecule. In a first step, the lysis efficiency is
assessed only by studying the release of a dye previously loaded into the cells
(calcein) and no further molecular investigation is reported using the resulting
lysate. Later, the authors report the integration of the alkaline lysis strategy in a

more complete microfluidic platform that combines cell culture and on-line lysis. In addition to this, they investigate the effect of electrochemical lysis on proteins (p53 and horseradish peroxidase (HRP)) and DNA: protein and plasmid solutions are simply flushed through the device. The functionality of the proteins is checked afterwards by immunodetection and enzymatic assays, respectively, and plasmid DNA is used for PCR amplification. The exposure to electrochemical lysis is found to affect the enzymatic activity of HRP (decrease of 20% activity after exposure to 2.5 V) while p53 and DNA remain unaffected.

12.3.4 Electrical Lysis

A very popular cell lysis protocol in a microfluidic format is electrical lysis.[46,54–56] For this technique the creation of pores in the membrane proceeds *via* the alignment of water dipole and the subsequent formation of water files that span the membrane, as occurs in the process of electroporation.[57] At a certain level, pores are too numerous or too large and do not close anymore, and the membrane is irreversibly damaged: the cell is lysed. The time scale for the lysis lies in the ms range. The applied electrical signal is chosen to exceed the trans-membrane potential of 60 mV, so that pores appear in the cytoplasmic membrane. In general, pore formation is observed as soon as the potential across the membrane exceeds 0.2–1.5 V, as indicated by Lu *et al.*,[55] and when a high enough amplitude is used, cell lysis is observed (irreversibility of the pores).[54] Upon irreversible breakdown of the membrane, surrounding medium can enter the cell by osmosis and the cell membrane becomes disrupted as the result of the extensive swelling of the cell. Electrical lysis still depends on the shape and size of the cell as well as on the membrane composition, and different cell types exhibit different responses to the application of the same electric field.

At the microscale, manipulating and shaping the electric field is easy by simply adding microstructures to make it locally higher for instance (Figure 12.7).[54,55] Subsequently, at this scale, the local generation of a high electric field requires lower voltages than in a conventional set-up *e.g.*, a few volts instead of kV, and subsequently the power consumption is decreased by six orders of magnitude. This not only reduces the risk of bubble formation as a consequence of water electrolysis, but makes the technique of electrical lysis highly attractive in a microfluidic format.

Many platforms are found in the literature on the topic of electrical lysis of cells in a microfluidic format for the lysis of both mammalian and yeast cells at the level of a cell population or that of individual cells. We will discuss here three examples of these platforms where the lysis is performed at the single cell level and is coupled to an on-line separation step of the cell content by capillary electrophoresis.

McClain *et al.* demonstrate electrical cell lysis in a cross-shape microfluidic device followed by on-line electrophoretic-based separation of the cell con-tent.[24] Cells are focused as a file in the center of the microfluidic channel, a single cell is isolated at the channel intersection and an electric signal is applied

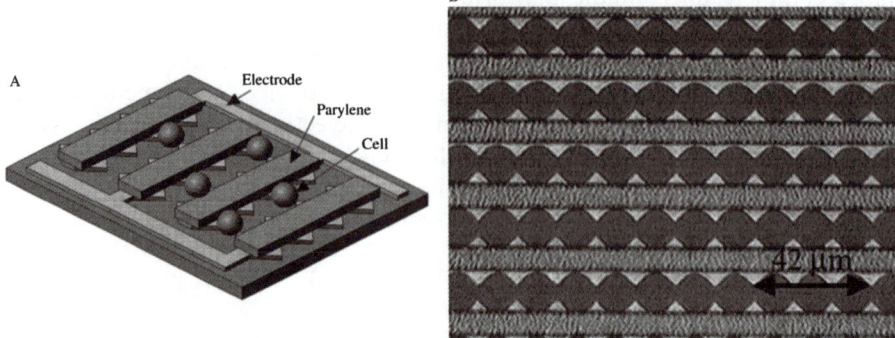

Figure 12.7   Electrical lysis. (A) Schematic view of a cell lysis device that includes
tooth-shaped electrodes for local enhancement of the electric field. A cell
placed at the tip of a tooth experiences a higher electric field and is
readily lysed. (B) Picture of the corresponding device. Reprinted, with
permission, from Lee and Tai.[54]

to trigger cell lysis, as illustrated in Figure 12.8. The separation is achieved
within 2.2 s, and the detection performed 3 mm downstream to the lysis point.
In a first instance, the authors use a DC voltage, but this gives rise to the
formation of gas bubbles in the reservoir and Joule heating; when a lower
electric field is used, these problems disappear but the signal is too low for
cell lysis. Therefore, they use an AC field with a DC offset (50% duty cycle
and 450 V cm^{-1} peak-to-peak square waves and 675 V cm^{-1} DC offset) to
solve this, and reach good lysis performance while alleviating bubble formation
and Joule heating. As soon as lysis is initiated the buffer solution can enter the
cell inducing biochemical reactions to terminate upon dilution of the cell
content. Subsequently, this fast lysis protocol enables the study of molecules
involved in fast processes in the cell, such as reactions occurring in signaling
cascades that can induce concentration changes by one order of magnitude
within less than 1 s.

Gao *et al.* propose another approach to achieve a similar goal, *i.e.* the
coupling of cell lysis to the on-line analysis of the intracellular content using a
CE separation.[23] Once a single cell has been isolated at the entry of the
separation channel, the voltage for the CE separation is applied in the main
channel. They use an alkaline buffer (pH 9.2 borate buffer) for the separation.
Subsequently, the voltage required to lyse cells is lower and the cell is simply
lysed when entering the separation channel and being subjected to the
separation voltage. Cells lysis is observed within 40 ms when applying an
electric field of 280 V cm^{-1} for the separation. A key issue in this setting is a
good mixing efficiency of the separation buffer and the cell solution to ensure
fast lysis of the cell. For that purpose, a single cell is first docked in the middle
of the separation channel and kept there for 15 s so that the buffer around the
cell is refreshed; this docking step enables to optimize the lysis-separation
coupled protocol, otherwise the cell would be progressively lysed during the

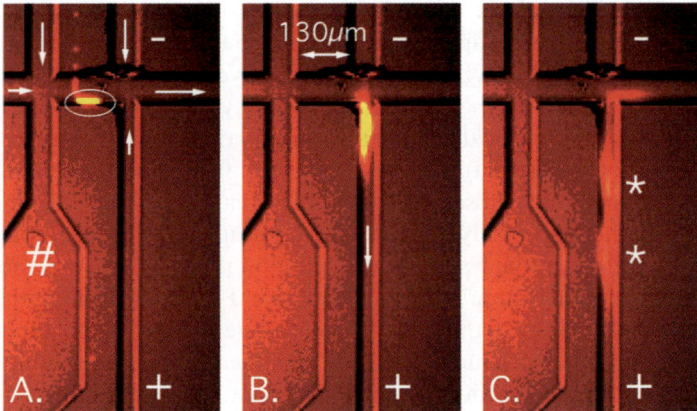

Figure 12.8   Chemically assisted electrical lysis. CCD images of cell lysis. (A) A Jurkat
cell loaded with calcein AM (within white oval) is being hydro-
dynamically transported to the lysis intersection. Arrows depict the
direction of fluid flow in the channels. (B) The cell has encountered the
electric field and is lysed. The fluorescently labeled content is injected into
the separation channel and migrates toward the anode. The arrow
depicts the direction of lysate migration in the channel. (C) Separation of
two of the fluorescently labeled components (marked by asterisks) loaded
in the cell in the separation channel. Reprinted, with permission, from
McClain *et al.*[24]

separation and the separation plug ill-defined, bringing peak broadening into
the electropherogram.

A last example of electrical lysis has been presented by Allbritton's group,
using microfabricated structures for isolation and culture of single cells.[44] The
device consists of a microwell array; every well being equipped with integrated
transparent ITO-based electrodes which are independently electrically addressable while the second electrode for cell lysis is placed above the microwell.
Modeling studies have shown the homogeneity of the electric field in the whole
cell trap, so that the position of the cell in the trapping well does not affect the
lysis performance. Lysis is achieved upon application of a 10 ms pulse of a
100 V electrical signal between the two electrodes. Lysis of rat basophilic leu-
kemic (RBL) cells is observed within 33–66 ms, as determined by the loss in
fluorescence intensity inside the cells that had been previously loaded with a
fluorescent probe. However, the strong conditions used for the lysis lead to the
formation of bubbles in the microwells that do not interfere according to the
authors with cell lysis or with cell analysis which is performed on-line after lysis.
For the analysis, the cellular content or lysate is sucked in a capillary placed
above the well, and separated by capillary electrophoresis inside the capillary.
Analysis is only demonstrated here by looking at the electrophoretic separation
of two fluorescent probes loaded in the cells before lysis, and no real biological
analysis is shown. This promising approach could easily be implemented in an

integrated platform, and the use of closer electrodes should allow for using lower voltage and thereby suppress the issue of bubble formation.

As already mentioned, electrical lysis is attractive in a microfluidic format as it gives a high control on the process of lysis. Lysis can only been triggered locally in the system, by local application or enhancement of the electric field with help of electrode structures. Lysis is much faster than when using chemicals for instance and subsequently enables real-time studies of biological processes. On other aspects, lysis can easily be coupled to on-line separation of biomolecules present in the lysate, also relying on the use of an electric field, as illustrated by the three examples discussed above. Finally, by tuning the electric field which is applied, only the plasma membrane is affected by the electrical signal and intracellular organelles whose trans-membrane potential is higher[55] (*e.g.* ∼ 160 mV for the mitochondria) remain intact. This selectivity is not found for other lysis techniques such as chemical, thermal, mechanical, and acoustic lysis whereby all membranes are simultaneously destructed. However, electrodes must be added in the microsystem. Their integration in the device gives a better control for the local application of the electric field, although it adds a level of complexity to the device fabrication. As seen in the work of McClain *et al.*, for example, the applied electric field can cause hydrolysis of water,[24] the appearance of bubbles in the solution, and these bubbles can interfere with the process of lysis. One possibility to circumvent bubble formation is the use of an AC voltage.[55] Similarly, Joule heating can become another issue if electrodes are not located close enough to each other and this is notably promoted when a buffer with a high ionic strength is used. Subsequently, the range of buffer compositions is limited with electrical lysis.

12.3.5 Mechanical Lysis

An approach which is specific to microfluidics is the mechanical lysis of cells. Thereby, pores occur in the cell membrane or it is torn as a result of the use of sharp structures or a high shear. Two examples of mechanical lysis of cells are found in the literature.[58,59] While they both belong to the class of mechanical protocols, they employed two different basic principles to rupture the membrane. One relies on the use of integrated sharp nanostructures or nanoknives that cut the membrane,[58] and the other on the combined used of spherical particles and centrifugation (CD-shaped system).[59]

The first report on mechanical and reagent-less lysis comes from Lee's group at the University of Berkeley.[58] The device includes narrow channels provided with nanoknives on their walls, as shown in Figure 12.9. Cells are pushed in these narrow channels with the help of a high flow-rate (pressure-driven flow), squeezed against the nanostructures and progressively lysed. The lysis efficiency is found to depend on the chosen flow-rate applied to force the cells into the narrow channels, and it involves shear and frictional forces that enable to rupture the membrane. After lysis, the amount of protein material available in the resulting lysate is determined by quantitative UV measurements; the

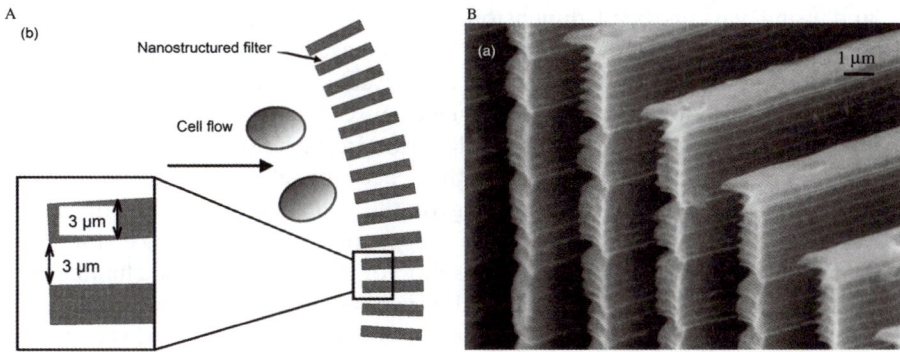

Figure 12.9 Mechanical lysis of cells using nanostructures (A) Schematic of the principle of mechanical lysis: cells are pushed in a semicircular array of narrow channels whose walls have been modified to present nanoknives. (B) SEM picture of the nanoknives. Sharp protrusions are clearly seen as orthogonal scallops meet at corners during the DRIE process. Distance between barbs is ~0.34 mm and radius of curvature of tips is below 25 nm. Reprinted, with permission, from Di Carlo *et al.*[58]

extraction yield represents 4.8% of the total amount of proteins contained in the cells and 7.5% of the hemoglobin content (*vs.* 1.9 and 3.2% in the absence of nanobarbs) when using a flow rate of $300\,\mu L\,min^{-1}$. This mechanical approach offers a number of advantages compared to other microfluidic-based cell lysis. The lysis protocol itself is much simpler and is *a priori* applicable for any class of biomolecules, and no purification step must be implemented to remove unwanted detergents before cell analysis. Moreover, there is no need for complicated microsystem design with multiple inlets and outlets to introduce various solutions, including a smart mixing structure or integrated electrodes. The lysing nanostructures are produced within the system; they are fabricated using a modified DRIE step to enhance the scallop effect found with the Bosch process, and to create nanobarb structures instead of a flat wall. However, the performance of the systems must still be improved as the retrieval yield of biological material is still very low and the principle has not been demonstrated yet at the single cell level.

The other approach found in the category of mechanical lysis has been reported by the group of Marc Madou.[59] Their strategy relies on the combined use of a CD-shaped device, a centrifugation system and spherical particles added to the cell suspension to tear their membrane. The solution containing the cells–particles mixture is introduced in an annular chamber included in the CD-shaped device. Rotation of the device results in the creation of a centrifugal force in the chamber, and the particles and cells are pushed against the external wall of the chamber. To force collisions between particles and cells and to prevent the particles to stick onto the chamber walls, the CD is sequentially spun forwards and backwards. Cell lysis is tested on three cell lines – yeast (*Saccharomyces cerevisiae*), bacteria (*E. coli*), and mammalian cells (CHO-K1)

– and using various bead dimensions depending on the cell type, from 10^6 (mammalian cells and *E. coli*) up to 600 µm (yeast cells) bead size. Thereafter, the authors look at DNA extraction after lysis and quantify the amount of extracted DNA: they reach 65% efficiency in comparison with conventional lysis. On other aspects, theoretical explanations are proposed for the process of lysis. Two phenomena are involved in the process of cell lysis, collision between cells and particles as well as shearing of the cell membrane. Identified key parameters for the lysis performance are, in order of importance: the bead density, the angular velocity, and the solid volume faction in the chamber.

This second protocol for mechanical lysis of cells in a microfluidic format is straightforward and highly attractive. As before, the cell membrane is damaged simply by the combined effect of the flow (centrifugal force arising in a CD device) and particles added to the cell solution. This strategy is notably suitable in the context of this particular project of development of an integrated CD platform for complete blood analysis: no additional structures are required for cell lysis and the lysis can easily be coupled to other analytical steps. The issue of pumping of liquids is solved by the use of a circular device which is spun. However, as before, the reported extraction yield is low compared to other approaches. Besides, care must be taken with the shear force which is exerted to lyse cells as similar shear flows can damage biomolecules, and are used for instance for DNA fragmentation.

12.3.6 Alternative Mechanical Lysis: Acoustic Lysis

A last possible protocol for cell lysis, which has also been applied in a microfluidic setting, is acoustic lysis. Here, ultrasonic waves are employed to generate localized areas of high pressure. The induced cavitation phenomena can be used to shear a cell and retrieve the intracellular content. Zhang and Jin report acoustic-based lysis of cells in a microfluidic system with the assistance of a digitonin pretreatment of the cells.[60,61] After injection of a digitonin buffer in a capillary containing the cell solution, a 3 s ultrasonication step is used, then the cell content is separated on-line in the capillary. The digitonin pre-treatment enables the lysis process to be shortened from 50 s to a few seconds, as longer ultrasonication causes the proteins to denaturate due to Joule heating of the solution.

With this technique, the solution is subsequently significantly heated and there is a risk as well of protein denaturation. Besides, the cell content diffuses extensively so that the detection of biological material is hindered downstream to the point of lysis. Last, the whole protocol is not straightforward to implement in a microfluidic format.

12.3.7 Optical Lysis

Optical lysis can be regarded as a group of protocols where cells are lysed upon indirect action of a laser light. Lysis occurs through either the generation of plasma or a bubble in solution, the creation of high shear forces on the cell membrane, or a local strong increase in the temperature of the solution (thermal lysis). In the first case, the energy of the laser light is absorbed directly by the solution while in the second case particles are added to absorb the light and dissipate energy (heat) locally in the solution.

In the first case, a high energy pulsed laser (*e.g.* 532 nm, pico- to nanosecond range pulses) is employed and highly focused in the solution with the help of a high numerical aperture (NA) objective.[62,63] The focused laser beam deposits energy which is absorbed by the solution and this gives rise to plasma formation, followed by the production of a propagating shockwave and subsequently to a cavitation bubble in the liquid. The cavitation bubble expands in microseconds and quickly collapses (Figure 12.10): this leads to the appearance of two strong and opposite flows in the vicinity of the point of focus of the laser. This shear stress induced by the flow can locally be employed to tear a cell's membrane and for lysis, and the effect of the laser light can eventually be limited to one part of the cell with a subcellular positioning of the laser. This technique enables very fast cell lysis, in the microsecond time range,[64] and is easily coupled with on-line analysis using single cell capillary electrophoresis (SCCE).[65] It is also worth noticing that, with this method, the cellular content is first dispersed as a consequence of the expansion of the cavitation bubble but that it is relocalized to a small region upon collapse of the bubble, preventing extensive dilution and loss of material.[64]

The authors investigate the applicability of the laser-mediated lysis technique in a PDMS microfluidic device, and have recently coupled it to an on-chip electrophoretic separation of the cell content.[64,65] The analysis is focused on signaling kinases using enzyme-specific fluorescent reporters, and a sampling efficiency of ~60% is reported. Laser-assisted lysis presents a series of advantages with regards to a microfluidic setting; it is easily implementable in a microfluidic platform as long as one material is transparent to the laser light, and basically only one optical access for the laser is required. No chemicals must be added, no care taken with the design of the microsystem and the addition of specific (*e.g.* mixing) structures, and in principle this lysis technique applies for the analysis of any biomolecule present in the cell as the cell content is not affected by the laser irradiation and other associated phenomena (cavitation). With regards to the last point, the authors report a lack of photodegradative effects in the cell when using this lysis technique. With regards to other aspects, the lysis is very fast (microsecond range); this is notably of interest for the analysis of cellular compounds which can be altered on a subsecond timescale, as for enzymatic reactions involved in signaling pathways. However, the persistence of bubbles in the liquid after the laser pulse may be detrimental to the cells, and non adherent cells may rather be displaced than sheared by the bubble-induced flow.[66] Besides, another limitation of the technique is found when working with PDMS systems: a large part of the laser energy is loss and used to deform the PDMS.[64,65]

In the second indirect optical lysis technique, the energy of the laser is absorbed by gold particles added to the solution and that specifically bind to the cell surface.[67] Here, the laser irradiation ultimately leads to the creation,

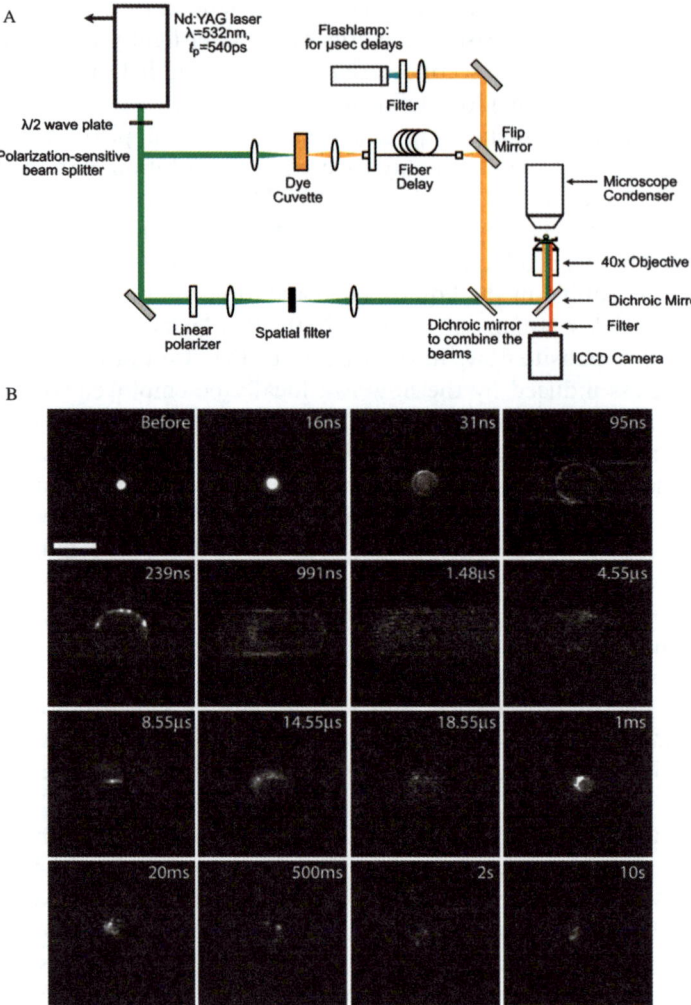

Figure 12.10   Optical lysis of a single cell in a PDMS microchannel with the help of a highly focused laser pulse. (A) Experiment set-up for time-resolved fluorescence imaging. (B) Fluorescent cell lysis dynamics inside the microfluidic chip. Fluorescent images of the laser-microbeam cell lysis process inside the microfluidic channel on time scales spanning nine orders of magnitude from 10 ns to 10 s. Scale bar = 50 μm. Reprinted, with permission, from Quinto-Su *et al.*[64]

locally, of a very high temperature that in turn induces the lysis of the cell anchored to the illuminated gold nanoparticle. This particular technique has been applied for the lysis of pathogens in an integrated microfluidic platform whose purpose is the isolation of pathogens from blood samples followed by their genetic analysis.[67] The integrated approach for the whole protocol compares favorably to a conventional approach with a similar DNA extraction yield while using a 12 min one-step process (microfluidic) versus a 30 min 10-step process (conventional approach).

12.3.8 Conclusion on Cell Lysis

A large variety of techniques, as for cell trapping, is described in the literature for cell lysis in a microfluidic system. Some are derived from conventional techniques while a number of specific techniques to have emerged these last years as microfluidic-based lysis methodologies.

As mentioned in the introduction to this section, three main types of issues are worth mentioning for cell lysis. The first is the time scale of the lysis process, which varies considerably depending on the technique, and this should be taken into account in the design of experiments depending of the molecules of interest and if the latter are likely to be changed (expression level, modifications) inside the cell in time scale of a few seconds.

The second issue is the damage the lysis process may cause to the cell or to the biomolecules to be analyzed; for instance, thermal or chemical lysis may lead to protein denaturation and induce further damage. The last issue is the compatibility of the lysis technique with a microfluidic setting, and the facility to implement it in a microfluidic system.

On other aspects, this book's topic is single cell analysis protocols, of which cell lysis is the second step. Consequently, interest is focused on lysis techniques which apply at the single cell level, and not every lysis principle discussed here has been demonstrated at the single cell level or is even suitable in the context of a single cell analysis protocol. Chemical lysis is conceivable at the single cell level and has already been included in a microfluidic single cell analysis protocol. Similarly, electrochemical lysis (*i.e.* using alkaline conditions) would be applicable while it has not been employed yet in such an experimental setting. Electrical lysis is more specific to a microscale format, and has been used for both cell population and single cell lysis. Mechanical lysis, in the way it has been reported so far, is not very likely to be extended to a single cell context, but the principle of shearing a cell membrane is attractive in a single cell analysis protocol as long as membrane debris does not pollute the analysis. Finally, cell lysis with the help of a focused laser beam has been shown in a microfluidic format. However, the variant of the technique where magnetic beads absorb the laser energy presents the same limitation as the use of magnetic beads for cell trapping, which is to be restricted to population applications.

12.4 Conclusion

In this chapter we have presented various conceivable techniques for, both cell, manipulation/trapping and cell lysis. We have discussed the pros and cons of the different techniques for both the purpose of single cell analysis and their implementation in a microfluidic device. The next step consists of examining possible trapping-lysis sequences in the context of a single cell analysis protocol. Firstly, the two techniques chosen for the successive trapping and isolation of a single cell and lysis must be compatible with each other. Secondly, the selected sequence should enable single cell analysis and especially give little or no dispersion of the resulting lysate for optimized analysis conditions.

From this viewpoint and when considering the characteristics of the different (trapping and lysis) techniques, in Table 12.1 we present all possible combinations for the cell trapping–lysis sequence in the context of a microfluidic single cell analysis protocol, and scored them in terms of suitability. From Table 12.1, the most promising approaches to our opinion $(+ + +)$ are detailed in the following paragraphs.

Confining a single cell in a small volume of liquid, which is defined by a series of valves or as a droplet, is highly attractive. The cell is lysed in this confined space and the lysate material is not lost due to its dispersion. Additionally, this strategy is easily scalable for parallel single cell analysis. Cell lysis can alternatively be performed by addition of chemicals (chemical or alkaline lysis) or by application of a high voltage pulse (electrical lysis). The choice of the lysis protocol is dictated by the targeted analysis technique and the layout of the microdevice. While chemical lysis is easy to implement, it remains a slow process and changes in the biomolecule expression or processing patterns may occur during the lysis. An electrical protocol is preferred for fast lysis of the cells, but this implies the addition of electrodes in the device and leads to complex microsystem layout and fabrication procedures. The last option consists of using alkaline conditions; again, electrodes must be added in the microchip device as most of the microchip materials cannot withstand the highly basic environment required to hydrolyze the cell membrane.

The second best option is to isolate a single cell in a dielectrophoretic (DEP) trap and to lyse it using the same electrodes as for the trapping. The time scale for cell lysis is much faster and better suitable for obtaining a reliable snapshot of a cell's content, and only the cytoplasmic membrane can selectively be hydrolyzed. Still, as before, electrodes must be integrated in the microfluidic device. Scaling-up is possible but more limited than when using droplets. For example, a row of DEP traps can be created and separately connected to individual channels for single cell analysis. Alternatively, the same DEP trap connected to a microfluidic channels can successively be used for different cells.

In a third approach, which has already been demonstrated, a cell is isolated by hydrodynamic focusing and subsequently lysed using high voltage (HV), chemicals or a combination of both. The main issue here is the timing between the lysis and the analysis to prevent loss of material by dispersion in the system. As before, scaling-up implies either the use of an array of individual channels for parallel single cell analysis or the successive use of the same channel.

Finally, an approach which is not fully integrated (yet) in a microfluidic platform starts with cell arraying – for example, in microwells – and cell lysis is carried out using an electrical or optical protocol. In this last case, capillary tubing must be brought above the well or place for cell immobilization to retrieve the cell content for its on-line analysis. The actual configuration (cell

Table 12.1 Summary of the different cell trapping and lysis techniques and of the possible trapping–lysis sequences.

		Thermal lysis	Chemical lysis	Alkaline lysis	Electr. lysis	Mech. lysis	Optical lysis
Mechanical trapping	Zero-dim.	NO / +	YES / + +	NO / +	NO / + +	N.A.	NO[d]
	One-dim.	NO+ / +	Under invest / +	NO / +	NO / + +	N.A.	NO / + +
	Two-dim.[a]	NO / –	NO / +	NO / +	YES / +	N.A.	NO
Electrical trapping	EP[a]	NO	NO	NO	NO	N.A.	NO
	DEP[a]	NO / –	NO / +	NO / +	YES / + + + / +[c]	N.A.	NO / +
Microfluidic trapping	Hydrodynamic	NO / –	YES / +	YES / +	YES / + +	NO / +	NO[d]
	Droplet	NO / –	YES / +	NO / +	?? / + + +	N.A.	NO[d]
Alternative trapping techniques	Chemical[a]	NO	NO / –	NO / –	NO / +	N.A.	NO
	Opt. tweezers	NO / –	NO / +	NO / +	NO / + +	NO? / ?	N.A
	Non conv. Opt. methods	N.A.	N.A.	N.A.	N.A.	N.A.	N.A.
	Magnetic	YES[b] / No single cell	NO / No single cell	NO / No single cell	NO / No single cell	N.A.	YES[b] / No single cell

A number of combinations have already been reported in the literature and are characterized by a "YES" while other combinations have not been described yet, noted as "NO". The suitability of the trapping–lysis sequence is also assessed as "–", and "+".

[a]If cells are immobilized as a 2D array, they must be independently released for a single cell analysis protocol; this requires active trapping and individual addressing of the trapping sites. Alternatively, a piece of capillary tubing can be brought above the cell to collect the resulting lysate of individual cells.
[b]Indirect optical lysis. Bacterial lysis caused by a temperature increase following the absorption of a laser light by small magnetic particles.
[c]If DEP trapping at the single cell level.
[d]Expected problems caused by the deposition of the laser energy in the liquid (linked to the resulting shock wave): perturbation of the hydrodynamic flow or of the droplets, or opening of the valves.

analysis in external capillary tubing) is easily suitable for large-scale trapping but the analysis is limited to one cell as long as integrated microchannels do not replace the external capillary tubing.

As a conclusion, even if several trapping–lysis combinations are suitable in the present context of a single cell lysis protocol, other parameters must be taken into account for choosing the best trapping–lysis sequence. The analysis technique or the biomolecules to be analyzed will obviously influence the choice of the lysis technique, as discussed earlier in the chapter. The ease of implementation of the trapping–lysis sequence in a microfluidic setting should not be neglected. Lastly, a few sequences may be preferred if scaling-up of the analysis is likely to occur, as mentioned above.

References

1. H. K. Wu, A. Wheeler and R. N. Zare, *Proc. Natl. Acad. Sci. U. S. A.*, 2004, **101**, 12809–12813.
2. J. W. Hong, V. Studer, G. Hang, W. F. Anderson and S. R. Quake, *Nat. Biotechnol.*, 2004, **22**, 435–439.
3. A. R. Wheeler, W. R. Throndset, R. J. Whelan, A. M. Leach, R. N. Zare, Y. H. Liao, K. Farrell, I. D. Manger and A. Daridon, *Anal. Chem.*, 2003, **75**, 3581–3586.
4. J. S. Marcus, W. F. Anderson and S. R. Quake, *Anal. Chem.*, 2006, **78**, 3084–3089.
5. N. Bontoux, L. Dauphinot, T. Vitalis, V. Studer, Y. Chen, J. Rossier and M. C. Potier, *Lab Chip*, 2008, **8**, 443–450.
6. Z. Sadani, B. Wacogne, C. Pieralli, C. Roux and T. Gharbi, *Sens. Actuators A-Phys.*, 2005, **121**, 364–372.
7. A. Valero, J. N. Post, J. W. van Nieuwkasteele, P. M. ter Braak, W. Kruijer and A. van den Berg, *Lab Chip*, 2008, **8**, 62–67.
8. M. Khine, A. Lau, C. Ionescu-Zanetti, J. Seo and L. P. Lee, *Lab Chip*, 2005, **5**, 38–43.
9. C. Ionescu-Zanetti, R. M. Shaw, J. G. Seo, Y. N. Jan, L. Y. Jan and L. P. Lee, *Proc. Natl. Acad. Sci. U. S. A.*, 2005, **102**, 9112–9117.
10. S. Le Gac, D. Wijnperle, H. de Boer, W. Meulemann, E. T. Carlen and A. Van den Berg, *Proceedings of the 13th International Conference on Miniaturized Systems for Chemistry and Life Sciences*, 2009, Chemical and Biological Microsystem Society, San Diego, CA, USA.
11. D. Di Carlo, N. Aghdam and L. P. Lee, *Anal. Chem.*, 2006, **78**, 4925–4930.
12. A. M. Skelley, O. Kirak, H. Suh, R. Jaenisch and J. Voldman, *Nat. Methods*, 2009, **6**, 147–152.
13. J. R. Rettig and A. Folch, *Anal. Chem.*, 2005, **77**, 5628–5634.
14. J. R. Kovac and J. Voldman, *Anal. Chem.*, 2007, **79**, 9321–9330.
15. J. Dragavon, T. Molter, C. Young, T. Strovas, S. McQuaide, M. Holl, M. Zhang, B. Cookson, A. Jen, M. Lidstrom, D. Meldrum and L. Burgess, *J. R. Soc. Interface*, 2008, **5**, S151–S159.

16. M. Ozkan, T. Pisanic, J. Scheel, C. Barlow, S. Esener and S. N. Bhatia, *Langmuir*, 2003, **19**, 1532–1538.
17. N. M. Toriello, E. S. Douglas and R. A. Mathies, *Anal. Chem.*, 2005, **77**, 6935–6941.
18. B. M. Taff and J. Voldman, *Anal. Chem.*, 2005, **77**, 7976–7983.
19. J. Godin, C. H. Chen, S. H. Cho, W. Qiao, F. Tsai and Y. H. Lo, *J. Biophoton.*, 2008, **1**, 355–376.
20. M. Yamada and M. Seki, *Lab Chip*, 2005, **5**, 1233–1239.
21. D. Di Carlo, D. Irimia, R. G. Tompkins and M. Toner, *Proc. Natl. Acad. Sci. U. S. A.*, 2007, **104**, 18892–18897.
22. L. R. Huang, E. C. Cox, R. H. Austin and J. C. Sturm, *Science*, 2004, **304**, 987–990.
23. J. Gao, X. F. Yin and Z. L. Fang, *Lab Chip*, 2004, **4**, 47–52.
24. M. A. McClain, C. T. Culbertson, S. C. Jacobson, N. L. Allbritton, C. E. Sims and J. M. Ramsey, *Anal. Chem.*, 2003, **75**, 5646–5655.
25. L. Shui, S. Pennathur, J. C. T. Eijkel and A. van den Berg, *Lab Chip*, 2008, **8**, 1010–1014.
26. H. Song and R. F. Ismagilov, *J. Am. Chem. Soc.*, 2003, **125**, 14613–14619.
27. L. Mazutis, J. C. Baret, P. Treacy, Y. Skhiri, A. F. Araghi, M. Ryckelynck, V. Taly and A. D. Griffiths, *Lab Chip*, 2009, **9**, 2902–2908.
28. J. Clausell-Tormos, D. Lieber, J. C. Baret, A. El-Harrak, O. J. Miller, L. Frenz, J. Blouwolff, K. J. Humphry, S. Koster, H. Duan, C. Holtze, D. A. Weitz, A. D. Griffiths and C. A. Merten, *Chem. Biol.*, 2008, **15**, 427–437.
29. A. Huebner, M. Srisa-Art, D. Holt, C. Abell, F. Hollfelder, A. J. Demello and J. B. Edel, *Chem. Commun.*, 2007, 1218–1220.
30. Y. H. Zhan, J. Wang, N. Bao and C. Lu, *Anal. Chem.*, 2009, **81**, 2027–2031.
31. J. Q. Boedicker, L. Li, T. R. Kline and R. F. Ismagilov, *Lab Chip*, 2008, **8**, 1265–1272.
32. C. S. Chen, M. Mrksich, S. Huang, G. M. Whitesides and D. E. Ingber, *Science*, 1997, **276**, 1425–1428.
33. C. S. Chen, M. Mrksich, S. Huang, G. M. Whitesides and D. E. Ingber, *Biotechnol. Prog.*, 1998, **14**, 356–363.
34. M. Thery and M. Bornens, *Curr. Opin. Cell Biol.*, 2006, **18**, 648–657.
35. M. Thery, V. Racine, A. Pepin, M. Piel, Y. Chen, J. B. Sibarita and M. Bornens, *Nat. Cell Biol.*, 2005, **7**, U947–U929.
36. G. H. Underhill and S. N. Bhatia, *Curr. Opin. Chem. Biol.*, 2007, **11**, 357–366.
37. C. Piggee, *Anal. Chem.*, 2009, **81**, 16–19.
38. J. E. Curtis, B. A. Koss and D. G. Grier, *Opt. Commun.*, 2002, **207**, 169–175.
39. G. T. Salazar, Y. L. Wang, G. Young, M. Bachman, C. E. Sims, G. P. Li and N. L. Allbritton, *Anal. Chem.*, 2007, **79**, 682–687.
40. H. Shadpour, C. E. Sims, R. J. Thresher and N. L. Allbritton, *Cytometry Pt. A*, 2009, **75A**, 121–129.
41. S. Bronzeau and N. Pamme, *Anal. Chim. Acta*, 2008, **609**, 105–112.

42. A. E. Saliba, L. Saias, J. Salamero, V. Fraisier, J. Y. Pierga, F. C. Bidard, C. Mathiot, P. Vielh, F. Farace and J. L. Viovy, *Anticancer Res.*, 2008, **28**, 704.

43. J. D. Adams, U. Kim and H. T. Soh, *Proc. Natl. Acad. Sci. USA.*, 2008, **105**, 18165–18170.

44. P. J. Marc, C. E. Sims, M. Bachman, G. P. Li and N. L. Allbritton, *Lab Chip*, 2008, **8**, 710–716.

45. L. C. Waters, S. C. Jacobson, N. Kroutchinina, J. Khandurina, R. S. Foote and J. M. Ramsey, *Anal. Chem.*, 1998, **70**, 158–162.

46. X. Chen, D. F. Cui and C. C. Liu, *Electrophoresis*, 2008, **29**, 1844–1851.

47. G. Ocvirk, H. Salimi-Moosavi, R. J. Szarka, E. A. Arriaga, P. E. Andersson, R. Smith, N. J. Dovichi and D. J. Harrison, *Proc. IEEE*, 2004, **92**, 115–125.

48. D. Irimia, R. G. Tompkins and M. Toner, *Anal. Chem.*, 2004, **76**, 6137–6143.

49. H. C. Birnboim and J. Doly, *Nucleic Acids Res.*, 1979, **7**, 1513–1523.

50. S. T. L. Harrison, J. S. Dennis and H. A. Chase, *Bioseparation*, 1991, **2**, 95–105.

51. S. T. L. Harrison, *Biotechnol. Adv.*, 1991, **9**, 217–240.

52. D. Di Carlo, C. Ionescu-Zanetti, Y. Zhang, P. Hung and L. P. Lee, *Lab Chip*, 2005, **5**, 171–178.

53. J. T. Nevill, R. Cooper, M. Dueck, D. N. Breslauer and L. P. Lee, *Lab Chip*, 2007, **7**, 1689–1695.

54. S. W. Lee and Y. C. Tai, *Sens. Actuators A-Phys.*, 1999, **73**, 74–79.

55. H. Lu, M. A. Schmidt and K. F. Jensen, *Lab Chip*, 2005, **5**, 23–29.

56. F. T. Han, Y. Wang, C. E. Sims, M. Bachman, R. S. Chang, G. P. Li and N. L. Allbritton, *Anal. Chem.*, 2003, **75**, 3688–3696.

57. D. P. Tieleman, *BMC Biochem.*, 2004, **5**, 10.

58. D. Di Carlo, K. H. Jeong and L. P. Lee, *Lab Chip*, 2003, **3**, 287–291.

59. J. Kim, S. H. Jang, G. Y. Jia, J. V. Zoval, N. A. Da Silva and M. J. Madou, *Lab Chip*, 2004, **4**, 516–522.

60. H. Zhang and W. R. Jin, *Electrophoresis*, 2004, **25**, 1090–1095.

61. H. Zhang and W. R. Jin, *J. Chromatogr. A*, 2006, **1104**, 346–351.

62. J. S. Soughayer, T. Krasieva, S. C. Jacobson, J. M. Ramsey, B. J. Tromberg and N. L. Allbritton, *Anal. Chem.*, 2000, **72**, 1342–1347.

63. C. E. Sims and N. L. Allbritton, *Lab Chip*, 2007, **7**, 423–440.

64. P. A. Quinto-Su, H. H. Lai, H. H. Yoon, C. E. Sims, N. L. Allbritton and V. Venugopalan, *Lab Chip*, 2008, **8**, 408–414.

65. H. H. Lai, P. A. Quinto-Su, C. E. Sims, M. Bachman, G. P. Li, V. Venugopalan and N. L. Allbritton, *J. R. Soc. Interface*, 2008, **5**, S113–S121.

66. S. Le Gac, E. Zwaan, A. van den Berg and C. D. Ohl, *Lab Chip*, 2007, **7**, 1666–1672.

67. Y. K. Cho, J. G. Lee, J. M. Park, B. S. Lee, Y. Lee and C. Ko, *Lab Chip*, 2007, **7**, 565–573.

CHAPTER 13

DNA Analysis in Microfluidic Devices and their Application to Single Cell Analysis

YANN MARCY AND ANGÉLIQUE LE BRAS

Genewave, 172 rue de Charonne, 75011 Paris, France

Abstract

The main developments of microfluidic systems for genetic analyses are aimed at the miniaturization of all biological operations from sampling to detection, an operation named by Manz Micro Total Analysis System (μTAS) 20 years ago.[1] Since then, intense efforts have been made to integrate on a single device all the steps of sample analysis. This integration proved highly complex and has hindered the development of hands-off systems with sample-in answer-out capabilities, apart from a handful of examples in the literature and only one commercially available device (Cepheid).[2-4]

Nevertheless, these efforts have not been wasted since we have witnessed successful uses of microfluidic systems for certain dedicated tasks, either as a preparative step or as a detection mean, with major improvements over conventional molecular biology. Some of these devices have now become part of the standard equipment of biology labs, replacing old technologies.

We will review here the state-of-the-art of microfluidic systems for genetic analyses and analyze their ability to reproduce biological operations and even go beyond, by achieving superior performances either in speed, consumption or efficiency, especially in the case of single cell analysis.

RSC Nanoscience & Nanotechnology No. 15
Unravelling Single Cell Genomics: Micro and Nanotools
Edited by Nathalie Bontoux, Luce Dauphinot and Marie-Claude Potier
© Royal Society of Chemistry 2010
Published by the Royal Society of Chemistry, www.rsc.org

13.1 Amplification on a Chip

Even if single molecule detection is now attained in laboratories, it usually needs equipment and some requirements that are not fully compatible with most microfluidic set-ups. Therefore, most of the time, an amplification step is needed to detect genetic material in microfluidic devices.

13.1.1 Polymerase Chain Reaction

In the vast majority of cases, Polymerase Chain Reaction (PCR) is used to amplify DNA in microfluidic chips as it is the most widely used technique in molecular biology, and also, it is possible to follow the reaction in real time, allowing for precise quantification. To perform a PCR reaction, reagents must be cycled between two to three different temperatures. Different strategies have been used to do so, *e.g.* using a stationary chamber and applying temperature cycles to it or moving the reagents over different temperature zones.

13.1.1.1 Stationary PCR

13.1.1.1.1 Geometry. In the first case, the microfluidic reaction vessel is usually a flat chamber (instead of being a tube) where the reagents are introduced and kept stationary while the temperature is cycled. Miniaturization of PCR in micro-chambers has been widely described in the literature.[5] Various materials have been used: silicon, glass, epoxy resins, silicone, and plastics.[6] The choice of material is usually dictated by the microfabrication technique commonly used in the laboratory where the study is performed, but it should be noted that this choice will impact the efficiency of the PCR reaction. First of all, materials must sustain not only the denaturation temperature of 95 °C but multiple cycles of heating/cooling. If this property is within reach using silicon or glass, plastics are more sensitive to heat. Furthermore, since the microfabricated chips are often an assembly of multiple layers, differences in dilatation coefficients between layers induce strain that the assembly must sustain. To date, a vast number of PCR chambers have been manufactured in silicon. Indeed, the first microfabrication techniques were directly derived from the semiconductor industry and silicon has very good heat transfer properties. Nowadays, silicon has been replaced by plastics and silicone materials as a result of the emergence of fast prototyping techniques in research and production.

13.1.1.1.2 Material and Surface Treatment. The choice of material is also critical due to the very high surface-to-volume ratio in microfluidic devices. Indeed, surfaces must be inert enough not to induce unspecific adsorption, mostly of the polymerase, and must not release PCR inhibitors, such as the solvents used, into the matrix. Various surface passivation strategies have been applied to circumvent this problem either through coating of the surface with silicon oxide (for use with silicon) or through covalent chemistry using silanes, or by adding substances such as bovine serum albumin (BSA) or polyethylene glycol (PEG) to the reaction mix, a technique called dynamic passivation.[6,7]

13.1.1.1.3 Heating. One of the greatest advantages of using micro-chambers for PCR is their small thermal mass. Indeed, as the heating modules are miniaturized, cycling times can be reduced to a few seconds, enabling amplification in a few minutes. Most of the apparatus used to heat microchambers is either thin film heaters or Peltier effect thermoelectric devices. Thin film heaters are usually made from metal, ideally platinum, and are placed directly beneath the device or directly manufactured inside it (Figure 13.1).[8] In this case, a temperature sensor can also be integrated with the heater. It is clear that the closer the heater/sensor is to the chamber, the better the temperature of the PCR reaction is controlled, allowing for a better reaction efficiency. Along with this control, the proximity of the heating system and its small size allow for a very small thermal mass and heating rates of several degrees centigrade per second. Cooling is also required during PCR and if a small thermal mass allows faster cooling, a cooling device can sometimes be used.[9] To this end, the use of a thermoelectric (TE) device allows for both heating and cooling, although it is only recently that TE devices have reached a sufficient resistance to cycle along with a strong temperature difference. Various non-contact heating techniques have also been demonstrated to perform PCR in microfluidic devices using hot air, micro-waves, induction or IR irradiation. The latter uses a lamp and presents the strong advantage of simplifying the assembly of the microchip, avoiding the integration of heaters and sensors.[10]

13.1.1.2 Circulating PCR

Instead of having the reagents in a closed chamber cycled in temperature, other strategies have been used to displace the reagents through different zones of fixed temperature. The first example of non-stationary PCR is the continuous flow PCR reported by Kopp *et al.*[11] In this study, PCR reagents are flushed into a serpentine-shaped channel that wanders over three different temperature zones. By adjusting the flow velocity to the geometry, the time the reagents are exposed at a given temperature can be set. The limitation on the number of PCR cycles imposed by the design was circumvented by Liu *et al.* who designed a rotary device in PDMS with integrated valves to pump the reagents over three temperature zones controlled by sputted electrodes beneath the chip.[12]

Moving liquids in microchannels are subject to Poiseuille flow: the velocity at the surface is null and this creates a parabolic flow profile. This results, in turn, in Taylor dispersion and analytes in the centre do not experience the same amplification as do those on the side of the channel.[13] A good way to circumvent this problem is to use droplets of reagents in oil, a technique called droplet micro-fluidics. The generation of droplets in a high throughput fashion has generated further interest and the splitting,[14,15] combination and mixing of droplets on-chip

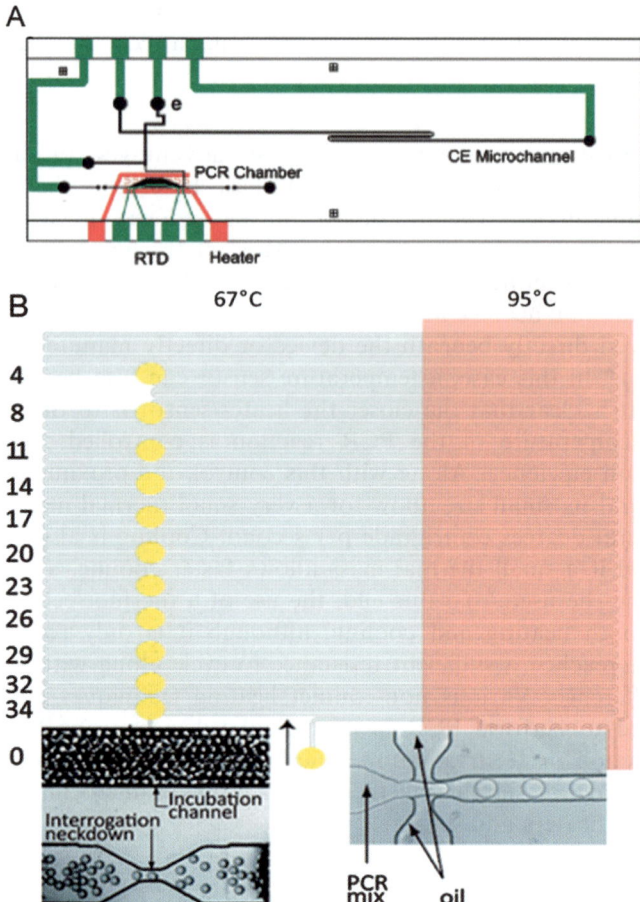

Figure 13.1 Two geometries of PCR amplification and their respective detection technique. (A) Stationary PCR chamber integrated with capillary electrophoresis (CE) in a microchip. (Reprinted, with permission, from Lagally *et al.*[42] © 2004 American Chemical Society.) The glass microchannels are indicated in black, the temperature sensor and microfabricated electrodes are in green, and the heater (located on the backside of the device) is shown in red. (B) Continuous droplet flow PCR microchip (Reprinted, with permission, from Kiss *et al.*[19] © 2008 American Chemical Society.) Schematic of the overall flow configuration. The regions highlighted in yellow correspond to the interrogation neckdowns, and the corresponding cycle numbers are noted on the left. The nozzle is highlighted in red. Bottom left: Optical image of uniform droplets in the downstream channel and flowing through one of the neckdowns. Bottom right: Optical image of droplet generation at the nozzle.

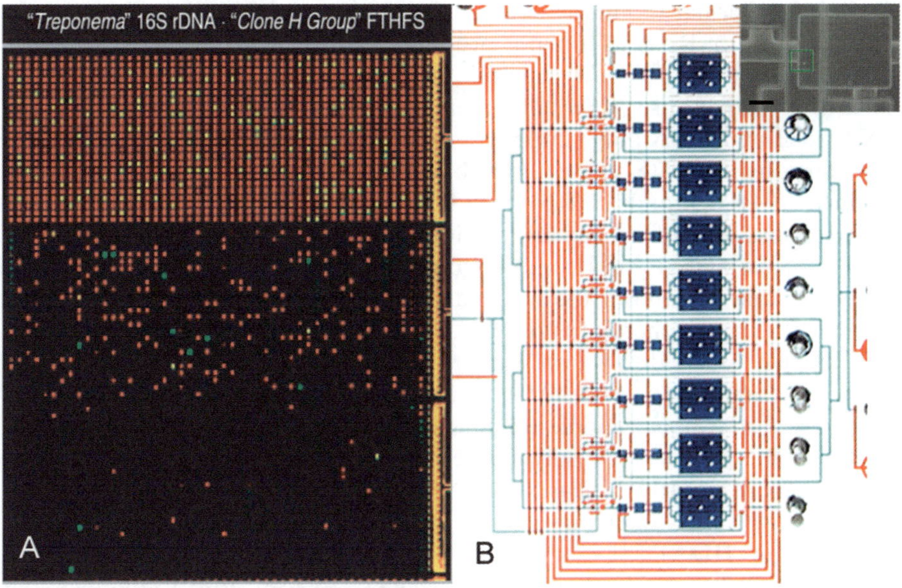

Figure 13.2 Single cell genetic analyses in microfluidics. (A) Microfluidic digital PCR. Three dilutions of termite hingut extract were introduced into an array of chambers. 16S Taqman probes targeted to the *Treponema* genus are in red fluorescence and FTHFS in green.[40] (B) Single cell whole genome amplification device. Left: micrograph of the chip; the different chambers and channel are in blue. Single *E. coli* cells (green fluorescence) are diverted into 3 nL chambers.[29]

can now be achieved.[16,17] Droplet microfluidics has become an active field especially for genetic analyses. Chabert *et al.* have presented a microdroplet platform in which a train of 0.5 µL droplets is generated by suction into a dual shaped capillary flow over a three zone heating cylinder, generating 3000 PCR reactions a day.[18] Moving to picoliters, Kiss *et al.* have introduced a microfluidic system that forms droplets of PCR/sample mix in oil with a microfluidic nozzle.[19] Those droplets move over two temperature zones for 34 cycles. By inserting restrictions at the end of a large channel (Figure 13.2), the authors succeeded in detecting single droplets at around 500 droplets per second, a full reaction lasting 30 min. This device was used to perform high-throughput digital amplification (a technique explained later).

13.1.2 Isothermal Techniques

Cycling microchips rapidly in various temperatures is not an easy task. Also, it must be noted that cycling between 55 and 95 °C is not a physiological working condition to replicate DNA. In humans, for example, most enzymes work at 37 °C and nature has devised a way to denature DNA without boiling bodies.

Using a combination of enzymes, isothermal techniques have then been developed and are now about to spread into molecular biology laboratories.

The first isothermal techniques were inspired by the retroviral replication system and produce RNA as a major product. Named "self-substained sequence replication" (3SR),[20] and nucleic acid sequence based amplification (NASBA),[21] they use three different enzymes, *i.e.* reverse transcriptase, RNAase H and RNA polymerase to produce both DNA and RNA templates at an exponential rate, even faster than PCR since a T7 RNA polymerase produces 10–100 strands of RNA per DNA molecule.

The second class of isothermal reactions is based on the strand displacement activity of certain enzymes coming from phage or bacteria. This strand displacement activity allows the enzyme to open up the DNA on its way to replicate it. There is therefore no need to heat the DNA at 95 °C to denature it. This activity was used to set up site-directed amplification with loop-mediated isothermal amplification (LAMP),[22] rolling cycle amplification of circular templates,[23] or whole genome amplification using the high processivity and fidelity of the phage phi29 polymerase named multiple displacement amplification.[24] More recently, helicases have been used to open DNA strands upfront polymerases, as it is in eukaryotes, to produce isothermal helicase-dependent amplification (HDA).[25]

Microfluidics engineers have moved quickly to use those techniques and apply them on chip. Real-time NASBA has been performed in parallel nanoliter chambers and proved to be as efficient as the macroscopic version to detect human papillomavirus (HPV) sequences.[26,27] LAMP was also used in microfluidic systems monitored by optical detection[28] to amplify hepatitis B virus DNA. Eventually, multiple displacement amplification was performed on chip to amplify the whole genome of single bacterial cells. This will be discussed later in this chapter (section 13.4).[29,30]

13.2 DNA Analysis

In the quest for iTAS, many detection devices have been integrated into microfluidic devices.

13.2.1 Real-time PCR Detection

The most straightforward detection of PCR products on a chip is to optically monitor the reaction either at the end or during the amplification, thus performing what is called real-time PCR. In this technique, intercalating dyes such as Sybr® Green or hybridization probes such as Taqman probes are used to monitor the amount of DNA replicated over time. This is usually done by fluorescence recording and requires an illumination source, at least two filters and a detector. In many studies a microscope system, which has a large footprint, was used. In some systems, however, the optics were integrated into more compact devices. Using a light-emitting diode (LED) as a light source and a photodiode as a detector seems to be the way to integrate optical detection, as shown by Cady *et al.* where a silicon microchip integrating purification, real-time PCR and optical detection allowed successful detection of *Listeria monocytogenes*.[9]

13.2.2 Capillary Electrophoresis

Capillary electrophoresis (CE) is another largely used detection technique and a natural attribute to microchip devices because of its easy microfabrication. Under the influence of an electric field it allows separation of charged molecules by displacing them through a matrix. This separation scheme was coupled very early to PCR on chip but it still requires some optics to record the signal from the molecules. An example of such integration is shown in Figure 13.1.[31]

If real-time PCR allows for on-chip detection of amplified DNA, it is usually not possible to simultaneously detect more than four targets and that requires bulky optics with different optical filters for fluorescence. If it is needed to detect multiple sequences, then a microarray is more suited. Microarrays have actually been integrated in microchips downstream of a stationary PCR or of a continuous flow PCR.[3,32] Unless the device already contains a detection capability, an imaging system is needed for readout. Diffusion limits the hybridization on microarrays but the microfluidic device usually provides fluid handling capability to improve mixing.[33]

13.3 Why and When Smaller is Better

Even if fully integrated systems are not yet widespread in laboratories, microfluidic systems present a clear-cut edge over conventional molecular biology techniques in diverse applications and subsystems tend to be part of our everyday life.

First of all, size matters for the speed of analysis. In PCR systems, the small thermal mass of microfluidic devices allows detectable amplification down to 4 min compared to typically 1 h using conventional thermocyclers.[34] The size reduction also enhances speed and efficiency in capillary electrophoresis with separation times close to 1 min.[31] Actually, microfluidic capillary electrophoresis constitutes the largest commercial success of microfluidic systems as millions of microchips have been manufactured by Caliper Inc. and sold by Agilent Technologies.

Secondly, microfluidic systems have proven to be more efficient for low concentration amplification reactions because of their small volume. As long as an efficient surface passivation strategy is deployed, single molecule amplification is commonly attained in microfluidic systems.[8,19,35–37] Attaining single molecule (or single cell) amplification in conventional tubes is not such an easy task. Failing to amplify a single molecule usually materializes by an unspecific amplification of either contaminants or self-generated by-products such as primer–dimers at the expense of the target DNA. Having the same amount of starting material but in a smaller volume decreases the probability of sampling contaminants and increases the relative concentration of target molecules compared to by-products, thus helping to improve the specificity of the reaction. This ability has been used in microfluidic digital amplification for an absolute quantification of molecules.

Last, the small footprint of microfluidic systems allows for unprecedented throughput by integration of multiple assays in a single device. Spurgeon *et al.* have shown that 2304 (48×48) assays can be done on a single device and 96 CEs were performed in parallel by Simpson *et al.*[38,39]

The integration of reactions on a single platform is, of course, pleading for microfluidics as it would reduce the hands-on time and the possibility of contamination by eliminating transfer steps. On the other hand, using such small volumes becomes a drawback when processing diluted samples. If no concentration step is included, the probability of having the cell or the molecule of interest trapped in the device is fairly low and users will have to solve the difficult problem of "world-to-chip interface".

13.4 Applications of Microfluidic Single Cell Genetic Analysis in Microbial Ecology

Single cell analysis can be of interest when variations in genetic content appear in cell populations. These variations among populations can be found in cancer, especially when dealing with early metastases that have a distinct genetic signature from primary tumors or in microbiota where there is usually a high diversity in bacterial species. For those applications where single cell precision is required, microfluidic systems have had an edge over conventional techniques for several reasons: the ability to isolate single cells, the superior performance of small volume amplification, and sometimes a higher throughput. In microbial ecology, microfluidics has been instrumental in assigning a single gene or part of the genome of a single not yet cultivated bacterium.[30,40] Since only a small fraction of environmental bacteria are amenable to laboratory culture, most of the genetic sequence of those microorganisms are either unknown or obtained through sequencing of whole communities. At present, it is not possible to assign genes and therefore function to a specific cell. For example, in the hindgut of wood-eating termites, the identities of the organisms dominating the production of acetate had remained uncertain. The goal of Ottesen *et al.* was to identify which single species were carrying the FTHFS gene involved in this symbiosis.[40] To do so, the authors performed a digital PCR in a poly (dimethylsiloxane) (PDMS) microchip. The principle of digital PCR is to partition a reaction mix so that each reaction vessel contains on average one molecule or cell. All vessels undergo a PCR and the amount of probed molecule in the sample is quantified by recording the fluorescence of Taqman probes at the end of the reaction and counting the number of positive reactions. The microfluidic here brings its advantages of small volume sampling and high throughput. In the microfluidic system of Ottesen *et al.*,[40] a suspension of cells is introduced into a device consisting of an array of chambers separated by valves. The cells are loaded into the chambers through a single inlet then the valves are closed and the whole device is placed on a conventional thermal cycler. At the end of the reaction, the fluorescence of each chamber is measured. The authors performed a duplex amplification for both the gene of interest FTHFS and the ribosomal 16S gene which is commonly used in microbiology to identify species. Where the gene of interest was localized, the amplicon was extracted from the chamber and sequenced, enabling the identification of not yet cultivated *Treponema* species carrying this specific gene.

To go beyond single gene analysis, the same group devised a way to amplify the whole genome of single uncultivated bacteria using multiple displacement amplification (MDA).[29] Unlike in PCR, for which the Taq polymerase can sustain a cell thermolysis step, phi29 polymerase cannot withstand annealing temperature nor alkaline lysis. Integrating the MDA on chip therefore required the integration of multiple fluid handling steps between several chambers into a single microfluidic chip. Relying on Poisson statistics to have only one cell in each chamber meant that only a third of a microfluidic chip would eventually be of use. A way to direct cells to the chambers was then needed. A microfluidic device that allows the isolation and genome amplification of eight individual microbial cells was designed in PDMS using the microfluidic large-scale integration (MLSI) technique.[41] In this device, the cells were directed to a first chamber by timed opening and closing of valves then lysed by alkaline treatment in the first chamber. The resulting solution was neutralized into a second chamber and pushed by the reaction mix containing the polymerase into a third 50 nL chamber. Whole genome amplification was then performed and yielded about 10^6 to 10^7 copies of the genome of each single microbe (depending on its size). After retrieval and off-chip re-amplification, ten billion copies of the genome (about a microgram) were available for sequencing. Using this approach, a member of the phylum TM7, which had no cultivated or sequenced member, was partially sequenced and more than a thousand genes were obtained. If this microfluidic approach was proven to more evenly represent the genome than amplification in tube,[29] a full sequence was not attainable at this point. This is probably because some DNA was broken during the alkaline lysis or fluid handling steps and any discontinuity in sequence would be largely under-represented in the amplified DNA, as MDA relies on multiple passages of the processive phi29. Nevertheless, a single cell could yield a huge amount of information thanks to this technique.

13.5 Conclusion

Although the microfluidic "grail" of μTAS has not yet been reached, microfluidic systems have become part of a lab's everyday life as subsystems. Now with the rush for next generation sequencing, we are witnessing a big push for microfluidic preparative technologies either as single cell amplification or as a way to automate sample preparation, the biggest remaining lock to fully automated microfluidic systems.

References

1. A. Manz, N. Grabera and H. M. Widmera, *Sens. Actuators B1*, 1990, 244–248.
2. L. C. Waters, S. C. Jacobson, N. Kroutchinina, J. Khandurina, R. S. Foote and J. M. Ramsey, *Anal. Chem.*, 1998, **70**, 158–162.
3. R. H. Liu, J. N. Yang, R. Lenigk, J. Bonanno and P. Grodzinski, *Anal. Chem.*, 2004, **76**, 1824–1831.
4. C. J. Easley, J. M. Karlinsey, J. M. Bienvenue, L. A. Legendre, M. G. Roper, S. H. Feldman, M. A. Hughes, E. L. Hewlett, T. J. Merkel, J. P. Ferrance and J. P. Landers, *Proc. Natl. Acad. Sci. USA.*, 2006, **103**, 19272–19277.
5. C. Zhang, J. Xu, W. Ma and W. Zheng, *Biotechnol. Adv.*, 2006, **24**, 243–284.
6. P. A. Auroux, Y. Koc, A. deMello, A. Manz and P. J. Day, *Lab Chip*, 2004, **4**, 534–546.
7. C. Zhang and D. Xing, *Nucleic Acids Res.*, 2007, **35**, 4223–4237.
8. E. T. Lagally, I. Medintz and R. A. Mathies, *Anal. Chem.*, 2001, **73**, 565–570.
9. N. C. Cady, S. Stelick, M. V. Kunnavakkam and C. A. Batt, *Sens. Actuators B-Chem.*, 2005, **107**, 332–341.
10. M. G. Roper, C. J. Easley, L. A. Legendre, J. A. Humphrey and J. P. Landers, *Anal. Chem.*, 2007, **79**, 1294–1300.
11. M. U. Kopp, A. J. de Mello and A. Manz, *Science*, 1998, **280**, 1046–1048.
12. J. Liu, M. Enzelberger and S. Quake, *Electrophoresis*, 2002, **23**, 1531–1536.
13. L. Chen, J. West, P. A. Auroux, A. Manz and P. J. Day, *Anal. Chem.*, 2007, **79**, 9185–9190.
14. T. Thorsen, R. W. Roberts, F. H. Arnold and S. R. Quake, *Phys. Rev. Lett.*, 2001, **86**, 4163–4166.
15. D. R. Link, S. L. Anna, D. A. Weitz and H. A. Stone, *Phys. Rev. Lett.*, 2004, **92**, 054503.
16. D. R. Link, E. Grasland-Mongrain, A. Duri, F. Sarrazin, Z. Cheng, G. Cristobal, M. Marquez and D. A. Weitz, *Angew. Chem. Int. Ed. Engl.*, 2006, **45**, 2556–2560.
17. H. Song and R. F. Ismagilov, *J. Am. Chem. Soc.*, 2003, **125**, 14613–14619.
18. M. Chabert, K. D. Dorfman, P. de Cremoux, J. Roeraade and J. L. Viovy, *Anal. Chem.*, 2006, **78**, 7722–7728.
19. M. M. Kiss, L. Ortoleva-Donnelly, N. R. Beer, J. Warner, C. G. Bailey, B. W. Colston, J. M. Rothberg, D. R. Link and J. H. Leamon, *Anal. Chem.*, 2008, **80**(23), 8975–8990.
20. J. C. Guatelli, K. M. Whitfield, D. Y. Kwoh, K. J. Barringer, D. D. Richman and T. R. Gingeras, *Proc. Natl. Acad. Sci. U. S. A.*, 1990, **87**, 7797.
21. J. Compton, *Nature*, 1991, **350**, 91–92.
22. T. Notomi, H. Okayama, H. Masubuchi, T. Yonekawa, K. Watanabe, N. Amino and T. Hase, *Nucleic Acids Res.*, 2000, **28**, E63.

23. G. T. Walker, M. S. Fraiser, J. L. Schram, M. C. Little, J. G. Nadeau and D. P. Malinowski, *Nucleic Acids Res.*, 1992, **20**, 1691–1696.
24. F. B. Dean, S. Hosono, L. Fang, X. Wu, A. F. Faruqi, P. Bray-Ward, Z. Sun, Q. Zong, Y. Du, J. Du, M. Driscoll, W. Song, S. F. Kingsmore, M. Egholm and R. S. Lasken, *Proc. Natl. Acad. Sci. U.S.A.*, 2002, **99**, 5261–5266.
25. M. Vincent, Y. Xu and H. Kong, *EMBO Rep.*, 2004, **5**, 795–800.
26. A. Gulliksen, L. Solli, F. Karlsen, H. Rogne, E. Hovig, T. Nordstrom and R. Sirevag, *Anal. Chem.*, 2004, **76**, 9–14.
27. A. Gulliksen, L. A. Solli, K. S. Drese, O. Sorensen, F. Karlsen, H. Rogne, E. Hovig and R. Sirevag, *Lab Chip*, 2005, **5**, 416–420.
28. S. Y. Lee, C. N. Lee, H. Mark, D. R. Meldrum and C. W. Lin, *Sens. Actuators B-Chem.*, 2007, **127**, 598–605.
29. Y. Marcy, T. Ishoey, R. S. Lasken, T. B. Stockwell, B. P. Walenz, A. L. Halpern, K. Y. Beeson, S. M. Goldberg and S. R. Quake, *PLoS Genet.*, 2007, **3**, 1702–1708.
30. Y. Marcy, C. Ouverney, E. M. Bik, T. Losekann, N. Ivanova, H. G. Martin, E. Szeto, D. Platt, P. Hugenholtz, D. A. Relman and S. R. Quake, *Proc. Natl. Acad. Sci. U. S. A.*, 2007, **104**, 11889–11894.
31. A. T. Woolley, D. Hadley, P. Landre, A. J. deMello, R. A. Mathies and M. A. Northrup, *Anal. Chem.*, 1996, **68**, 4081–4086.
32. M. Hashimoto, F. Barany and S. A. Soper, *Biosens. Bioelectron.*, 2006, **21**, 1915–1923.
33. J. G. Lee, K. H. Cheong, N. Huh, S. Kim, J. W. Choi and C. Ko, *Lab Chip*, 2006, **6**, 886–895.
34. B. C. Giordano, J. Ferrance, S. Swedberg, A. F. Huhmer and J. P. Landers, *Anal. Biochem.*, 2001, **291**, 124–132.
35. N. R. Beer, B. J. Hindson, E. K. Wheeler, S. B. Hall, K. A. Rose, I. M. Kennedy and B. W. Colston, *Anal. Chem.*, 2007, **79**, 8471–8475.
36. R. Dettloff, E. Yang, A. Rulison, A. Chow and J. Farinas, *Anal. Chem.*, 2008, **80**, 4208–4213.
37. Y. Schaerli, R. C. Wootton, T. Robinson, V. Stein, C. Dunsby, M. A. Neil, P. M. French, A. J. Demello, C. Abell and F. Hollfelder, *Anal. Chem.*, 2009, **81**, 302–306.
38. S. L. Spurgeon, R. C. Jones and R. Ramakrishnan, *PLoS One*, 2008, **3**, e1662.
39. P. C. Simpson, D. Roach, A. T. Woolley, T. Thorsen, R. Johnston, G. F. Sensabaugh and R. A. Mathies, *Proc. Natl. Acad. Sci. U. S. A.*, 1998, **95**, 2256–2261.
40. E. A. Ottesen, J. W. Hong, S. R. Quake and J. R. Leadbetter, *Science*, 2006, **314**, 1464–1467.
41. T. Thorsen, S. J. Maerkl and S. R. Quake, *Science*, 2002, **298**, 580–584.
42. E. T. Lagally, J. R. Scherer, R. G. Blazej, N. M. Toriello, B. A. Diep, M. Ramchandani, G. F. Sensabaugh, L. W. Riley and R. A. Mathies, *Anal. Chem.*, 2004, **76**, 3162–3170.

CHAPTER 14

Gene Expression Analysis on Microchips

MAX CHABERT

RHODIA Lab of the Future, F-33608 Pessac Cedex, France

Abstract

Integration of the whole operations enabling gene expression analysis on microchips is a rather recent advance in the – rather young – world of microfluidics. Here, we summarize the gradual evolutions that led from the very first "millifluidic" integrated chip for RNA analysis to more sophisticated and sensitive devices. We show how the developments of an initially applied problematic (creating portable integrated devices for *e.g.* point of care analysis) are finally used for studies of fundamental importance in cell biology. Two main families of devices are found in the literature: those integrating multiple steps to carry out RNA reverse transcription and DNA amplification separately, and those that take advantage of the recent developments of biology chemicals to perform RNA analysis in a single step. The latter are generally highly sensitive, and single cell RNA analysis has been made possible in very low volume handling systems. Nevertheless, in contrast to some other microfluidics biological applications, there remains room for progress in miniaturized gene expression analysis. Further efforts are still needed to provide reliable tools for use in applied biology as well as in fundamental research.

RSC Nanoscience & Nanotechnology No. 15
Unravelling Single Cell Genomics: Micro and Nanotools
Edited by Nathalie Bontoux, Luce Dauphinot and Marie-Claude Potier
© Royal Society of Chemistry 2010
Published by the Royal Society of Chemistry, www.rsc.org

14.1 Introduction

Analyzing gene expression, *i.e.* measuring the level of given RNAs in a sample of interest, has a broad range of applications, from fundamental biological studies to pathogen detection and clinical diagnosis. The key step in RNA analysis is to amplify the initial number of molecules to obtain a measurable (generally optical) signal. This can either be done by reverse transcription of RNA into cDNA (complementary DNA) followed by polymerase chain reaction (RT-PCR), either by direct amplification of RNA. In both cases, combining these methods with microfluidic technology not only opens the way to high-throughput screening and point-of-care applications, but also gives access to previously hardly measurable signals. In this perspective, although most development efforts are directed towards portable devices for biological analysis, some systems are also dedicated to more fundamental laboratory studies. From an applicative point of view, detecting RNA in small samples is generally easier than working directly on DNA, as the initial number of target molecules is relatively high when compared to a single DNA molecule per cell. From the molecular biology point of view, the low volumes used in microfluidics make it possible to take advantage of concentration effects to increase the yield of RNA transcription and circumvent artifacts such as preferential amplification of most concentrated RNAs.[1] It also permits the analysis of single cells, thus allowing a study of the variability in gene expression at the "biological unit of life" level.[2]

Integration of the whole protocol permitting RNA analysis on a single chip arrived late and remains rather rare. Indeed, although the very first microfluidic PCR dates back to 1994,[3] the first example of RNA analysis using microfluidics was only reported in 2000.[4] This is mainly because of the complex and multiple steps involved. For instance, one common problem encountered with RT-PCR is that reagents from the RT interfere with the PCR, leading to a low fidelity and a low amplification yield. In the first versions of microfluidic RT-PCR systems, RT products therefore have to be purified, or at least diluted before encountering PCR. Alternatively, nested PCR can be used: the RT products are amplified by two successive PCRs with two different primer sets, thus increasing reaction specificity. Whatever the solution, many technological bottlenecks exist that hinder the integration of these multiple steps on a microchip. Interestingly, what triggered the development of more and more RNA analysis chips mainly came from the biochemistry side, with the appearance of new reaction mixes allowing for one step RT-PCR reactions.

In this chapter, we first describe microfluidic devices permitting RNA analysis, with various levels of integration. Characteristics of these devices are summarized in Table 14.1. A brief description of relevant systems dedicated to cDNA analysis after off-chip reverse transcription is then carried out. We finally focus on devices that take advantages of microfluidics to go towards single cell RNA analysis. This is certainly the domain where, besides technological development, microfluidics offers a unique route to a better understanding of the fundamental mechanisms underlying the role of RNA in cell life and death.

Table 14.1 Main characteristics of typical microfluidic systems developed for gene expression analysis.

Reference	Geometry	Temperature control	Detection	Sensitivity (starting RNAs)	Genes analyzed at a time	Remarks
Anderson et al. (2000)	Polycarbonate chambers Hydrophobic vents Microliter volume	4-zones external heating block	Hybridization Benchtop scanner	500	1	Nested PCR
Obeid et al. (2003)	Continuous flow glass chip Pressure pumping Microliter volume	4 zones external copper block	Off-chip gel electrophoresis	Over 10^6	1	Continuous flow. Two working modes
Liao et al. (2005)	Two PDMS chamber MSL built On-chip micropump Microliter volume	Integrated heaters and sensors	Off-chip gel electrophoresis	1000	1	Designed to be portable
Marcus et al. (2006)	72 parallel reactions chambers built (MSL) 500 pL volumes	External heating block	Benchtop scanner	50	9	Uses one step RT-PCR kit
Toriello et al. (2006)	2 parallel reactions chambers and electrophoresis channels 380 nL volumes	Integrated sensors and heaters	Benchtop scanner after on-chip gel electrophoresis	11	2×2	Uses one step RT-PCR kit Possibly more genes analyzed using multiplex PCR

Reference	Device	Heating	Detection	Sensitivity		Comments
Kagaila *et al.* (2008).	Chambers separated by MSL valves. 600 nL volume	Integrated sensors and heaters	Laser induced fluorescence after gel electrophoresis	?	1	Totally portable instrument Unknown sensitivity
Lee *et al.* (2008)	COC chambers + pinch valves Microliter volume	80 W IR lamp	Chemoluminescence + photodiode	?	1	Unknown sensitivity Detection with no need for excitation source
Beer *et al* (2008).	Static droplets in a hydrophobized glass chip 70 pL droplets	Flexible resistive elements and thermocouple	Fluorescence microscope	1	1	Does not take fully advantage of droplets microfluidics
Pipper *et al.* (2007)	100 nL droplet actuated by magnetic forces	Integrated sensors and heaters	Fluorescence microscope	1	1	Fast analysis (28 min total)
Dimov *et al* (2008)	PDMS chamber and purification column	Peltier device	Fluorescence microscope	100 cells equiv.	1	NASBA isothermal amplification

14.2 Multi-step Microfluidic RT-PCR

Interestingly, the first example of RNA analysis on a microchip is already a highly integrated device developed by Affymetrix[4] (Figure 14.1). This device uses sequential operations in multiple chambers to extract RNA from serum samples, do the reverse transcription and PCR amplification and finally analyze products by hybridization (including fragmentation and labeling). Here, nested PCR is used to circumvent the drawbacks of one-step RT-PCR in terms of specificity. The whole operation is performed in a polycarbonate chip held between heating elements and pneumatic controls, and hybridization is quantified using a custom scanner. The system used is "millifluidic" rather than microfluidic and all steps are performed using volumes in the tens of microliters range, thus corresponding to classical manual benchtop reactions. A detection limit of 500 starting RNA copies was reached when analyzing the mutation of one gene of the HIV virus.

The need for two independent steps for RT and PCR in RNA analysis has also been addressed by Liao *et al.*[5] using two independent chambers for RT and PCR, connected by an actuation channel. The device is a hybrid poly (dimethylsiloxane) (PDMS)/glass system, with integrated heaters and temperature sensors. Pumps and valves are also included on-chip using multi-layer soft lithography (MSL). The operations are carried out sequentially: RNA is

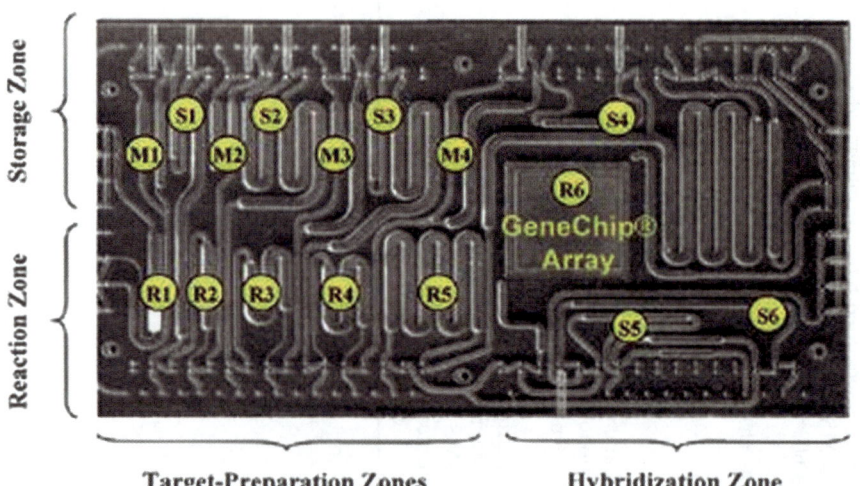

Figure 14.1 First integrated microsystem for RNA analysis by RT-PCR. The device is 8×40×70 mm. Reagents are stored in S zones, reactions are carried out in R zones, intermediate products are stored in M zones. Temperatures of the storage, reaction and hybridization zones are controlled independently by external elements. Reprinted, with permission, from Anderson *et al.*[4]

reverse transcribed in a $10\,\mu L$ reaction chamber, and $2\,\mu L$ of the synthesized cDNA is then transported in the PCR reaction chamber for direct amplification. The use of integrated peristaltic pumps allows precise metering of the sample and reagents volumes. The whole device is conceived to be portable but, unfortunately, amplicon analysis is done off-line by gel electrophoresis. This leads to a rather high detection limit (1000 starting copies) for the two RNA viruses used in this study (around 400 base pairs (bp) – amplicons). In 2007 the same group also presented a refinement of their device where the virus are captured, separated from the surrounding biological medium and the RNA extracted on-chip using antibody-coated paramagnetic beads.[6]

Lee *et al.* recently presented a fully integrated system for HIV detection by RT-PCR in view of point of care applications.[7] RT and PCR are carried out in two different chambers separated by an original system of pinch valves. The device is constructed in cyclic olefin copolymer (COC), and volumes are in the microliter range. Heating is provided by an $80\,W$ infrared (IR) lamp. PCR primers are tagged with biotin and synthesized DNA is detected using a sandwich chemoluminescent assay (no excitation is needed). The authors give no information on their system sensitivity.

The only example of continuous flow microfluidic RT-PCR can be found in Obeid *et al.*[8] Reverse transcription and PCR are performed on a single microfluidic chip. The system is fabricated in silanized borosilicate glass. The resulting products are analyzed off-chip using gel electrophoresis. This RT-PCR system can work in a semi- continuous or totally continuous flow mode. In the first mode, RNA reverse transcription is accomplished in a closed chamber, and continuous flow PCR is carried out after this first static step. In the second mode, RNA reverse transcription is also carried out in continuous flow, the resulting products being directly mixed with the PCR reagents and subjected to PCR. In both cases, to circumvent the negative interaction of RT products with the PCR reaction, the former are diluted ten times in the PCR mix, and reverse transcriptase is inactivated at $95\,°C$. Only one RNA sequence is analyzed in each run. Multiple samples can be sequentially injected and analyzed in the system provided that a water wash is performed. However, given the sensitivity of the PCR reaction, contamination between successive samples is likely. The detection limit is high (2.5×10^6 starting RNA molecules). Although very powerful in numerous microfluidic applications, continuous flow is not necessarily well-suited for carrying out RT-PCR due to the inherent sequential aspect of this reaction. Indeed, no other attempts to use this technique can be found in the literature.

In all the above examples, reactions are performed in volumes close to those used in benchtop experiments, therefore preventing any applications to single cell analysis. These devices are nevertheless well-suited for point-of-care applications, for example, where an average signal is used for diagnostics. In contrast, new devices appeared that work with a simplified amplification protocol on much smaller volumes. The latter open the way toward single-cell analysis applications.

14.3   One-step Microfluidic RNA Analysis

RT-PCR in a 72 parallel chambers microfluidic matrix built using the MSL technology has been demonstrated in Steve Quake's group.[9] The device is constructed in PDMS. RT-PCR is performed in a single step with no need for post-RT dilution or nested PCR. This is made possible by the commercialization of new reagents allowing one step RT-PCR, although the specificity and yield of the reactions carried out with the latter is sometimes thought to be lower than in classical two steps RT-PCR. Heating and endpoint detection are done using external instruments. A detection limit of about 50 total RNA copies was reached for the detection of a 240 bp portion of the β-actin gene. Note that this sensitivity, linked to the concentration effects obtained in the 500 pL reactors, corresponds to what is needed for single cell analysis.

Using the same kind of one-step RT-PCR kit, a sensitivity of 11 starting total RNA molecules was reached on a device coupling RT-PCR in 380 nL chambers with integrated capillary electrophoresis (CE).[10] A hybrid glass/PDMS design allows the integration of heaters, sensors and electrodes. Detection of separated products is carried out on an external scanner. The first multiplex RT-PCR on a microfluidic chip was demonstrated on this device. This was used to quantify the relative abundance of two transcripts in a mixed *Escherichia coli* population and to perform splice variant analysis of the BRCA1 suppressing gene RBBP8 (RNA extracted from breast tumor tissue). In an effort to make this type of integrated system totally portable at an affordable price, Kaigala *et al.* also developed an on-chip RT-PCR-CE system coupled with a reduced dimensions electronic box for fluidic and thermal controls, high voltage supply, laser excitation and fluorescence detection using a CCD camera.[11] Analysis is carried out in 600 nL volumes. The proof of concept was demonstrated on β2M gene of KMS-34 cell line. The sensitivity of the system still has to be assessed.

Due to the important surface-to-volume ratio in microfluidics, surface state of the devices can have a dramatic influence on the yield of biochemical reactions.[3] This influence can be minimized when working in biphasic microfluidics, and that combined with the particularly small volumes attainable with this configuration makes sensitivity of RT-PCR in microdroplets very high. One-step RT-PCR in microdroplets was recently demonstrated by Beer *et al.*[12] Picoliter droplets containing the reagents and samples for RT-PCR are generated in mineral oil at a T-junction. The flow is then stopped, and reverse transcription followed by PCR are carried out on chip using external resistive heating. Detection is performed by laser-induced fluorescence. A high sensitivity was achieved with the detection of a single copy of MS-2 virion in 70 pL droplets. About 80 droplets are analyzed at a time, but these droplets are all the same. Therefore, this device does not take advantage of the usefulness of droplet microfluidics for creating and manipulating independent microreactors. An alternative biphasic approach consists in using 100 nL droplets actuated on a planar surface using magnetic forces[13] and was used for detecting H5N1 virus in throat swab samples by RT-PCR. The super-paramagnetic particles permitting actuation are also used for solid phase extraction of the viral RNA, allowing for a 500× pre-concentration. RT-PCR is carried out in one step. The device is fully integrated and provides single copy sensitivity with a 28 min analysis time. Going further from microfluidic technologies, but still in micro-droplets, RT-PCR was demonstrated using the principle of *in vitro* compart-mentalization (IVC)[14] (see Chapter 8 for more information on this technique). A single copy of RNA could also be detected.

Finally, RNA can also be directly amplified by nucleic acid sequence-based amplification (NASBA). This reaction uses three enzymes and allows expo-nential amplification of RNA in isothermal conditions. Only one example of complete integration of this powerful tool can be found.[15] It was used to detect *E. coli* bacteria. A silica beads purification column is integrated on-chip to extract RNA directly from cell lysate. NASBA is carried out isothermally and reaction products are detected in real-time using a fluorescence microscope (RNA is tagged by molecular beacons). The authors demonstrate a limit of detection below 100 cells in 2 µL volumes. Note, however, that tmRNA (pre-sent in all bacteria), rather than mRNA, is amplified and that the copy number per cell of the former is much higher than that of the latter. At present, inte-gration of NASBA in microfluidics is poorly developed, but should become a method of choice for point-of-care applications because of an open intellectual property policy, its simple one-step protocol and its isothermal aspect. Indeed, although heating and cooling can be very efficient due to the high surface-to-volume ratio in microsystems, controlling the multiple temperatures involved in RT-PCR reactions is a hindrance to the development of cost- and energy-effective integrated devices. While off-chip heating generally remains energy consuming due to the thermal inertia of the external elements, integrating resistances and sensors in a device is expensive, thus impeding the development of disposable point of care microchips. The most promising technology to solve this paradigm might be the use of external heating and sensing by optical means, *e.g.* laser diodes.[16]

14.4 Microfluidic cDNA Analysis

As seen above, systems enabling micro total analysis of RNA are rare. In contrast, microfluidic systems for cDNA analysis are numerous, and describing them all is beyond the scope of this review. In this part we will only focus on some relevant examples – potentially amenable to routine applications – that illustrate the various microfluidic technologies used for cDNA amplification by PCR. Three main approaches hold promises for microfluidic high-throughput amplification of DNA: microarray PCR coupled with robotics, microdroplet platforms, and microfluidic matrices (using MSL technology).

High-throughput PCR of cDNA has been performed in 200 nL volumes using a microfabricated microtiter plate (MTP) coupled with robotic nano-dispensing.[17] The system was demonstrated to work reproducibly for the analysis of four transcript level in three mouse tissues, with a sensitivity of five starting DNA molecules. Using this method, steps of reverse transcription are performed classically. The sample and reagents are then manipulated using either piezo or valve dispensing, depending on the desired volumes. Particular attention must be given to the control of evaporation. Temperature cycling is provided by an external heating block. Product detection is accomplished in real-time by live-fluorescence imaging of the MTP. The obtained results correlate well with what is observed in classical RT-PCR. Morrison *et al.*[18] demonstrated RT-PCR using 48 sub-arrays of 64 wells, thereby carrying out 64 identical reactions in 33 nL volumes. Samples and reagents are dispensed manually. Heating and detection are achieved in a custom thermal cycler. A close agreement was found between the developed device and conventional instruments, with a 24-fold higher throughput.

Another approach for high throughput analysis of cDNA consists of using circulating microdroplets.[19,20] Thermal cycling is achieved by circulation in three temperature zones. An automated platform enables the analysis of 3000 samples per day in microliter droplets.[21] Droplet formation and reagent mixing, as well as endpoint fluorescence detection of the amplified products are performed automatically. The fluidic system is designed to avoid any contamination between neighboring droplets. Although the detection limit is rather high, this approach proposes a solution to the classical "world-to-chip" problem and opens a new way toward high throughput analysis in microdroplets.

A microfluidic matrix for quantitative PCR was used to analyze 45 genes expression in 18 tissues with replicates.[22] RNA transcription and DNA pre-amplification were performed off-chip. Using a microfluidic PCR chip allows for a much higher throughput than conventional methods, as 2300 simultaneous analyses can be carried out with fewer pipetting steps than to set up a classical 96-well plate. Pneumatic control of integrated valves, thermal cycling and product detection are done using an external platform specifically developed to accommodate these microchips (Fluidigm). An excellent concordance was demonstrated between the results obtained using the microchip and those using conventional real-time PCR. This is an important point in order to bring microfluidics technology to a broad audience (*e.g.* physicians). Using a microfluidic matrix, quantitative digital PCR was also demonstrated to be a powerful tool for gene expression studies.[23] Thermal cycling and optical detection are carried out using external elements. Five blood progenitor populations were isolated as single cells using fluorescence activated cell sorting, and reverse transcribed using classical protocols. The transcripts were then cycled on the digital PCR chip (7.5 μL of transcripts sampled in 1200 chambers). The digital PCR chip enables a high level of dilution, so that each compartment statistically contains one or no DNA template. A Poisson analysis of the amplification results gives the initial number of DNA molecules contained in a given sample, and thereby allows one to back out the initial number of RNA molecules in the single cells. A clear advantage of this digital PCR method is an absolute quantification of transcript number with no need for internal reference. The biases associated with classical RT are not avoided but this technique allowed for the profiling of transcription factors at the single cell level with a relatively high throughput, and might therefore help bring new biological insight into problematics, such as cell differentiation.

14.5 Single Cell RNA Analysis

Microfluidic volumes are perfectly suited for studying complex biological processes at the single cell level. This holds for gene expression analysis, and as seen in section 14.3, microfluidic devices allowing the analysis of the equivalent RNA of a single cell exist. These devices have been further developed to integrate the capture, the extraction and the purification of the content of a single cell.

Following their first demonstration of RT-PCR in a 72 microchamber array, Marcus *et al.*[24] built a device permitting the capture of a single cell, its lysis, mRNA purification and cDNA synthesis. The operations can be carried out in four parallel chambers built using MSL. Purification following cell lysis is performed by mRNA capture using an on-chip constructed affinity column.

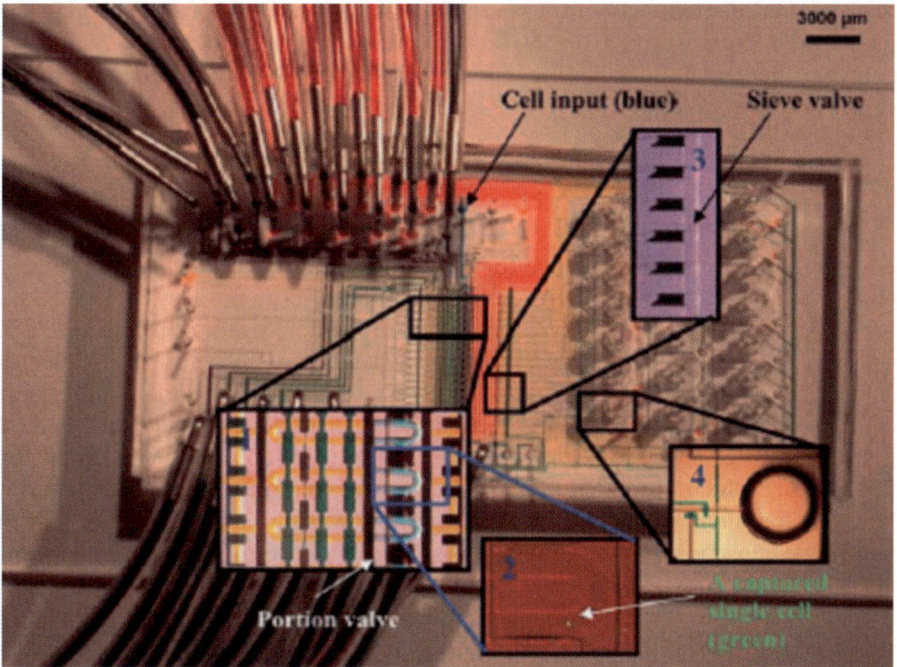

Figure 14.2   Example of a multi-chamber microfluidic chip for RT-PCR built using MSL. This device enables one to work independently on 20 single cells, from RNA extraction to reverse transcription into cDNA. Insets 1 and 2 show modules for single cell capture and lysis (flow channels are blue). Inset 3 shows purification columns formed *in situ*. Inset 4 is the collection chamber for cDNA-labeled beads. Reprinted, with permission, from Zhong *et al.*[25]

cDNA synthesis is performed using primers grafted to the affinity column beads. The resulting products are analyzed by conventional real-time PCR. The device was used to study the expression of the highly expressed GAPDH gene in NIH/3T3 cells, at the single but also at multiple cell level. A device of the same category was used to determine the absolute number of mRNA of three genes in a single human embryonic stem cell[25] (Figure 14.2). In both cases, PCR amplification of synthesized cDNA is made off-chip, which might be detrimental to reproducibility.

Toriello *et al.*[26] further developed their integrated platform[10] to include cell capture and lysis. Cell capture is carried out using DNA recognition: cells of interest are surface-labeled using a 20-base single strand oligonucleotide and captured on a gold electrode labeled with the complementary sequence. The device was used to study GAPDH mRNA silencing in single Jurkat T-cells.

Taking the next step in terms of complexity, Bontoux *et al.* designed a microfluidic device aimed at analyzing the whole transcriptome of a single cell on a microfluidic chip.[1] This device will be described in more detail in chapter 16. Besides an important technological effort, this device enabled the authors to bring a new biological insight into the complex domain of neurosciences thanks to an unprecedented yield of the reverse transcription. This is rare enough to be highlighted. Interestingly, the PCR step had to be performed off-chip to circumvent detrimental crowding effects associated with a too high cDNA concentration.

14.6 Conclusion

Although of utmost importance for the understanding of fundamental biological mechanisms as well as for developing portable analysis systems, "RNA microchips" are still under development. Portable devices for applications such as point-of-care analysis exist, but some work remains to be done in order to achieve fully integrated and portable systems. From a more fundamental point of view, examples where RNA analysis in microfluidics could bring a real insight into a given biological process are still rare. Many reasons account for this, among which are inherent difficulties in the manipulation of RNA and a lack of capability to reliably integrate multiple washing and purification steps in a single, closed microfluidic device. As pointed out in this chapter, the recent progress in microfluidic systems allowing complete gene expression analysis in these devices is due to improvements in molecular biology reagents, permitting single step RT-PCR analysis, rather than to the development of new microfluidic tools. On the other hand, microfluidic cDNA analysis is now a mature and powerful technology that will hopefully be used as a high-throughput routine technique in the near future. As its development was largely posterior, and in view of the increasing quantity of innovative work carried out in this field, RNA analysis should follow the same trend in the years to come.

Gene expression is closely related to protein abundance in cells, but a high level of mRNAs in a given cell does not necessarily mean a high level of protein expression. From the cellular biology point of view, there is therefore immense interest in looking directly at protein expression rather than at the intermediate mRNA. Many microfluidic devices have thus been developed that take advantage of concentration effects in small volumes to target directly protein expression in single cells (see, for example, Cai *et al.*[27]). These systems represent an indispensable complement to RNA analysis microdevices, and are described in more detail in the following chapters.

Acknowledgement

The author thanks Dr. H. Bertrand and Professor K.D. Dorfman for useful comments on this manuscript.

References

1. N. Bontoux, L. Dauphinot, T. Vitalis, V. Studer, Y. Chen, J. Rossier and M. C. Potier, *Lab Chip*, 2008, **8**, 443.
2. J. M. Raser and E. K. O'Shea, *Science*, 2005, **309**, 2010.
3. P. Wilding, M. A. Shoffner and L. J. Kricka, *Clin. Chem.*, 1994, **40**, 1815.
4. R. C. Anderson, X. Su, G. J. Bogdan and J. Fenton, *Nucleic Acids Res.*, 2000, **28**, e60.
5. C.-S. Liao, G.-B. Lee, H.-S. Liu, T.-M. Hsieh and C.-H. Luo, *Nucleic Acids Res.*, 2005, **33**, e156.
6. K.-Y. Lien, W.-C. Lee, H.-Y. Lei and G.-B. Lee, *Biosens. Bioelectron.*, 2007, **22**, 1739.
7. S. H. Lee, S. W. Kim, J. Y. Kang and C. H. Ahn, *Lab Chip*, 2008, **8**, 2121.
8. P. J. Obeid, T. K. Christopoulos, H. J. Crabtree and C. J. Backhouse, *Anal. Chem.*, 2003, **75**, 288.
9. J. S. Marcus, W. F. Anderson and S. R. Quake, *Anal. Chem.*, 2006, **78**, 956.
10. N. M. Toriello, C. N. Liu and R. A. Mathies, *Anal. Chem.*, 2006, **78**, 7997.
11. G. V. Kaigala, V. N. Hoang, A. Stickel, J. Lauzon, D. Manage, L. M. Pilarski and C. J. Backhouse, *Analyst*, 2008, **133**, 331.
12. N. R. Beer, E. K. Wheeler, L. Lee-Houghton, N. Watkins, S. Nasarabadi, N. Hebert, P. Leung, D. W. Arnold, C. G. Bailey and B. W. Colston, *Anal. Chem.*, 2008, **80**, 1854.
13. J. Pipper, M. Inoue, L. F. P. Ng, P. Neuzil, Y. Zhang and L. Novak, *Nat. Med.*, 2007, **13**, 1259.
14. M. Nakano, N. Nakai, H. Kurita, J. Komatsu, K. Takashima, S. Katsura and A. Mizuno, *J. Biosci. Bioeng.*, 2005, **99**, 293.
15. I. K. Dimov, J. L. Garcia-Cordero, J. O'Grady, C. R. Poulsen, C. Viguier, L. Kent, P. Daly, B. Lincoln, M. Maher, R. O'Kennedy, T. J. Smith, A. J. Ricco and L. P. Lee, *Lab Chip*, 2008, **8**, 2071.

16. H. Kim, S. Dixit, C. J. Green and G. W. Faris, *Opt. Express*, 2009, **17**, 218.
17. A. Dahl, M. Sultan, A. Jung, R. Schwartz, M. Lange, M. Steinwand, K. Livak, H. Lehrach and L. Nyarsik, *Biomed. Microdev.*, 2007, **9**, 307.
18. T. Morrison, J. Hurley, J. Garcia, K. Yoder, A. Katz, D. Roberts, J. Cho, T. Kanigan, S. E. Ilyin, D. Horowitz, J. M. Dixon and C. J. H. Brenan, *Nucl. Acids Res.*, 2006, **34**, e123.
19. M. Curcio and J. Roeraade, *Anal. Chem.*, 2003, **75**, 1.
20. K. D. Dorfman, M. Chabert, J.-H. Codarbox, G. Rousseau, P. de Cremoux and J.-L. Viovy, *Anal. Chem.*, 2005, **77**, 3700.
21. M. Chabert, K. D. Dorfman, P. de Cremoux, J. Roeraade and J. L. Viovy, *Anal. Chem.*, 2006, **78**, 7722.
22. S. L. Spurgeon, R. C. Jones and R. Ramakrishnan, *PLoS. ONE*, 2008, e1662.
23. L. Warren, D. Bryder, I. L. Weissman and S. R. Quake, *Proc. Natl Acad. Sci. U. S. A.*, 2006, **103**, 17807.
24. J. S. Marcus, W. F. Anderson and S. R. Quake, *Anal. Chem.*, 2006, **78**, 3084.
25. J. F. Zhong, Y. Chen, J. S. Marcus, A. Scherer, S. R. Quake, C. R. Taylor and L. P. Weiner, *Lab Chip*, 2008, **8**, 68.
26. N. M. Toriello, E. S. Douglas, N. Thaitrong, S. C. Hsiao, M. B. Francis, C. R. Bertozzi and R. A. Mathies, *Proc. Natl Acad. Sci. U. S. A.*, 2008, **105**, 20173.
27. L. Cai, N. Friedman and X. S. Xie, *Nature*, 2006, **440**, 358.
28. S. H. Lee, S. W. Kim, J. Y. Kang and C.-H. Ahn, *Lab Chip*, 2008, **8**, 2121.

CHAPTER 15
Analysis of Proteins at the Single Cell Level

SÉVERINE LE GAC

BIOS the Lab-on-a-Chip group, University of Twente, P.O. Box 217, 7500 AE Enschede, The Netherlands

Abstract

Proteins bring a consequent additional level of information in comparison with nucleic acids on a cell's state as proteins are dynamically processed and chemically modified in the cell as a function of the cell life. Yet, the analysis of proteins is challenging as no amplification step is possible as is the case for nucleic acids, and another difficulty lies in the dynamic range of protein expression in a single sample (*e.g.* a single cell). While the same challenges are still found for microfluidic-based analysis of proteins, microsystems bring about enhanced analytical performance and novel analysis opportunities. This is illustrated here for two different strategies that can be adopted for protein analysis in a chip format. A first strategy consists of transposing the standard proteomic protocol in miniaturized analytical tools, and this provides a number of advantages and enhancement for the analysis: an overall improvement is expected when using smaller systems whose capacity matches better the size of the samples; sample manipulation is minimized when using LOC technology, and this goes together with a decrease in sample loss and contamination; enhanced analytical performance in terms of analysis time and detection sensitivity is ensured by micro- and nano-scale features; last, the use of microfabricated structures guarantees higher analysis reproducibility. In a second strategy, the analysis is actually performed at the single cell level. This strategy does not enable protein mapping anymore, but the investigation focuses on

RSC Nanoscience & Nanotechnology No. 15
Unravelling Single Cell Genomics: Micro and Nanotools
Edited by Nathalie Bontoux, Luce Dauphinot and Marie-Claude Potier
© Royal Society of Chemistry 2010
Published by the Royal Society of Chemistry, www.rsc.org

given proteins (a single protein of a small number thereof) which are specifically targeted. For that purpose, innovative microfluidic-based protocols have been developed, and we classify them in three categories of fully destructive, partially invasive and non invasive protocols. Ongoing developments in the area of nanotechnology would enable truly protein mapping at the single cell level, with the use of nanofabricated tools in a LOC platform

15.1   Introduction

After the effervescence caused by the deciphering of the human genome sequence and DNA analysis, proteins have been acknowledged as the carriers of further and more interesting information. Proteins have a central role in cells and organisms and their importance is notably reflected by their name as the word *protein* derives from the Greek word *protos* meaning "primary". Proteins are the key actors in a cell's life while nucleic acids passively encode the genetic information. Protein profiling, or the identification of all proteins expressed in one sample, brings about valuable information on the state of cells or an organism. In particular, the expression level and post-translational modification patterns of proteins in a sample are of great interest to detect and understand diseases. For instance, protein (de)-phosphorylation is a central process in cellular intracellular signaling for conveying information within a cell. Changes in phosphorylation patterns are often the signature of the changes in a cell's state and its entry in a disease state. Even if protein profiling is mostly carried out on small-scale samples, it has not been demonstrated yet at the single cell level. However, populations are heterogeneous, and heterogeneities found in a cell population are indicative of the development of diseases such as cancer and can therefore be exploited for early detection.

15.1.1   Protein Analysis: The Challenge

While protein profiling is essential, it raises a number of challenges compared notably to the analysis of nucleic acids, especially at the single cell level. A first challenge lies in the protein dynamic ranges: there are more than six orders of magnitude in the protein expression level between low-abundant and high-abundant proteins in a cell, which makes it difficult to develop a protocol enabling the simultaneous analysis of all proteins in a cell. Secondly, proteins of interest are low-abundant and present in a cell as traces (few copies per cell); they are subsequently often lost during sample processing. A third challenge derives from the great variety of physical and chemical properties proteins may possess. Nucleic acids share the same chemical and physical nature so that a universal protocol applies for the processing and detection of all nucleic acids. In addition to this, DNA can be amplified to yield sufficient amount of samples for analysis and their detection relies on the use of highly specific dyes. On the contrary, proteins are all different, a universal analysis protocol is excluded, they cannot be amplified and there is no such specific and universal dye for fluorescence-based detection of proteins.

If identified proteins (biomarkers) are to be analyzed, specific antibodies are developed and used for the capture of the targeted proteins; this is often followed by a fluorescent labeling step with the help of a second antibody and a secondary antibody coupled to a fluorophore unless a label-free technique is used for the detection (*e.g.* surface plasmon resonance (SPR), cantilever-based detection). This implies the development of specific antibodies for every protein of interest and this approach is limited to the analysis of a low number of proteins.

Therefore, as introduced in chapter 10, protein profiling relies on the use of another universal and label-free strategy that, theoretically, enables the analysis and detection of any protein in a complex sample: the proteomics approach. The key steps identified in the proteomic approach are: (1) the lysis of cells, when working with cell samples, (2) the isolation of the material contained in the lysate and the eventual extraction of the protein content, (3) the separation of the proteins using 2D gel electrophoresis, (4) the digestion of the proteins using specific enzymes after appropriate chemical processing, (5) the purification and concentration of the peptide mixture and their eventual separation, followed by (6) the detection and analysis of the peptides using mass spectrometry techniques. To circumvent some limitations of the conventional proteomics protocol and in particular the two-dimensional (2D) gel separation step, an alternative approach, the shot-gun approach, has been introduced. In this approach, protein extraction is directly followed by their digestion (and chemical processing), and the resulting peptide mixture undergoes a purification/concentration step followed by two liquid chromatography (LC) separation steps on cation exchange and hydrophobic phases respectively and, finally, mass spectrometry (MS) analysis. In both approaches, the use of MS as a detection technique at the end of the protocol provides extra information on the analyzed proteins, such as their molecular weight (MW) and structure, and this information is confronted with data contained in a database to identify the proteins analyzed.

However, these time-consuming and labor intensive protocols do not lend themselves well to high-throughput analysis and the extensive manipulation of samples often results in contamination and partial loss of the sample. In addition to this, certain classes of proteins are lost during the analysis; these include proteins in low abundance as well as basic and hydrophobic membrane proteins.

15.1.2 Why Microfluidics?

As a consequence, the field of protein analysis needs new analytical tools that (1) exhibit enhanced performance (higher reliability, higher sensitivity), (2) limit human intervention for sample manipulation, and (3) enable automated and faster analysis.

Lab-on-a-chip (LOC) devices are good candidates in that context, for the numerous advantages they bring about for biological analysis. These advantages can be divided in three categories: (1) the gain in analysis quality and sensitivity, (2) performance improvement by the high level of integration of sample treatment and preparation, and (3) the suitability of such microfabricated devices for high-throughput analysis. The amount of liquid conventionally handled in these LOC devices matches the size of crude biological samples; consequently, there is no need to dilute the samples for their analysis to finally concentrate the molecules of interest at the end of the analysis as often occurs using conventional techniques. Microfluidics is compatible with the realization of complex platforms that integrate on a few square centimeters a whole analytical process consisting of a series of operations and, eventually, a number of independent devices in parallel for simultaneous analysis of different samples. A first essential consequence of this level of integration is a decreased manipulation of sample, limiting sample loss and contamination, resulting thereby in faster and more sensitive analysis. This gain in sensitivity is all the more enhanced with the use of microfabricated tools. For mass spectrometry analysis, this gain is accounted for, for electrospray ionization (ESI)-MS, by the improved ionization efficiency which is reached by decreasing the size of the ionization source, while matrix-assisted laser desorption ionization (MALDI) benefits from both the miniaturization of the spots of analytes and the use of a microfabricated interface for sample deposition, as explained later in this chapter. The miniaturization of the analysis may also result in a greater tolerance to sample contamination.[1] On other aspects, microfluidics is currently widely used for cellular investigation; subsequently, it is ultimately possible to realize a multi-step platform that includes preliminary steps of cell culture and eventual treatment, followed by the analysis of identified cells and their on-line analysis. In that context, the microfluidic platform can include smart structures such as valves to easily control flow and manipulate objects such as single cells.[2] A second advantage of integrated platforms is the possible automation of the analysis as a result of their compatibility with the use of robots that fully control the analysis. Finally, LOC technology enables low-cost analysis. Not only is the overall consumption in chemicals, solvents, and sample decreased but the actual price of every device is lower as a result of large-scale production. Besides, for biological analysis, single-use devices are preferred, suppressing thereby the need for device regeneration as currently occurs for conventional analytical tools. This trend is particularly stressed by both the increasing amount of LOC devices fabricated using polymer material[3] and emerging new fabrication technologies that does not require any access to a dedicated environment such as cleanrooms.

15.1.3 Microfluidics and Protein Analysis

Protein analysis using LOC devices follows two general trends: (1) a proteomic approach and (2) innovative approaches.

A first and general trend consists of miniaturizing – or transposing – the conventional proteomics (and shot-gun) approach(es) in a microfluidic format towards enhanced analysis performance.[4] This implies the development of individual microfluidic tools to carry out the successive steps found in the proteomic protocol. In that context the implementation on a microfluidic format of the steps of protein digestion, protein (2D) separation, peptide separation, sample purification and concentration, and coupling of the microsystem to mass spectrometry techniques (ESI-MS and MALDI-MS) have been demonstrated. In more integrated approaches, a succession of steps have been combined into a microfluidic platform, such as protein digestion followed by the purification/concentration of the resulting peptide mixture, sample purification/concentration or separation and on-line analysis by MS, for instance. An ultimate hope for this miniaturization approach is to be able to reach the sensitivity required for single cell analysis but the analysis has only been illustrated on small samples containing much more material than a single cell.

In contrast, the second approach relies on the development of innovative protocols for the actual analysis of proteins at the single cell level. In that case, MS techniques are not used as the analysis focuses on targeted and well-identified proteins while in the other proteomic-like approach all proteins are analyzed. These innovative protocols include, among others, cell lysis followed by the isolation of the protein of interest, the permeabilization of the cell membrane for the selective release of targeted proteins, the analysis of membrane proteins or intracellular measurements of targeted proteins. The goal of the analysis in that case is more specific than with the proteomics approach. Selected applications are cancer detection using surface biomarkers, and the elucidation of intracellular signaling pathways.

In this chapter, we will review what has been done so far for the two trends found for protein analysis in a microfluidic format. The first "proteomics" trend will be presented in two sections dedicated to microfluidic development for analysis using ESI-MS and MALDI-MS, respectively. The third part will present the "innovative trend" through a selection of new protocols actually aiming at protein analysis at the single cell level. We will finish with a "conclusion and perspectives" section, presenting a vision for the future of protein analysis using microfluidic devices in combination with nanoscale structures for actual single protein detection.

15.2 Electrospray Ionization Mass Spectrometry

Electrospray ionization (ESI) is performed on a continuous flow of liquid,[5] and for that reason, sample purification or separation using nanoLC columns is already often coupled to on-line ESI-MS analysis.[6] As a consequence, the LOC device for protein sample analysis can easily be placed upstream to the mass spectrometer as a nanoLC column would be. It is worth mentioning that the use of microfabricated structures could help establish a lower flow rate that would be beneficial to the ionization process and the analysis. A key aspect for this coupling is a reliable and zero dead volume connection between the microsystem and the ionization source, and for smaller systems, the dead volumes are larger!

In this ESI-MS section, we will first successively present different modules that have been developed for protein analysis in a microfluidic format, the interface with the mass spectrometer and the modules for processing the samples, for protein digestion, purification/concentration and separation. Following this, several integrated systems that combine a series of operations will be described.

15.2.1   Connections and Coupling

A key module of the LOC device is the interface with the mass spectrometer, and this mostly includes the ionization source for the analysis. Not only must the coupling be optimized in terms of dead volume, but the characteristics of the interface will dictate the quality and sensitivity of the analysis by mass spectrometry, and a source with smaller dimensions translates into a gain in the ionization efficiency and sensitivity of the detection as explained by a number of parameters.[7] First, smaller source apertures give rise to the generation of smaller droplets, which favors the processes underlying the formation of ions in the gas phase. Besides, the charge concentration in the droplets varies as $r^{-3/2}$ and is consequently higher for smaller droplets, as reflected by the following equation:

$$\frac{q}{V} = 3\left(\frac{\varepsilon_0 \gamma}{2r^3}\right)^{\frac{1}{2}}$$

where q/V represents the charge volume concentration in a droplet, r is the radius of a droplet, and γ the surface tension. As a result, the ionization yield increases, and typically shifts from 10^{-9} to 10^{-4} for conventional ESI ($100\,\mu m$ *i.d.*) and nanoESI (sources of $< 10\,\mu m$ *i.d.*), respectively. Another particular consequence is a decrease in ion suppression phenomena, which translates into a higher tolerance to the presence of salts or other contamination.[1] This is of great interest for the technique of ESI where the presence of salts usually completely hinders the process of ionization. All together, providing the source is small enough, 100% of the molecules are eventually ionized, enter the mass spectrometer and are actually analyzed.

Several strategies have been employed for coupling a microsystem to a mass spectrometer.[8] A first strategy employs an external coupling with a conventional ionization source which is connected to the outlet port of the microsystem with the help of capillary tubing.[9] This configuration does not bring about any improvement in the ionization performance as a conventional interface is employed and the manual connection to the capillary appears as a limitation for the mass production of integrated systems. In addition, this approach presents two main issues. Firstly, a dead volume is created at the junction between the channel outlet and the capillary tubing, possibly degrading the quality of the separation.[10] Secondly, the use of glue to maintain the capillary in the channel often translates into the appearance of chemical noise in the mass spectra due to the dissolution of the glue in the solvent used for the analysis.[11]

A second strategy consists of suppressing the ionization source by creating the electrospray directly from the edge of the microsystem, at the channel outlet.[12] However, without any sharp structure, the liquid tends to spread on the LOC side, preventing thereby the formation of an electrospray. This has been improved with a hydrophobic treatment of the system edge to alleviate any spreading phenomena or a pneumatic assistance to spray formation.[13]

A preferred approach relies on the integration of the ionization source onto the microsystem and its fabrication using microtechnology techniques. Not only is the coupling optimized but also the fabricated ionization sources are more robust and more reproducible than standard capillary-based sources and they are directly fabricated at the same time as the channels in the microsystem. Two classes of microfabricated and integrated sources are found: ionization sources mimicking conventional capillary sources and planar sources that are fabricated at the end of the analysis channel.[14,15] For the first category of microfabricated sources, two main examples, made from polymer[15] or silicon,[14] are reported in the literature. They are fabricated as multi-nozzle chips but have not been integrated as part of a LOC device. They are more used as stand-alone sources as enhanced alternatives to capillary sources and their geometry is a limiting factor for the generation of low-micron-scale sources. The silicon-based sources are commercially available (Advion Biosciences, Ithaca, NY) and they are routinely used with a robotic interface that makes the junction with 96-well plates. More examples are found for the other category of fabricated sources with various geometrical characteristics and using a great variety of materials such as parylene,[16] poly(dimethylsiloxane) (PDMS),[17–19] SU-8,[20–24] cyclo-olefin,[25] polyethylene terephthalate (PET),[26,27] polyimide,[28] and poly (methyl methacrylate) (PMMA).[29] Most of these microfabricated sources have apertures of 10–50 µm, and while they exhibit enhanced performance compared to conventional capillary, they are still not compatible with protein profiling at the single cell level. Of interest in that prospective is the work of Arscott and Le Gac who have reported low-micron sized ionization sources fabricated from polycrystalline silicon and compatible with single molecule detection, also in purely aqueous solutions (Figure 15.1).[30]

15.2.2 Sample Processing: Purification and Digestion

Sample processing includes several steps that have been illustrated in a microfluidic format: sample separation, purification/concentration, and protein digestion. For these different steps a common requirement is found, the creation a large surface area in the microchannel which is subsequently functionalized as a stationary phase in the former case or with immobilized enzymes for the latter case. This is usually achieved with the help of porous beads which are

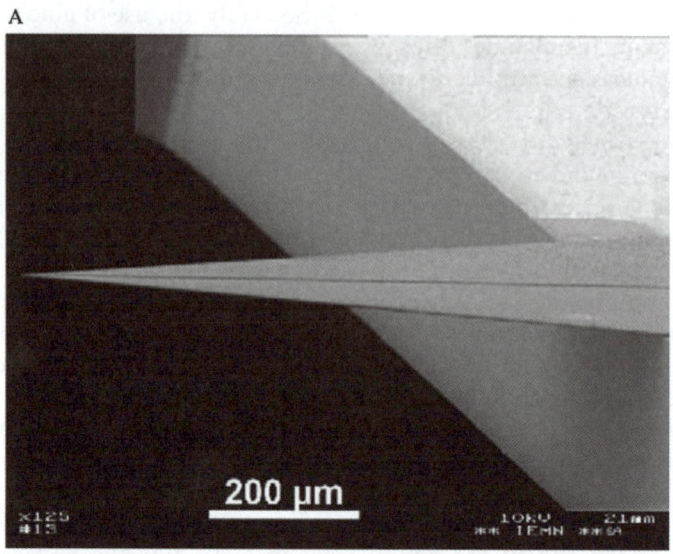

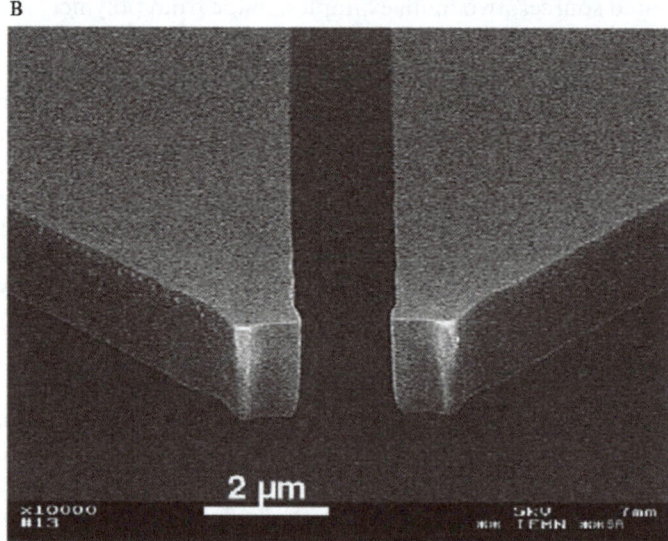

Figure 15.1 Example of microfabricated nanoESI ionization sources made from polycrystalline silicon. (A) Lateral view of a free standing 800 nm length nanoESI-MS structure (SEM picture); (B) Front view of the source opening for a source of 2 μm height and 1.8 μm opening width. Reprinted from [30] with permission.

packed, but in the case of a closed microchannel bead packing is ill-controlled and implies the addition of structures for local packing of the beads. Therefore, two other approaches are alternatively adopted in a microsystem for both the purposes of sample purification/concentration and enzymatic digestion: the *in situ* polymerization of a porous phase (monolith),[31,32] or the micro-structuration of the channels with a dense network of pillars.[33] These different strategies to create a large surface area in a microchannel are illustrated in Figure 15.2.

15.2.2.1 Protein Digestion

Most of the microfluidic approaches for the enzymatic digestion of proteins rely on the immobilization of the enzyme on a solid support; solely on the channel walls or after creation of a large surface area.[34] In any case, the digestion proved to be more efficient in a microfluidic channel than in a microtube. The reaction is also faster (few minutes against several hours) as a result of the shorter distance molecules must travel to reach the enzymes and the high surface-to-volume ratio. The enzymatic turnover is higher and autolysis phenomena are suppressed as well, and consequently the issue of digest contamination.

For instance, trypsin has been immobilized on beads subsequently packed in a microchannel with the help of a dam structure to produce a flow-through microreactor.[35] The sample is pumped through the column *via* the application of an electric field, the resulting digest separated in a CE column and analyzed on-line by MS with a capillary-based coupling. Protein digestion takes place in 3–6 min to give 71% and 92% sequence coverage for BSA and cytochrome *c*, respectively. In another example, the enzyme is covalently immobilized on the channel walls which have previously been made porous to increase the surface area of exchange.[34] The immobilized enzymes showed a 170-fold increase in their proteolytic activity and the digestion of lysozyme proceeds in 1–3 min, *i.e.* 200–1000 times faster than in solution to yield 90% sequence coverage. Methacrylate-based monoliths have also been reported for the same purpose,[36] and the digestion of eight proteins demonstrated in less than 1 min. Figure 15.3 shows some digestion results obtained in a microsystem, using a macroporous monolithic phase for enzyme anchoring, the digest analysis being performed off-line using MALDI-MS techniques. Last, an original approach has recently been reported by Nissila *et al.* where the digestion solution is introduced in an open system, let for evaporation and the digest desorbed by the analysis solvent and analyzed on-line by ESI-MS. More examples of digestion microreactors can be found in the recent review of Krenkova and Foret[37] but all of them are restricted to the mere digestion step and do not include any extra chemical processing steps of the sample as does occur in a conventional approach.

15.2.2.2 Sample Purification and Concentration

Crude samples present a high level of contamination such as high concentrations of salt, urea, and surfactants that hinder not only their processing (HPLC separation) but also their ionization by inducing ion suppression phenomena.

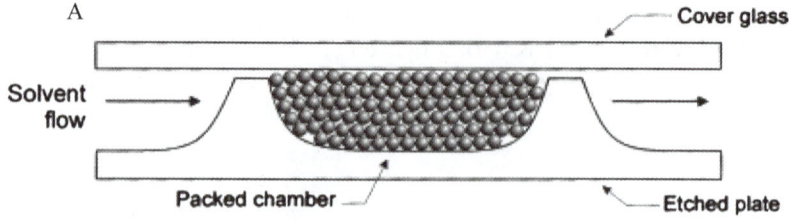

A

Cover glass

Solvent flow

Packed chamber

Etched plate

B

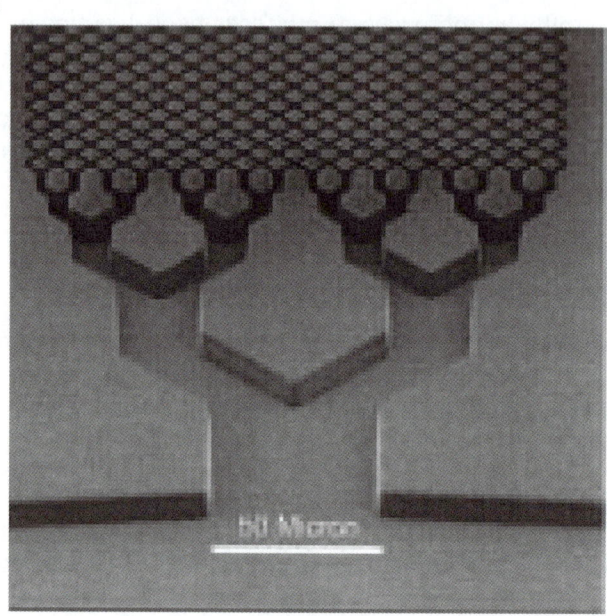

C

Figure 15.2

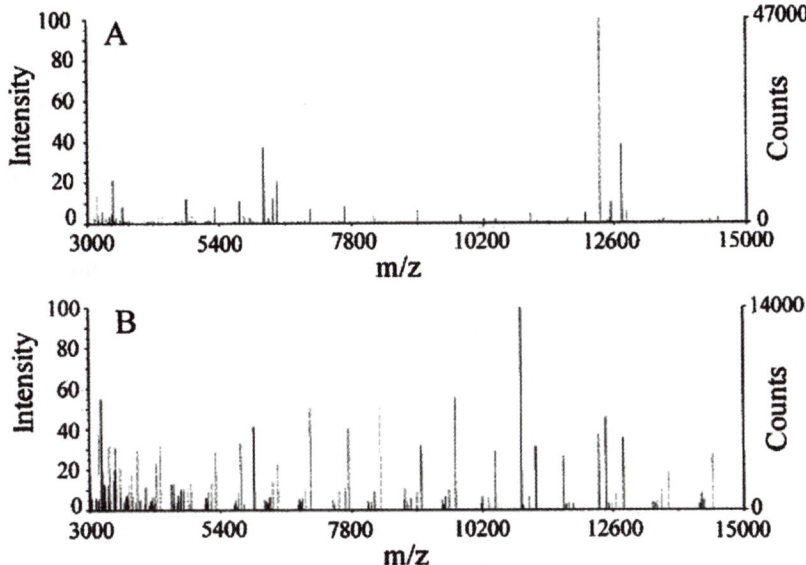

Figure 15.3 On-chip protein digestion using a macroporous monolithic support for the immobilization of the digestion enzyme (trypsin). MALDI mass spectra of casein ($10\,\mu g\,mL^{-1}$) obtained off-line, without digestion (top panel) or after digestion in the microreactor (flow of $26.5\,cm\,min^{-1}$). Reprinted from [36] with permission.

Different approaches have been proposed in a microfluidic format to clean samples before their actual MS analysis.

In a first approach a dialysis membrane is employed to filter out compounds with a low (salts) or high (cellular matrix) molecular weight and simultaneously concentrate compounds of interest. For instance, Xiang *et al.* developed a polycarbonate system that includes two cellulose membranes with a low and a high molecular cut-off (8 and 50 kDa), for the elimination of low and high MW contamination. A 20-fold increase in the signal-to-noise ratio was observed for *Escherichia coli* cellular lysates.[38]

Figure 15.2 Different options for creating a high surface area stationary phase in a microchannel for both sample cleaning and processing (digestion) (A). Packing of beads in the portion of a channel. (B) Microstructuration of the channel using microfabrication technique; SEM image of the inlet of a silicon-based microfabricated column for sample cleaning. (C) Polymer-based monolithic phase that is prepared *in situ*, and restricted to a certain portion of the channel when using a photopolymerization process. Scanning electron microscopy (SEM) photograph of a section of capillary tubing which contains a monolithic phase; enlarged view of the monolith morphology in inset. Reprinted from [40], [41] and [31], respectively, with permission.

A second approach relies on the use of a solid support as already evoked for the digestion step, to create a solid phase extraction (SPE) module. A hydrophobic phase is mostly employed for the removal of any hydrophilic contamination and analytes are eluted in a small amount of polar solvent for concentration purposes. As for the digestion the solid phase can be packed beads,[39,40] in channel micromachined structures,[41,42] or a porous polymer phase can be prepared *in situ*.[31,43]

Oleschuk *et al.* report a 500-fold sample concentration of a dye solution using reverse phase (RP) beads packed in a cavity machined in a microsystem (see Figure 15.2A).[40] The same device was employed for separation purposes using CEC, but not on protein samples. Le Gac and co-workers report on the purification of 1 pmol of a model peptidic sample spiked with a high amount of contamination (20% salts) on a monolithic phase polymerized in a microchannel (see Figure 15.2C), using both off-line and on-line MS analysis.[31]

15.2.2.3 Peptide Separation

Sample purification is often not sufficient when working with complex mixtures, and a separation step must be added. The same solid supports as for sample purification are employed as stationary phases while the module is made longer for separation purposes. Two options are conceivable for fluid actuation, either using electro-osmosis (capillary electrochromatography (CEC))[40,44,45] or hydrodynamic pumping (LC).[28,46] While the operation of a CEC module in a LOC device is easier, the separation faster and the flow profile flat, this still implies the addition of charged groups on the stationary phase to support the flow and the device presents a higher sensibility to the solvent composition. On other aspects, nanoLC columns suffer from a high pressure drop and the smaller the channel the higher the hydraulic resistance as it varies as r^4 for a circular channel with a radius r.

Ceriotti *et al.* describe a PDMS chip for CEC separations where RP particles are packed in a tapered structure to form a 4-cm long column; the system performance is illustrated on a mixture of amino acids stained with fluorescein isothiocyanate (FITC).[39] He *et al.* introduced the concept of microfabricated columns[41] (see Figure 15.2B) to alleviate issues raised by bead packing in a channel. A 4.5 cm column was structured with an array of posts of $5\,\mu m \times 5\,\mu m$ and $1.5\,\mu m$ spacing which were modified with a hydrophobic coating, and the system was tested for the separation of small organic dyes.

15.2.3 Integrated Systems

Ultimately, these different modules of sample preparation and separation are combined on a single microsystem together with an interface with ESI-MS; this eliminates the connections and manipulation between the different steps of the analysis process as well as any dead volume issue if the ionization source is also integrated on the chip.

Two monolithic phases have been prepared together to combine the digestion and purification/concentration steps.[47] For that purpose, one part of a hydrophobic monolithic phase is locally activated with the help of UV light to immobilize reactive groups that enable later anchoring of trypsin to yield a digestion microreactor. The combined device is coupled to MS for on-line analysis of the purified peptide mixture. Higher sequence coverage is reported after addition of the SPE module; 80% and 60% for 400 and 40 pmol myoglobin, respectively.

A fully integrated device is commercially available at Agilent.[28,48] The polyimide system includes an enrichment column, a 4.5-cm separation column, both based on conventional packed beads, and an integrated ESI source, and fluids are actuated with the help of a nanoLC rotary valve. Sub-femtomolar detection sensitivity is reported for the RP separation of a 20-fmol BSA digest sample at a flow rate of 100–400 nL min^{-1}, which is comparable to state-of-the-art analysis, but still with minimal human intervention. The performance of the system has been further investigated by Ghitun *et al.* by varying the hydrophobicity of the phase used for the enrichment and the separation, and by applying it for phosphoproteome profiling.[49] In that case, an ion metal affinity chromatography (IMAC) phase which is specific to phosphorylated species is used in the enrichment column.

Xie *et al.* reported a similar device[46] that includes not only a separation column (1.2 cm length) based on packed beads and an integrated ESI interface but also internal electrodes for electrolysis-based electrochemical pumping of the solutions[50] and a mixer for gradient elution. The integrated platform is tested for the analysis of a BSA digest (0.6 pmol) at a flow rate of 80 nL min^{-1} and the lower sequence coverage observed is explained by a loss of very hydrophobic species that were not eluted from the column.

Another example of interest is the LC-MS chip of Lazar and co-workers developed for biomarker discovery and differential protein profiling between two cell samples.[51] The chip includes a sample pre-concentrator and a LC column based on packed particles and an original internal electro-osmotic pumping/valving system based on shallow microchannels. The ESI tip is inserted in the outlet channel with a zero dead volume connection. Sample analysis is demonstrated on 0.1–1 μg of tryptic extracts of MCF-7 breast cancer cells, and 40–50 proteins were identified as potential biomarkers.

15.3 MALDI-MS

In the case of MALDI-MS, the ionization is performed on solid crystals of samples and samples are manipulated as microliter-sized droplets. As a consequence, a number of LOC devices for MALDI-MS analysis are adapted to this droplet format of the samples and the LOC device is developed on the MALDI target itself using appropriate surface chemistry.

Microsystems for MALDI-MS analysis can be divided into three groups: (i) microfabricated MALDI target plates, (ii) microsystems for off-line preparation of the samples and working with a continuous flow of liquid and (iii) microsystems that integrated both the sample preparation steps and the MALDI targets, with on-line or off-line sample analysis and we will successively present these three strategies.

15.3.1 Microfabricated MALDI Targets

The first category is comprised of micromachined MALDI targets that aim at concentrating and eventually cleaning/purifying samples on site using passive structures. One limitation in MALDI analysis lies in the fact that samples spread on the surface of the target upon deposition, and this results in a decrease of the analysis sensitivity. To circumvent this liquid spreading, MALDI targets can be patterned so as to decrease the spot size and thereby to yield local concentration of the analytes. Two main approaches have been reported.

The first consists of machining nanovials where samples are deposited. These vials have an inverted pyramid structure, at the bottom of which the sample is concentrated.[52] Such systems have first been fabricated in silicon,[52] and more recently their realization has been demonstrated in a polymer material.[53] The analysis time is also decreased as the experimenter does not need to search for the sample on a large spot of crystal.

A second approach relies on chemical patterning of the surface. Typically, the whole surface of the MALDI target is made hydrophobic *via* appropriate coating, (*i.e.* a monolayer or the use of a hydrophobic material) and small hydrophilic spots are created therein.[54] Thereby, the sample is confined to the hydrophilic spots, and the smaller the spots, the higher the sample confinement. Both approaches give a sensitivity improvement down to the attomole range. Another major technological development in this direction is the implementation of the purification step on such a chemically patterned MALDI target.[55] For that purpose, the hydrophilic areas of the MALDI target are functionalized with a stationary phase aiming at retaining a specific class of compounds. These devices have especially been developed by the company Ciphergen and are now commercially available under the name of ProteinChip® devices.[55]

15.3.2 Off-line Sample Preparation

The second category of microsystems for MALDI-MS analysis focuses on the sample preparation steps before their deposition on a (conventional) MALDI target. MALDI-MS analysis is subsequently done off-line and not on the same device. Such a microsystem is comprised of two parts, one for the sample treatment (digestion, purification, separation) and the other for sample dispensing onto the MALDI target. The first part consists of microfluidic device working on a continuous flow of liquid. The second part aims, as before, at limiting sample spreading on the MALDI target; however, instead of structuring a MALDI target, a robotic interface is used to improve the deposition of the sample on a confined area on a (conventional) MALDI target. Sample dispensing has been described using various types of technologies: a spotting technology, piezoelectric actuation or by spraying the sample as a thin film. The spotting technology[56] borrows much from the interface which is used for coupling LC to MALDI-MS, but works here for smaller amounts of liquids. Another technique uses a piezoelectric dispenser[52] that enables continuous deposition of sample on a MALDI plate *via* sub-nanoliter droplets of liquid. As soon as the droplet reaches the surface, solvent evaporates so that the sample does not spread on the surface. This dispensing technique has been coupled to a microreactor for the tryptic digestion of proteins and to a nano-vial-based MALDI plate.[52] Figure 15.4 represents a complete microfabricated platform for MALDI-MS analysis that integrates the different modules described above. An original dispensing method developed by Murray's group is based on a rotating ball placed at the outlet of a microfluidic system where samples are separated;[57,58] the rotating ball makes the interface between the zone at atmospheric pressure where samples are prepared and the zone under vacuum where MALDI ionization takes place. The last reported technology to couple sample preparation steps with a continuous flow to MALDI-MS consists of spraying the sample into a uniform film onto the MALDI target surface.[59] This technology provides more homogeneous sample crystals, whose size is dictated by the distance between the spraying head and the MALDI target. Spots down to 170 µm diameter with 150 µm spacing have been realized

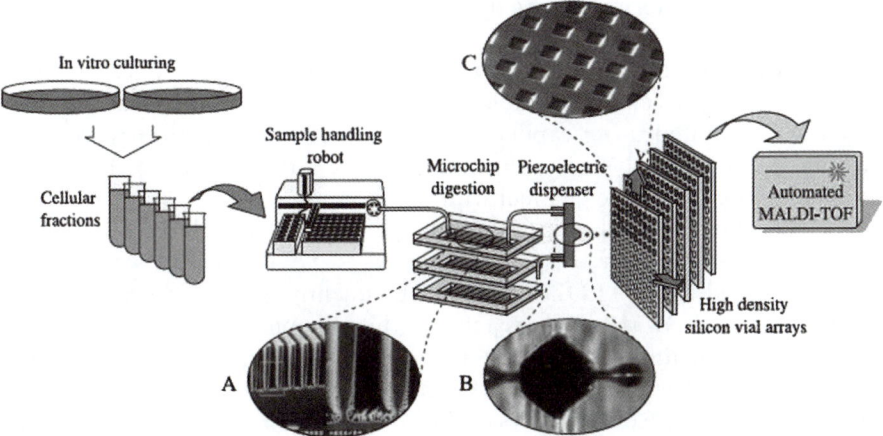

Figure 15.4 Complete microfluidic platform for MALDI-MS based analysis of proteins. From left to right: sample preparation and placing in a standard well plate; robot-based injection of the samples in a digestion microreactor, and automatic dispensing using a piezoelectric dispenser on a microfabricated MALDI plate that consists of a 2D array of inverted pyramids as nanovials, and that eventually contains a stationary phase for sample purification or selective trapping. Reprinted from [4] with permission.

using the spraying technique. Sample deposition has been demonstrated after digestion of protein samples inside the spraying device using immobilized trypsin. One major advantage of these devices is their ability to be multiplexed and to work in parallel without any risk of contamination from one sample to the other one.

15.3.3 Integrated Microsystems

The third and last group of microfluidic systems for MALDI analysis consists of integrated systems that include both steps of the analysis, *i.e.* sample preparation and the MALDI target. The advantage of these systems is that they are fully integrated and hardly need any external intervention for sample analysis. While for most of these devices sample preparation is carried out off-line (*i.e.* not in the mass spectrometer), one unique system works "under vacuum" and enables on-line sample treatment and analysis.

In this category, the first microfluidic device has been developed by Gyros AB and gave rise to a commercial product, the Gyrolab®, which is dedicated to biological sample analysis using affinity chromatography, and particularly targets the field of proteomics.[60] This device has the shape of a compact disk that includes a series of radial microfluidic networks with MALDI targets placed at its external edge. Sample preparation is performed off-line on a dedicated workstation, with a robot making the link between conventional 96-well plates and Gyrolab® devices that contain 96 independent analysis networks. Thereafter the device is cut and the MALDI target parts are placed in the mass spectrometer for the analysis.

Liu *et al.*[61] have developed an alternative system whose channels are not covered with a lid for accessibility to the laser. In those open microchannels, peptide and oligosaccharide samples to which a matrix solution has first been added are separated using capillary electrophoresis and dried in place under vacuum. MALDI analysis is done on the crystals formed in the channels by scanning along the whole channel with the laser.

Electrowetting on a dielectric (EWOD) is particularly suitable for MALDI-MS analysis, where liquids are displaced as droplets, *i.e.* the format which is conventionally used for MALDI analysis. Consequently, several integrated microsystems for MALDI-MS analysis rely on this pumping principle that enables sample treatment and mixing with the matrix solution. Wheeler *et al.* have reported EWOD-based microfluidic systems for MALDI-MS analysis. In a first system,[62] the principle of EWOD is exploited to displace a sample droplet, to mix it with a droplet of matrix solution and to place it on a Teflon-based MALDI target. They further improved their system by including a purification step of the sample;[63] a droplet of sample is displaced onto a Teflon-coated place where analytes and contamination adsorb. This spot is washed with deionized water to remove contamination, and later, matrix is added on the purified analytes. In both cases, the system is opened at the end of sample preparation and spots are dried under vacuum before the system is introduced in the mass spectrometer.

15.4 Innovative Approaches for Protein Analysis at the Single Cell Level

The second approach for analyzing proteins in a single cell does not aim at protein profiling as does the proteomic strategy but mostly focuses on one or a small amount of targeted proteins which are specifically detected and possibly quantified. This second approach finds application for the detection of diseases once relevant biomarkers have been identified, for example. In that context, it is essential that the analysis is done at the single cell level as tumors consist of heterogeneous populations of cells, with proliferating cells, malign cells, and quiescent cells. The pooled analysis of several cells does not reveal the specific properties and behavior of the individual cells and may hide cancerous cells. A second main application is the study of signaling pathways in a single cell and the activation of protein kinases involved in the intracellular communication. Here again, as different cells may respond differently to the same stimulus and as their response may not be synchronized, protein analysis at the single cell level is essential.

We review here a certain number of innovative protocols proposed in the literature for protein analysis along this "philosophy" and at the single cell level. These different examples can easily be classified in three categories of (1) fully invasive techniques with the lysis and destruction of the cell; (2) partially invasive techniques where the cell is transiently disturbed for the analysis but however kept intact and the analysis proceeds on living cells; and (3) non-invasive techniques where the cell is not (or little) affected by the analysis, again performed on a living system.

15.4.1 Invasive Analysis

The first category of protocols for protein analysis using innovative methodologies corresponds to invasive cellular analysis; these protocols involve a first step of cell destruction *via* its lysis to access the intracellular content and analyze targeted molecules in the resulting lysate.

We will present in this category two types of protocols; in a first protocol the cell lysis is followed by the on-line separation of compounds of interest by capillary electrophoresis, while in a second protocol cells are isolated in droplets and chemically stimulated therein to study signaling pathways. In this second approach, protein analysis is not included in the microfluidic device but performed off-line using the individual cell lysate.

15.4.1.1 Single Cell Capillary Electrophoresis

Capillary electrophoresis-based analysis at the single cell has now been carried out for almost two decades. In this approach a cell must be introduced at the entry of a conventional column for CE analysis, lysed *in situ* and its lysate analyzed on-line. Still, this conventional protocol presents a limited throughput up to a few tenths of cells a day as a result of this tedious manipulation and the time required for both the analysis and the column washing steps. As a consequence, this protocol is upgraded and combined with microfabricated structures, or even implemented in a microfluidic platform for easier and faster analysis.

A number of articles in the literature describe single cell CE-based analysis. In these articles, different lysis protocols are used to optimize the speed of the lysis step, not to disturb biological processes inside the cell and "freeze" enzymatic processes and to limit the loss in biological material by diffusion phenomena. The separation step and its coupling to cell lysis is the object of investigation to alleviate peak broadening issues due to differential mobility of analytes. Last, different detection protocols are proposed, depending on the targeted molecules (dyes or proteins), with or without signal amplification. In general, the protocol requires a pre-treatment step of the cells to introduce a fluorescent compound which is later detected during the analysis.

State-of-the-art methodology for single cell capillary electrophoresis has recently been reviewed elsewhere.[64,65] These reviews notably discuss recent improvements brought to the technique for sampling, separation and detection, and cover various fields of application of the technique, among which proteomics and subcellular analysis.

A first idea to increase the analysis throughput for single cell capillary electrophoresis is to optimize the coupling between the cell lysis and the CE column, and this can be done by using a microfabricated interface. In general, the cell is grown on a surface or placed in a solution and the CE column must be placed manually in close vicinity to the cell to suck the resulting lysate upon application of suitable voltage. However, this imposes the need to locate the cell to be analyzed before this procedure is started, and thereby limits the throughput of the analysis. In that context, Marc et al. propose placing and culturing cells in microfabricated structures to facilitate this first step of cell localization, and they integrate an electrode in every microwell for in situ and fast lysis of the cell.[66] With this improved integrated system, cells are lysed electrically within 33–66 ms, thereby limiting the risk for loss of biological material by diffusion, and the authors reach a throughput of ~ 120 cells h^{-1}. The system is notably illustrated for the investigation of intracellular signaling pathways and for measurement of kinase activities. For that purpose, cells are previously loaded with a fluorescent substrate for a given kinase, and after lysis, the latter substrate is separated by CE and quantified to yield information on the activity level of the kinase in the cell. The technique is easily amenable to multiplexed analysis, as demonstrated in another article from the same group.

Fully microfluidic approaches for single cell analysis by CE have first been restricted to the analysis of small fluorescent molecules previously introduced into the cells, and the protocol focuses on the implementation of the full analytical process on a single device. In that respect McClain et al. use a slightly modified CE cross-shape microfluidic system for single cell analysis.[67] A Jurkat cell is introduced from a side reservoir and focused close to the channel cross-section where they are lysed electrically using a 75 Hz square wave. The same signal is employed for the injection and the analysis of the lysate, and the separation of dyes (carboxy fluorescein diacetate, Oregon green carboxylic acid diacetate, calcein) previously loaded into the cell. Cell lysis is achieved in less than 33 ms, which is fast enough not to affect any cellular process and to limit diffusion of the cell content, and the separation performed within 2.2 s, in a reproducible way. The overall throughput reported for this system is of 7–12 cells h^{-1}. Moreover, the authors were able to detect differences in the metabolism of the Oregon green diacetate dyes by individual cells using this platform. The main issue mentioned here is the adsorption of cell debris on the channel walls that lead to a degradation of the separation performance.

Gao *et al.* describe a similar platform for single cell analysis that integrates an injection, cell lysis and CE separation. The lysis is done here electrically using the same voltage as for the sample injection and the CE separation under alkaline conditions (pH = 9.2), and a single cell is isolated at the cross-section between the loading and analysis channels with the help of a combination of a hydrostatic pressure and voltages. However, the authors slightly modify their injection protocol with the addition of a docking step of the cell for ∼15 s before the lysis procedure, and to ensure thus good homogeneity of the buffer and reproducible lysis conditions. The integrated platform is illustrated with the lysis of erythrocytes and the investigation of the metabolism of individual cells by looking at the concentration of gluthatione (GSH) previously derivatized in the cells using naphthalene-2,3-dicorboxaldehyde (NDA) to make its detection possible. A comparable analysis throughput of 15 cells h^{-1} is reported.

A more sophisticated platform was proposed by Wu *et al.* for single cell isolation, eventual stimulation, lysis, and derivatization of molecules of interest using well-calibrated amounts of chemicals prior to on-chip CE analysis.[68] The system includes a set of novel three-state valves based on PDMS material which are easily controllable using an extra fluidic layer and that define a 70 pL reaction chamber where a single cell is easily isolated and subjected to a series of chemical treatments. Chemical plugs are precisely and reproducibly metered using the combination of valves and an on-chip picopipette. Downstream to the reaction chamber a long channel is found for the on-line separation of the cell lyate using micellar electrokinetic chromatography (MEKC). The system is demonstrated for the separation of amino acids in individual Jurkat cells, and the staining agent for the amino acids is added to the lysis buffer for simultaneous and efficient cell lysis and analyte derivatization. Nonetheless, the analysis is contaminated with a broad peak corresponding to cellular debris and the process would benefit from the addition of purification step between the cell lysis and the MEKC analysis.

Protein analysis using single cell lysate has also been carried out in similar integrated microfluidic platform, and the detection is done either indirectly using a fluorescent substrate specific for the protein enzyme to be analyzed or directly. Of interest in that respect are the studies by Ocvirk *et al.*[69] (indirect measurements), Huang *et al.*[70] and Ros and co-workers[71] (direct measurements). Interestingly, the different reports found in the literature for true protein analysis do not require specific pre-treatment of the cells and the introduction of fluorescent species to make their detection possible after the separation.

Ocvirk *et al.* report a microfluidic system for protein analysis at the single cell level, with an indirect analysis of the protein enzyme *via* the determination of the amount in a fluorescent degradation substrate specific to the enzyme.[69] The system presents a standard Y-shape configuration for on-stream mixing of the cell solution with a buffer containing both the lysis agent and the enzymatic substrate, and the fluorescent product is detected ~6 mm downstream to the Y junction. Cells are lysed chemically with the help of a detergent (SDS or Triton 100X). The authors look at the activity of a key-enzyme for intracellular sugar degradation, β-galactosidase whose deficiency is correlated with various diseases. They use fluorescein-di-β-D-galactopyranoside as enzymatic substrate which is degraded into the fluorescent product fluorescein mono-β-D-galactopyranoside. Their platform integrates the whole β-galactosidase assay without preliminary incubation steps as would happen when using the fluorescence activated cell sorting (FACS) technique; it brings about enhanced and faster mixing performance for both the cell lysis and the enzymatic reaction; and it enables to decrease reagent consumption. This microfluidic protocol is reported as more robust, and can as well be automated for continuous analysis of a cell population.

Direct measurements of proteins at the single cell level by protein counting have also more recently been described by Zare's group.[70] Their platform resembles that of Wu *et al.* in that it includes a capture chamber for a single cell where lysis and antibody-based labeling of specific proteins are simultaneously done. After refreshing the buffer with a separation buffer, the resulting mixture is injected in the separation channel and single proteins are counted in a low-detection volume and using cylindrical optics. This novel and highly sensitive methodology is firstly applied on the quantification of a human trans-membrane protein, β_2 adrenergic receptor expressed in an insect cell line (SF9). Secondly, the authors consider a more relevant biological application of single cell analysis, the degradation of phycobiliprotein complexes in individual cyanobacteria *Synechococcus* when the cell environment is depleted from essential macronutrients such as nitrogen. Interestingly, the analysis of three individual cells show a heterogeneous degradation pattern of the phycobiliprotein complexes, and one cell notably shows incomplete proteolysis of its phycobilisome. Such a cell accounts for less than 5% of a whole population, and this is detectable by using a single cell analysis approach while the information would be lost using a conventional analysis strategy at the level of a whole population.

A last example we present here for single cell analysis has been reported by Ros and co-workers.[71,72] They aim at detecting a single protein in single cell, after cell lysis and electrophoretic separation of the lysate, using a label-free detection technique. This UV-LIF detection exploits the intrinsic fluorescence of the amino acids tryptophan, tyrosine, and phenylalanine, and has yielded a possible limit in the picomolar range. In their first article, Chao and Ros use PDMS microsystems for this purpose,[72] and subsequently switch to quartz to reduce the background fluorescence,[71] and demonstrate a gain of one order of

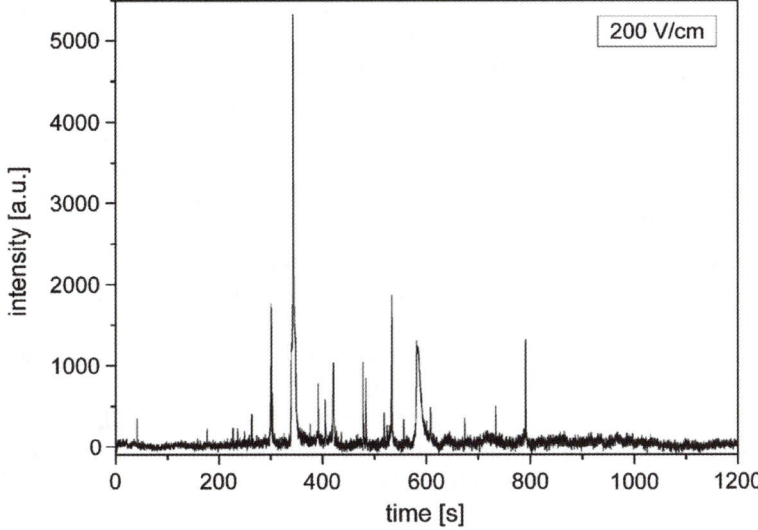

Figure 15.5 Single cell capillary electrophoresis. Baseline-corrected electropherogram obtained from a single Sf9 insect cells in a microfluidic quartz chip using native UV-LIF detection (266 nm). Reprinted from [71] with permission.

magnitude in the detection limit to reach a low nanomolar sensitivity when using tryptophan samples. Following these optimization experiments using simple amino acid solutions, they test their devices on a mixture of three native standard proteins, α-chymotrypsogen A, catalase, ovalbumin, each introduced at a concentration of 100 μg mL^{-1} (or around 1 μM). Again, they achieve good separation performance within 2 min. Finally, the protocol is applied to the analysis of single Sf9 insect cells. The cells are first resuspended in the separation buffer before their introduction in the chip. A cell is selected and placed at the intersection position using optical tweezers. Subsequently, a short pulse of electric field (2000 V cm^{-1}, 50 ms) is applied for cell lysis, the separation performed using an electric field strength of 200 V cm^{-1} and detection done using the UV-LIF technique 5 mm after the injection point. With this method, 20 peaks were detected in the electropherogram obtained for the label-free analysis of a single Sf9 insect cell, as shown in Figure 15.5.

15.4.1.2 Droplet-based Approach

Besides this CE-based protocol, a droplet-based approach has been proposed to investigate the expression level of targeted proteins at the single cell level and to resolve fast transient responses in cell signalling networks.[73] Cells are isolated in small aqueous droplets to create a confined environment for selective single cell stimulation and analysis and for enhanced mixing efficiency. Droplets are manipulated as a gas–liquid segmented flow in a microfluidic platform that also includes heating and cooling systems below the fluidic chip to finely control the temperature of different zones. Droplets containing individual cells are first merged with droplets containing chemicals for the specific treatment and stimulation of the individual cells; the droplets flow in a first serpentine reactor for the action of the chemical stimulant. In a second stage, a lysis buffer is added to the individual droplet to yield a series of droplets containing the lysates of the individual cells, and analysis of the content of the droplets is carried out off-line. The incubation times for both the stimulation and lysis steps are defined by the lengths of the reaction channels and the flow-rate values and can finely be tuned to change the time duration for these cells. The platform is notably illustrated on the stimulation of Jurkat E6-1 cells and the resulting activation using α-CD3 of two protein kinases, extracellular regulated kinase (ERK) and Jun n-terminal kinase (JNK), which have been previously transfected in the cells as kinase–green fluorescent protein (GFP) constructs. This microfluidic protocol gives identical temporal activation patterns as observed using conventional techniques.

15.4.2   Partially Invasive Analysis

A second series of innovative protocols can be characterized as partially invasive. The analysis is performed on a living cell without a priori endangering its viability and typically relies on sampling of a small amount of molecules out of a living cell, either out of the cytoplasm by creating transient pores in the membranes or from the membranes with the help of "smart hydrophobic droplets".

15.4.2.1   Cellular Poration

A first approach still involves rupturing the cell membrane, but only in a temporary way. Transient pores are formed in the cell membrane, as would occur for cell transfection for instance, but here to retrieve biological material from the cytoplasm. Pores close again naturally after molecular sampling so that the viability of the cell is not endangered. Two protocols have been reported in the literature for the creation of transient pores in a membrane, with the help of an electric field (electroporation) or by using chemicals (chemical poration).

15.4.2.1.1   Electroporative Flow Cytometry.   While single cell electroporation (SCE) has widely been described in a microfluidic format, its aim mainly focuses on the introduction of foreign substances in cell in a highly controlled way. Still, the same principle has been employed by Bao *et al.* for an opposite purpose, *i.e.* the targeted retrieval of molecules localized close to the plasma membrane.[74,75] Their main goal is to elucidate intracellular signaling pathways by tracking the intracellular localization of signaling proteins following specific chemical treatment of the cell, as reflected by the name they gave to their technique "electroporative flow cytometry".

The microfluidic device borrows much from a SCE platform, and presents a channel with a narrow section where a high voltage is applied and is consequently enhanced (Figure 15.6A). Cells which are flowing in the constricted area see this locally enhanced electric field and become porated. The authors first demonstrate the power of the technique with the diffusion of small molecules (dyes) previously loaded into the cells. They show thereby that the release of molecules can finely be tuned by changing the electric signal applied on the cells; the higher the signal the larger the molecules which are sampled. Following this, they focus on a particular protein kinase, spleen tyrosine kinase (Syk), that is translocated to the cell membrane upon activation. Cells are first transfected to express a GFP–Syk construct, and this fluorescent protein can subsequently be tracked after cell poration. When activated, the protein is translocated from the nucleus to the cell membrane, where it is easily released in the extracellular environment after creation of pores in the membrane.[75] Figure 15.6B compares the release of small fluorescent dyes and a protein kinase out of the cells as a function of the applied electric field to induce cell poration.

15.4.2.1.2 Digitonin-based Permeabilization of a Cell Membrane. A second approach to permeabilize the cell membrane (in a transient way) relies on the use of specific pore-inducing chemicals. Digitonin for instance is known to bind to cholesterol molecules present in the membrane and to subsequently induce poration of the membrane.[76]

This phenomenon is employed by Olofsson *et al.* in a microfluidic platform to access the intracellular content at the single cell level.[77] They use the commercially available platform Dynaflow 16 from Celletricon AB for intracellular experimentation in single cells. The platform includes 16 parallel microchannels that end in an open chamber, enabling the generation of a 16-phase flow (Figure 15.7A). The cell is alternately moved between different flows of digitonin and buffer or dye solution with the help of a micropipette and a programmable scanning stage, for its sequential stimulation and analysis (Figure 15.7B). Here again, the level of permeabilization of the cell can be controlled by adjusting the exposure parameters to digitonin (time, concentration), the intracellular solution can be exchanged in less than 1 min after careful calibration of the protocol, and sub-cellular experimentation is conceivable if a cell is placed at the interface between two adjacent flows. The technique is first demonstrated on the loading of fluorescent molecules in a cell, and later applied to the activation of enzymatic reactions towards a better understanding of signaling pathways.

15.4.2.2 Probing the Plasma Membrane Using Smart Droplet Microtools

While the two former techniques apply for the analysis of molecules present in the cytoplasm of a cell, they do not enable the analysis of hydrophobic molecules such as membrane proteins which are difficult to solubilize and isolate. To overcome this limitation, Lanigan *et al.* propose a novel protocol for the

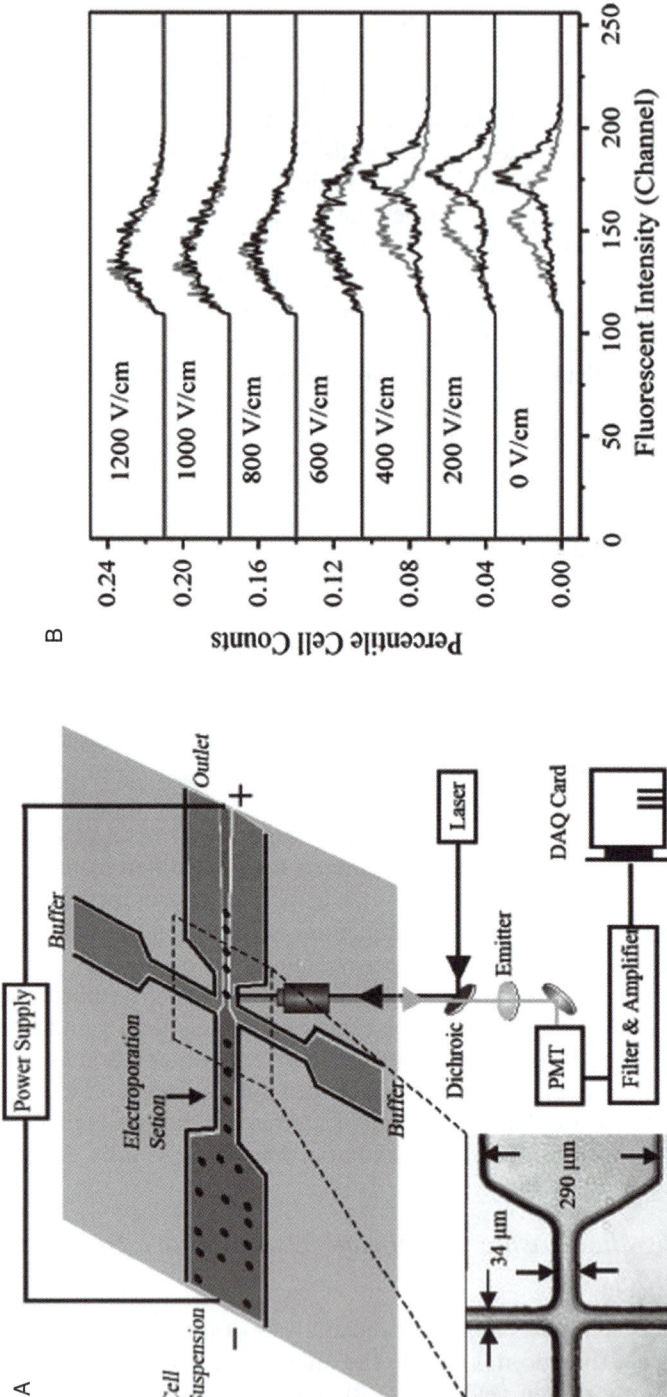

Figure 15.6 Electroporative flow cytometry. (A) Schematic of the microfluidic system for the electroporative release of intracellular materials at the single-cell level. The cells are focused a single-cell line in a narrow section of a channel where they are exposed to a high electric field for a certain period of time. Consequently, pores are formed in the cell membrane, and molecules can escape the cells to diffuse into the medium. (B) Comparison of the release of two fluorescent entities out of SykEGFP-DT40-Syk cells. Calcein (black) and a fusion protein between a protein kinase and GFP (SykEGFP) (grey) under different electric field intensities with the duration of 100 ms. Reprinted from [75] with permission.

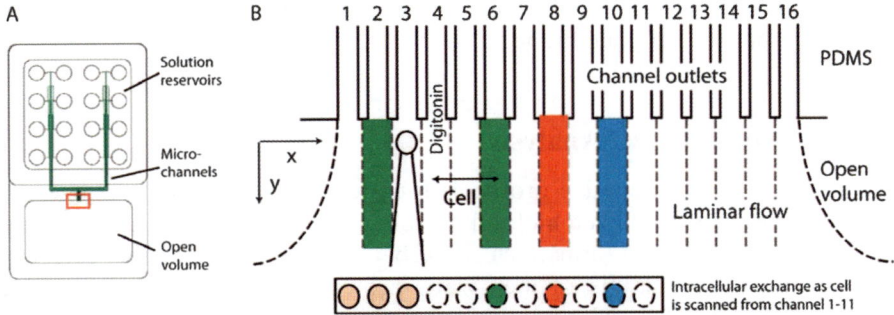

Figure 15.7 Accessing the intracellular content by chemical-inducing permeabilization of the cell membrane. (A) a microfluidic device that generates a 16-phase flow is used. (B) The system includes a series of 16 independent microchannels that exit in a large open volume where a 16-phase flow is created as a result of the laminar properties of the flow. A cell is manipulated with a micropipette and with the help of a x,y,z micropositioner between the 16 flows and exposed to different solutions. For instance, the cell is moved to a digitonin solution to be permeabilized, subsequently to buffer (washing) and to a fluorescein solution. Reprinted from [77] with permission.

selective isolation of membrane proteins using smart droplet microtools (SDMs) from biliary epithelial human colon carcinoma cells.[78,79] These droplets, 1–4 µm in size, are made from an organic solvent and coated with amphiphilic molecules to promote their solubilization in an aqueous solution. SDMs are brought into contact with a cell membrane with the help of optical tweezers, eventually fused with the cell membrane to allow membrane molecules to be transferred in the SDMs where they easily dissolve. The translocation of material into SDMs is visualized by using membrane green fluorescent proteins previously expressed in the cells to be analyzed.

In a first stage, the authors use SDMs based on hexadecane solvent and coated with phospholipid molecules found in natural membranes.[78] These SDMs are truly tethered to the cell membrane and act as reservoirs for the storage of plasma membrane material during a process the authors called "nanodigestion". Successful and selective transfer of biological material was observed using these SDMs and cells cultured in a conventional Petri dish. This protocol enables sampling and isolation of membrane proteins for further analysis and eventual quantification using fluorescence detection. In a second stage, the authors have modified their experimental set-up to perform large-scale sampling and use a microfluidic platform.[79] The microsystem includes two independent chambers for the SDMs and the cells, respectively. SDMs are still manipulated and brought into contact with GFP-expressing cells with the help of optical tweezers. After 60–300 s contact, the SDMs are removed slowly and in a controlled way with the optical tweezers and stored in the chip. The SDMs are here slightly modified; they are based on heptane and coated with Triton-100X so that they do not fuse anymore with the cell membrane. Still transfer of

GFP is still possible by diffusion phenomenon and the SDMs are easily retrieved from the cell membrane.

15.4.3 Non-invasive Analysis

The last group of innovative protocols for protein analysis at the single cell level are non-invasive. Given molecules/proteins are detected and quantified in living cells without any (apparent) perturbation of the cell. In this category are found (i) the analysis of membrane proteins that are easily accessible without affecting the cell and (ii) intracellular measurements of targeted molecules after the eventual introduction in the cell of specific probes such as gold nanoparticles.

15.4.3.1 Determination of the Expression Level of Targeted Membrane Proteins

As mentioned, membrane proteins are easily accessible as they are spanning the cell membrane and can be analyzed without lysing or permeabilizing the cell. One highly relevant application is the quantification of certain membrane receptors whose expression level can be used for the early detection of cancer, for instance.

15.4.3.1.1 Early Cancer Detection Using Two Microfluidic Protocols. Fitzpatrick *et al.* report two different microfluidic protocols for the analysis of membrane proteins at the single cell level and detect the over-expression of given biomarkers on a cell surface.[80] One protocol borrows much from conventional flow cytometry with the staining of proteins of interest and the fluorescent detection of cells positive to a given stain. However, their microfluidic protocol applies on a lower population level (few tenths of cells *versus* millions for flow cytometry). The second protocol relies on a flow retardation system or on the interactions of cells with a functionalized surface in the same way as employed for cell affinity chromatography; cells presenting the targeted antigens on their surface flow slowly in a channel functionalized with the corresponding antibody. Both protocols have been used for the analysis of integrin alpha5 and low abundant Her-2/neu in three breast cancer cell lines, MCF7, SK-BR-3, and MDA-MB-231 cells. It is worth mentioning that while the first protocol still implies preliminary preparation of the samples and staining, the second methodology can be directly applied on crude samples with no or minimal preparation, and is label-free. While these two platforms enable detection of membrane protein in a reliable and innovative manner, they do not really give quantitative information on the protein expression level on the cell surface and at the single cell level.

15.4.3.1.2 In-droplet Enzymatic Amplification. Another methodology was proposed by Joensson *et al.* to not only detect but also quantify the amount of low-abundant biomarkers located in the cell membrane, at almost the single cell level.[81] Their methodology should, in particular, yield enhanced detection sensitivity compared with flow cytometry and fluorescence techniques. Most of the time, biomarkers are expressed in a low abundance, below the critical amount that can be detected by fluorescence techniques. As a consequence, direct staining of the biomarkers of interest is not sufficient for their detection by conventional flow cytometry techniques. Here, the authors report on an enzymatic amplification method (Figure 15.8) to indirectly increase the fluorescent signal generated by the presence of a single protein on the membrane of U937 human monocytic cells. The biomarker (CCR5 or CD19) of interest is coupled to a β-galactosidase enzyme *via* a biotin–streptavidin interaction and the resulting functionalized cells are incubated for a given period of time in the presence of a fluorogenic substrate (fluorescein-di-β-galactopyranoside) for the enzyme. Besides, cells are isolated in droplets to confine the enzymatic reaction, enhance mixing conditions and make possible the quantification of the biomarker. After incubation, the fluorescence intensity in the individual droplets is measured and used to derive the amount of biomarkers expressed on the cell surface. The authors observe a linear increase of the fluorescent signal for a 5 h incubation period, and they claim a detection level down to 1.1 pM, or 22 molecules of β-galactosidase per 40 μm diameter droplet. Last, the authors propose a color-based coding of

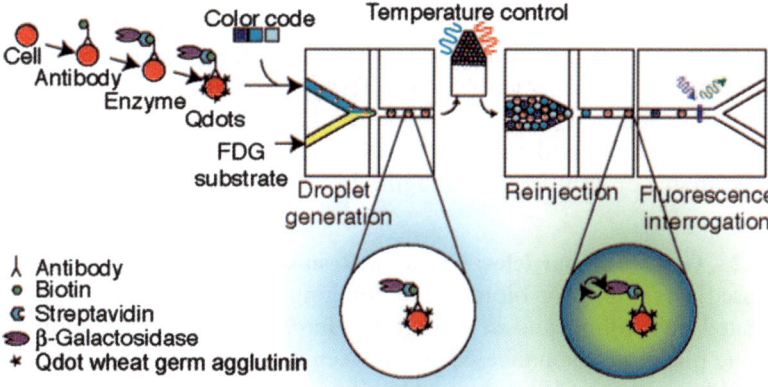

Figure 15.8 Principle of the in-droplet enzymatic amplification detection method. Cells are labeled with an antibody specific to a cell surface protein, and this antibody is coupled to an enzyme (beta-galactosidase) *via* a streptavidin biotin interaction and to a quantum dot wheat germ agglutinin (WGA) stain for visualization. Cells are subsequently encapsulated individually in droplets in a microfluidic chip with a sample-specific color code and a fluorogenic substrate (FDG or fluorescein-di-β-D-galactopyranoside) specific to the enzyme immobilized on the cell surface. After incubation to allow the enzymatic reaction take place, the fluorescence level is measured and analyzed for each droplet. Reprinted from [81] with permission.

the droplets for multiplexed analysis and simultaneous investigation of two different biomarkers, using a permanent dye (Alexa 405) at three various concentrations.

15.4.3.2 Intracellular Measurements of Targeted Biomolecules with the Eventual Help of Specific Probes

Alternatively, intracellular measurements enable to follow biological processes and track and quantify in real-time biomolecules in a living cell, and this eventually implies the introduction and targeting of nanoprotein in the cell. This category of measurements exploits the absorption properties of the biomolecules, the particles introduced in the cells or a combination thereof, as detailed in the coming paragraphs.

15.4.3.2.1 Thermal Lens Microscopy. Thermal lens microscopy (TLM) has been proposed as an attractive technique for the optical detection and quantification of biomolecules in living cells.[82] TLM relies on the absorption properties of biomolecules in the visible and UV light, and does not require any sample preparation as it is a label-free technique. This non-invasive technique can be used for dynamic and time-lapse imaging of biomolecules in living cells, and their distribution over time in a single cell. The technique has notably been applied in a microfluidic format for monitoring the cytochrome *c* distribution in cells undergoing apoptosis using a 532 nm laser. This example illustrates the spatial resolution and the sensitivity of the technique, as cytochrome c was imaged with a 1 μm lateral resolution and 10 zmol molecules could be detected in a single cell. As TLM is a label-free technique, it enables the detection and analysis of other molecules although the authors do not comment on multiplexed imaging in a single cell using this technique.

15.4.3.2.2 Gold Nanoparticles: Plasmon Resonance. Gold nanoparticles are widely used as sensors for biomolecule imaging, also in living cells, as well as for targeted cell killing and cell surgery. Here, the imaging of biomolecules rely on the plasmon resonance occur between adjacent nanoparticles. While a single gold nanoparticle weakly absorbs and emits light, the coupling between two identical nanoparticles results in a bright and strong signal whose characteristics depend on the dimensions of the nanoparticles. This phenomenon is exploited here; when two 20-nm gold nanoparticles are closely assembled, they emit around 620 nm and the bright red signal is easily detected with a high lateral resolution.[83]

The gold nanoparticles are first made multifunctional for both intracellular delivery and protein targeting, using a TAT-HA2 peptide and a specific antibody for binding to a protein of interest, respectively. While TAT is used for cellular delivery, the addition of HA2 is added for non-endosomal retention of the particles and suitable delivery. The authors demonstrate here the efficiency of their gold nanoparticle system for imaging actin rearrangement in live 3T3 fibroblasts, and in particular the retraction of filapodia and spreading of cells on a collagen coated surface.

The technique shows promising results with sharp and precise imaging and localization of the molecule and possible dynamic tracking of the molecule. As actin is relatively abundant in cells, the sensitivity of the technique is not discussed here.

15.4.3.2.3 Gold Nanoparticles: The PRET Technique. Of interest is recent report of Lee and co-workers on the use of gold nanoparticles for plasmonic-based imaging in HepG2 cells of targeted molecules by exploiting the absorption properties of both the nanoparticles and the biomolecules to be analyzed.[84] This technique described as plasmon resonance energy transfer (PRET) relies on the use of gold nanoparticles that emit in the frequency where a given biomolecule to be imaged absorbs (electronic transition energy). The presence of the biomolecules in the close vicinity of the Au NP gives rise to dips in the gold nanoparticles (GNPs) Rayleigh spectrum caused by the resonant plasmonic energy transfer to the biomolecules: these dips are used for both detection and quantification purposes, as illustrated in Figure 15.9. They successfully illustrate the power of the technique to follow a key-step in the apoptotic process; the intracellular release of cytochrome c out of mitochondria, with the help of 50 nm GNP functionalized with carboxylic acid moieties with which cytochrome c easily interacts. Limitations of the technique are the need to tune the GNPs so that their emission matches with the biomolecule absorption and to develop suitable surface chemistry for the biomolecule–GNP interactions; however, providing this, the technique is easily amenable to multiplexed detection and the simultaneous imaging of different intracellular probes.

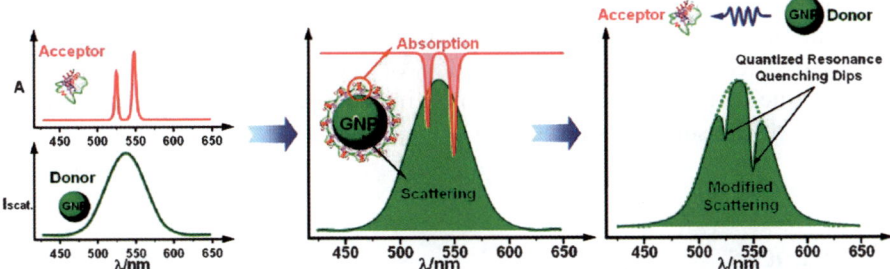

Figure 15.9 Principle of the plasmonic resonance energy transfer (PRET)-based molecular imaging technique. A gold nanoparticle is functionalized with given biomolecules whose electronic transition frequencies overlap with the plasmon resonance frequency of the gold nanoparticle. Selective energy transfer occurs between the two entities as a consequence of this overlap and this gives rise to a distinguishable spectral resonant quenching dips in the Rayleigh scattering spectrum of the particles. Reprinted from [84] with permission.

15.5   Conclusion and Perspectives

While first approach for single cell protein analysis can be seen as a transposition of the conventional proteomic protocol to a small cell population with minimal sample handling, emerging approaches tend to develop innovative protocols to look at the protein content of a single cell. In addition to this, these novel protocols are not always invasive as would the proteomic approach be; they can also apply on living cells. In the latter case, they focus on certain protein targets, and leave the cell alive and intact for other experimentation or use.

On other aspects, it should be noted that microfluidic tools are still not appropriate to analyze proteins at the single cell level; typical volumes handled with microfluidics remain mostly in the nanoliter range while a cell has a typical volume of 1 pL. Consequently, single protein analysis in a microfluidic format still reminds us of looking for a needle in a haystack!

Still, the current emergence of nanotechnology and the increasing amount of novel and performing nanoscale tools give a hope for single protein detection at the single level, also using a proteomics approach. These nanotools not only make possible the manipulation of subcellular amount of liquids, but also bring enhanced detection sensitivity down to the single molecule level. Among nanofluidics tools are found, for instance: (1) nanochannels presenting a capacity lower than a cell's volume and that have recently been applied for single molecule detection,[85] (2) nanofluidic pumps for sampling the cellular content such as the bubble-based nanofluidic pump[86] and attochemical atto-syringe,[87] and (3) droplets whose size can easily be tailored to create small reactors in which the cell content is slightly diluted. The "nano" toolbox also contains nanostuctures such as nanoparticles and nanowires[88] that enable single molecule detection. Raman nanoparticles can either be brought into cells for on-site analysis in living platforms or implemented on a surface for highly sensitive and label-free analysis of cell lysates. Alternatively, nanowires are suitable candidates for single molecules detection, and the biomolecule analysis relies in the targeted hybridization of identified proteins on antibodies and electrical detection. The implementation of these upcoming nanotools in a microfluidic format is a promising route in the near future for single protein detection at the single cell level.

References

1. R. Juraschek, T. Dulcks and M. Karas, *J. Am. Soc. Mass Spectrom.*, 1999, **10**, 300.
2. V. Studer, G. Hang, A. Pandolfi, M. Ortiz, W. F. Anderson and S. R. Quake, *J. Appl. Phys.*, 2004, **95**, 393.
3. H. Becker and C. Gartner, *Anal. Bioanal. Chem.*, 2008, **390**, 89.
4. T. Laurell, J. Nilsson and G. Marko-Varga, *TrAC, Trends Anal. Chem.*, 2001, **20**, 225.

5. J. B. Fenn, M. Mann, C. K. Meng, S. F. Wong and C. M. Whitehouse, *Science (Washington, D.C.)*, 1989, **246**, 64.
6. B. Mehlis and U. Kertscher, *Anal. Chim. Acta*, 1997, **352**, 71.
7. M. Wilm and M. Mann, *Anal. Chem.*, 1996, **68**, 1.
8. S. Koster and E. Verpoorte, *Lab Chip*, 2007, **7**, 1394.
9. D. Figeys, C. Lock, L. Taylor and R. Aebersold, *Rapid Commun. Mass Spectrom.*, 1998, **12**, 1435.
10. J. J. Li, P. Thibault, N. H. Bings, C. D. Skinner, C. Wang, C. Colyer and J. Harrison, *Anal. Chem.*, 1999, **71**, 3036.
11. D. M. Pinto, Y. B. Ning and D. Figeys, *Electrophoresis*, 2000, **21**, 181.
12. R. S. Ramsey and J. M. Ramsey, *Anal. Chem.*, 1997, **69**, 1174.
13. B. Zhang, H. Liu, B. L. Karger and F. Foret, *Anal. Chem.*, 1999, **71**, 3258.
14. G. A. Schultz, T. N. Corso, S. J. Prosser and S. Zhang, *Anal. Chem.*, 2000, **72**, 4058.
15. K. Tang, Y. Lin, D. W. Matson, T. Kim and R. D. Smith, *Anal. Chem.*, 2001, **73**, 1658.
16. C.-H. Yuan and J. Shiea, *Anal. Chem.*, 2001, **73**, 1080.
17. J. S. Kim and D. R. Knapp, *J. Chromatogr., A*, 2001, **924**, 137.
18. J. S. Kim and D. R. Knapp, *J. Am. Soc. Mass Spectrom.*, 2001, **12**, 463.
19. J.-S. Kim and D. R. Knapp, *Electrophoresis*, 2001, **22**, 3993.
20. S. Arscott, S. Le Gac, C. Druon, P. Tabourier and C. Rolando, *J. Micromech. Microeng.*, 2004, **14**, 310.
21. S. Arscott, S. Le Gac, C. Druon, P. Tabourier and C. Rolando, *Sens. Actuators, B Chem.*, 2004, **98**, 140.
22. S. Arscott, S. Le Gac and C. Rolando, *Sens. Actuators, B Chem.*, 2005, **B106**, 741.
23. S. Le Gac, S. Arscott and C. Rolando, *Electrophoresis*, 2003, **24**, 3640.
24. S. Le Gac, C. Cren-Olive, C. Rolando and S. Arscott, *J. Am. Soc. Mass Spectrom.*, 2004, **15**, 409.
25. J. Kameoka, R. Orth, B. Ilic, D. Czaplewski, T. Wachs and H. G. Craighead, *Anal. Chem*, 2002, **74**, 5897.
26. J. S. Rossier, N. Youhnovski, N. Lion, E. Damoc, S. Becker, F. Reymond, H. H. Girault and M. Przybylski, *Angew. Chem. Int. Ed.*, 2003, **42**, 54.
27. V. Gobry, J. Van Oostrum, M. Martinelli, T. C. Rohner, F. Reymond, J. S. Rossier and H. H. Girault, *Proteomics*, 2002, **2**, 1474.
28. N. F. Yin, K. Killeen, R. Brennen, D. Sobek, M. Werlich and T. V. van de Goor, *Anal. Chem.*, 2005, **77**, 527.
29. M. Schilling, W. Nigge, A. Rudzinski, A. Neyer and R. Hergenroder, *Lab Chip*, 2004, **4**, 220.
30. S. Arscott, S. Le Gac and C. Rolando, *Sens. Actuators, B Chem.*, 2005, **106**, 741.
31. J. Carlier, S. Arscott, V. Thomy, J. C. Camart, C. Cren-Olive and S. Le Gac, *J. Chromatogr. A*, 2005, **1071**, 213.
32. D. S. Peterson, T. Rohr, F. Svec and J. M. J. Frechet, *J. Proteome Res.*, 2002, **1**, 563.

33. B. He, J. Y. Ji and F. E. Regnier, *J. Chromatogr. A*, 1999, **853**, 257.
34. S. Ekstrom, P. Onnerfjord, J. Nilsson, M. Bengtsson, T. Laurell and G. Marko-Varga, *Anal. Chem.*, 2000, **72**, 286.
35. C. Wang, R. Oleschuk, F. Ouchen, J. J. Li, P. Thibault and D. J. Harrison, *Rapid Commun. Mass Spectrom.*, 2000, **14**, 1377.
36. D. S. Peterson, T. Rohr, F. Svec and J. M. J. Frechet, *Anal. Chem.*, 2002, **74**, 4081.
37. J. Krenkova and F. Foret, *Electrophoresis*, 2004, **25**, 3550.
38. F. Xiang, Y. H. Lin, J. Wen, D. W. Matson and R. D. Smith, *Anal. Chem.*, 1999, **71**, 1485.
39. L. Ceriotti, N. F. de Rooij and E. Verpoorte, *Anal. Chem.*, 2002, **74**, 639.
40. R. D. Oleschuk, L. L. Shultz-Lockyear, Y. B. Ning and D. J. Harrison, *Anal. Chem.*, 2000, **72**, 585.
41. B. He, L. Tan and F. Regnier, *Anal. Chem.*, 1999, **71**, 1464.
42. E. Mery, F. Ricoul, N. Sarrut, O. Constantin, G. Delapierre, J. Garin and F. Vinet, *Sens. Actuators, B Chem.*, 2008, **134**, 438.
43. J. Liu, K. W. Ro, R. Nayak and D. R. Knapp, *Int. J. Mass Spectrom.*, 2007, **259**, 65.
44. N. Gottschlich, S. C. Jacobson, C. T. Culbertson and J. M. Ramsey, *Anal. Chem.*, 2001, **73**, 2669.
45. A. B. Jemere, R. D. Oleschuk and D. J. Harrison, *Electrophoresis*, 2003, **24**, 3018.
46. J. Xie, Y. N. Miao, J. Shih, Y. C. Tai and T. D. Lee, *Anal. Chem.*, 2005, **77**, 6947.
47. D. S. Peterson, T. Rohr, F. Svec and J. M. J. Frechet, *Anal. Chem.*, 2003, **75**, 5328.
48. H. F. Yin and K. Killeen, *J. Sep. Sci.*, 2007, **30**, 1427.
49. M. Ghitun, E. Bonneil, M. H. Fortier, H. F. Yin, K. Killeen and P. Thibault, *J. Sep. Sci.*, 2006, **29**, 1539.
50. J. Xie, Y. N. Miao, J. Shih, Q. He, J. Liu, Y. C. Tai and T. D. Lee, *Anal. Chem.*, 2004, **76**, 3756.
51. J. M. Armenta, A. A. Dawoud and I. M. Lazar, *Electrophoresis*, 2009, **30**, 1145.
52. S. Ekstrom, D. Ericsson, P. Onnerfjord, M. Bengtsson, J. Nilsson, G. Marko-Varga and T. Laurell, *Anal. Chem.*, 2001, **73**, 214.
53. G. Marko-Varga, S. Ekstrom, G. Helldin, J. Nilsson and T. Laurell, *Electrophoresis*, 2001, **22**, 3978.
54. T. Redeby, J. Roeraade and A. Emmer, *Rapid Commun. Mass Spectrom.*, 2004, **18**, 1161.
55. N. Tang, P. Tornatore and S. R. Weinberger, *Mass Spectrom. Rev.*, 2004, **23**, 34.
56. E. Nagele and M. Vollmer, *Rapid Commun. Mass Spectrom.*, 2004, **18**, 3008.
57. H. K. Musyimi, J. Guy, D. A. Narcisse, S. A. Soper and K. K. Murray, *Electrophoresis*, 2005, **26**, 4703.
58. H. Orsnes, T. Graf, H. Degn and K. K. Murray, *Anal. Chem.*, 2000, **72**, 251.

59. Y. X. Wang, Y. Zhou, B. M. Balgley, J. W. Cooper, C. S. Lee and D. L. DeVoe, *Electrophoresis*, 2005, **26**, 3631.
60. M. Gustafsson, D. Hirschberg, C. Palmberg, H. Jornvall and T. Bergman, *Anal. Chem.*, 2004, **76**, 345.
61. J. Liu, K. Tseng, B. Garcia, C. B. Lebrilla, E. Mukerjee, S. Collins and R. Smith, *Anal. Chem.*, 2001, **73**, 2147.
62. A. R. Wheeler, H. Moon, C. J. Kim, J. A. Loo and R. L. Garrell, *Anal. Chem.*, 2004, **76**, 4833.
63. A. R. Wheeler, H. Moon, C. A. Bird, R. R. O. Loo, C. J. Kim, J. A. Loo and R. L. Garrell, *Anal. Chem.*, 2005, **77**, 534.
64. L. A. Woods, T. P. Roddy and A. G. Ewing, *Electrophoresis*, 2004, **25**, 1181.
65. I. G. Arcibal, M. F. Santillo and A. G. Ewing, *Anal. Bioanal. Chem.*, 2007, **387**, 51.
66. P. J. Marc, C. E. Sims, M. Bachman, G. P. Li and N. L. Allbritton, *Lab Chip*, 2008, **8**, 710.
67. M. A. McClain, C. T. Culbertson, S. C. Jacobson, N. L. Allbritton, C. E. Sims and J. M. Ramsey, *Anal. Chem.*, 2003, **75**, 5646.
68. H. K. Wu, A. Wheeler and R. N. Zare, *Proc. Natl Acad. Sci. U. S. A.*, 2004, **101**, 12809.
69. G. Ocvirk, H. Salimi-Moosavi, R. J. Szarka, E. A. Arriaga, P. E. Andersson, R. Smith, N. J. Dovichi and D. J. Harrison, *Proc. IEEE*, 2004, **92**, 115.
70. B. Huang, H. K. Wu, D. Bhaya, A. Grossman, S. Granier, B. K. Kobilka and R. N. Zare, *Science*, 2007, **315**, 81.
71. D. Greif, L. Galla, A. Ros and D. Anselmetti, *J. Chromatogr. A.*, 2008, **1206**, 83.
72. T. C. Chao and A. Ros, *J. R. Soc. Interface*, 2008, **5**, S139.
73. J. El-Ali, S. Gaudet, A. Gunther, P. K. Sorger and K. F. Jensen, *Anal. Chem.*, 2005, **77**, 3629.
74. N. Bao, J. Wang and C. Lu, *Electrophoresis*, 2008, **29**, 2939.
75. J. Wang, N. Bao, L. L. Paris, H. Y. Wang, R. L. Geahlen and C. Lu, *Anal. Chem.*, 2008, **80**, 1087.
76. M. J. H. Geelen, *Anal. Biochem.*, 2005, **347**, 1.
77. J. Olofsson, H. Bridle, A. Jesorka, I. Isaksson, S. Weber and O. Orwar, *Anal. Chem.*, 2009, **81**, 1810.
78. P. M. P. Lanigan, K. Chan, T. Ninkovic, R. H. Templer, P. M. W. French, A. J. de Mello, K. R. Willison, P. J. Parker, M. A. A. Neil, O. Ces and D. R. Klug, *J. R. Soc. Interface*, 2008, **5**, S161.
79. P. M. P. Lanigan, T. Ninkovic, K. Chan, A. J. de Mello, K. R. Willison, D. R. Klug, R. H. Templer, M. A. A. Neil and O. Ces, *Lab Chip*, 2009, **9**, 1096.
80. E. Fitzpatrick, S. McBride, J. Yavelow, S. Najmi, P. Zanzucchi and R. Wieder, *Clin. Chem.*, 2006, **52**, 1080.
81. H. N. Joensson, M. L. Samuels, E. R. Brouzes, M. Medkova, M. Uhlen, D. R. Link and H. Andersson-Svahn, *Angew. Chem. Int. Ed.*, 2009, **48**, 2518.

82. E. Tamaki, K. Sato, M. Tokeshi, K. Sato, M. Aihara and T. Kitamori, *Anal. Chem.*, 2002, **74**, 1560.
83. S. Kumar, N. Harrison, R. Richards-Kortum and K. Sokolov, *Nano Lett.*, 2007, **7**, 1338.
84. Y. H. Choi, T. Kang and L. P. Lee, *Nano Lett.*, 2009, **9**, 85.
85. H. T. Hoang, I. M. Segers-Nolten, J. W. Berenschot, M. J. de Boer, N. R. Tas, J. Haneveld and M. C. Elwenspoek, *J. Micromech. Microeng.*, 2009, **19**, 065017.
86. N. R. Tas, J. W. Berenschot, T. S. J. Lammerink, M. C. Elwenspoek and A. van den Berg, *Anal. Chem.*, 2002, **74**, 2224–2227.
87. F. O. Laforge, J. Carpino, S. A. Rotenberg and M. V. Mirkin, *PNAS*, 2007, **104**, 11895–11900.
88. S. Y. Chen, J. G. Bomer, W. G. van der Wiel, E. T. Carlen and A. van den Berg, *ACS Nano*, 2009, **3**, 3485–3492.

CHAPTER 16

A Concrete Case: A Microfluidic Device for Single Cell Whole Transcriptome Analysis

NATHALIE BONTOUX, LUCE DAUPHINOT AND MARIE-CLAUDE POTIER

CRICM, CNRS UMR7225, INSERM URMS975, UPMC Hôpital de la Pitié Salpétrière, Paris, France

Abstract

Single cell whole transcriptome analysis, *i.e.* the analysis of all the genes that are expressed by a cell at a given time and under given physiological or pathological conditions, constitutes a major challenge in understanding cellular diversity and the complexity of living organisms. Indeed, such analyses will be key in unravelling cellular regulatory networks and understanding cell growth, differentiation and migration mechanisms.[1] They are also of significant interest for diagnosis and could prove a very efficient tool to identify new therapeutic targets.[2]

With the recent development of DNA microarrays, the transcriptome, *i.e.* the expression of all the genes, can now be studied in a single experiment. However, current labelling and detection methods require a starting amount of total RNA of about 100 ng, which is around 10^4 times more than the content of a single cell. The sensitivity thus undoubtedly needs to be improved to achieve accurate single cell whole transcriptome analysis. In this context, microfluidic devices offer interesting perspectives since they enable studies to be performed at the pico or nanoliter scale.

RSC Nanoscience & Nanotechnology No. 15
Unravelling Single Cell Genomics: Micro and Nanotools
Edited by Nathalie Bontoux, Luce Dauphinot and Marie-Claude Potier
© Royal Society of Chemistry 2010
Published by the Royal Society of Chemistry, www.rsc.org

In this chapter, we will detail our microfluidic approach for whole gene profiling of single cells. We will briefly review the choice of protocols for single cell transcriptome amplification as well as materials and techniques that can be used to fabricate microfluidic devices. We will describe the integration of the reverse-transcription (RT) and polymerase chain reaction (PCR) steps on chip and then discuss how all the steps of the biological protocol can be integrated in a single lab-on-a-chip.

16.1   Introduction

Gene expression analysis always starts with the relatively low efficient reverse transcription (RT) of RNA into complementary DNA (cDNA), an essential step as unprocessed RNAs will not be further analysed. Using microarrays, gene profiling is now well established for samples that contain at least 100 000 cells (more than 1 μg of total RNA). But for smaller samples, assays remain extremely difficult to carry out and prior amplification of the genetic content is required. Because standard laboratory techniques are not adapted to very low volumes, they do not seem adapted to single cell analysis. Indeed, the 10 pg of total RNA contained initially in the picoliter volume of a single cell is currently analyzed in several microliters. This million-fold dilution of the cell content is likely to impact the efficiency of reactions.

In this context, microfluidics lab-on-a-chip seem a most promising tool for single cell assays.[3] Indeed, they allow reactions to be performed at the scale of the cell, thus at higher concentrations, and the integration of multiple reactions on a single microchip reduces the risks of contamination and the consumption of reagents.

Since 1993 and the first report of a microfluidic device for polymerase chain reaction (PCR),[4] lab-on-a-chip for nucleic acid analysis has been developed at an ever-increasing pace. Most of the effort has been targeted towards integrating PCR and capillary gel electrophoresis on chip, leading to significant development in the field of surface modification and characterization, as well as thermal cycling and temperature control of microchips. Very few examples of a lab-on-a-chip have been reported, however, for gene expression studies. Obeid *et al.* and Beer *et al.* are the only authors who integrated RT-PCR in, respectively, a continuous stream[5] or in droplets.[6] Quake and co-workers have reported consistently more elaborated devices, where cell capture, lysis, mRNA extraction, and reverse-transcription are integrated on-chip. Although this approach proved very sensitive, the major limitation remains in the number of genes analyzed at a time.[7,8] To our knowledge, the amplification protocols reported do not allow for whole transcriptome amplification.

We have developed a microfluidic approach for whole transcription analysis of single cells[9] integrating cell capture, lysis, and efficient reverse transcription in a 10 nL poly(dimethylsiloxane) (PDMS) rotary mixer,[10,11] followed by template-switching PCR (TS-PCR) amplification. This approach allows for more sensitive gene expression profiling of single cells.

16.2 Choice of Biological Protocol, Material and Fabrication Technique

16.2.1 Protocols for Single Cell Whole Transcriptome Analysis

Whole transcriptome analysis of single cells currently requires messenger RNAs (mRNAs) being first reverse transcribed into complementary DNAs (cDNAs). cDNAs are then amplified, either linearly by *in vitro* transcription (IVT),[12] exponentially by global PCR[13,14] or using a combination of those two techniques (PCR + IVT).[15,16] Gene expression profiles are obtained after hybridization on pangenomic microarrays and data analysis.

Several methods for whole transcriptome analysis of single cells on benchtop have been reported with various levels of success.[2,15,17–20] Among these, the protocol developed by Kurimoto and colleagues[16,18] seemed to be promising. Indeed, it was shown to be a good approach for single-cell gene profiling and allowed for the identification of new markers important for cell lineage.[21–24] However, looking carefully at this protocol,[16] we identified some negative points towards microfluidic integration: (1) it requires PCR amplification steps followed by IVT, (2) it is necessary to perform a purification step on agarose gel to cut the amplified cDNAs, and (3) the complete achievement of the protocol takes 6 days. Looking at the literature, template-switching PCR (TS-PCR)[13] amplification seemed to be another interesting technique for transcriptome analysis of single cells. It has indeed been designed to amplify full-length transcripts, is less time-consuming than IVT-based protocols and has been reported to give the highest true-positive discovery rate in differential expression assays.[25,26] With this method, all cDNAs are amplified in one step using one single primer. In our hands, it proved the most efficient and reproducible protocol and we thus decided to use this technique, although the risk of getting contaminations is higher when using a single primer for PCR.

16.2.2 Miniaturizing Reactions: Continuous Flows, Reaction Chambers or Droplet Micro-fluidic Reactions

Having selected the amplification method, the next step was to miniaturize and integrate in a lab-on-a-chip all the steps of the protocol, *i.e.* single cell capture, lysis, reverse transcription (RT), and PCR amplification.

In contrast to our macrofluidic environment, most flows in microchannels are laminar, *i.e.* parallel streamlines will remain parallel, and mixing of species is controlled by diffusion. For many applications such as biological reactions or kinetics studies, the reaction should not be diffusion-limited; fast homogenization of samples is thus required.

Microfluidic reactions can be performed in closed chambers or in a continuous flow mode. In the first case, an external source of power is necessary to induce mixing. In the flow mode, mixing can be achieved passively, either by splitting and recombining flows or by inducing chaotic stirring.[27] An alternative is to use microfluidic devices to encapsulate reagents in microdroplets and then manipulate those droplets as individual reactors. Although very interesting, this approach was too much in its early days to be an option when we started our project. We refer interested readers to Chapter 17 for more details on droplet microfluidics for single-cell analyses.

Choosing how to manipulate fluid flows and control reactions definitely stands out as one of the key steps in designing a microfluidic device. When several reactions need to be integrated or when reagents are limited, a reaction chamber may be preferable as it relies less on flow engineering, allowing better control of individual reactions. We decided to build our device for transcriptome profiling by integrating on a single microchip several microreactors, each dedicated to a given step of the protocol.

16.2.3 Choosing the Microchip Material

Choice of strategy for fluid flows engineering and choice of chip material and fabrication techniques are often related as flow control elements should easily be integrated.

Most of the early microfluidic devices were made of glass or silicon, using standard microfabrication techniques such as photolithography and etching. Although their fabrication is expensive, cumbersome and requires a cleanroom facility, glass and silicon lab-on-a-chips prove advantageous when there is a need for a well-characterized surface chemistry or when metallic or even optical parts should be integrated (*e.g.* for sensing or heating). They also withstand strong solvent and high pressure. But with the development of "non-conventional" fabrication techniques, in the mid 1990s, plastic materials became more popular. They indeed are better suited to prototyping or for single-use devices, as they are cheap and easily microfabricated.

For gene profiling, contamination from samples to samples is a key issue and single-use lab-on-a-chip should be preferred. Plastic materials seem the best option for proof-of-principle experiments as prototype fabrication will be straightforward, rapid, and inexpensive.

We chose to make our microfluidic device in poly(dimethylsiloxane) (PDMS). Solid PDMS elastomers are obtained after a liquid mixture of a pre-polymer and cross-linker is cured above the elastomer transition temperature. PDMS is transparent, permeable to gas and organic solvents, and, most importantly, biologically inert. Although PDMS surfaces are naturally hydrophobic, appropriate surface treatment can make them hydrophilic. Another advantage of PDMS devices is that active flow control elements such as monolithic valves and pumps can easily be integrated using multilayer soft-lithography.

16.2.4 Microchip Fabrication

Complex PDMS devices made of several layers of channels can be fabricated by multilayer soft-lithography.[28] The first step consists in designing and fabricating a mold for each channel layer. It requires a cleanroom facility and some lithography expertise. In this work, the optical masks were designed using Adobe Illustrator and were printed on transparencies using a high-resolution printer. Once the molds have been obtained, PDMS structures are fabricated within 1 day by cast-molding each layer and bonding those together. Fabrication cost is very low as material is inexpensive and molds can be reused over time. Active flow control elements such as monolithic valves and pumps can easily be integrated. For example, in a two-layer device, a thin membrane can be engineered at the intersection of two superimposed perpendicular channels (Figure 16.1). This membrane will deflect and act as a valve when correctly pressure-actuated.[28,29] Moreover, using a slightly more complex geometry, successful mixing can be achieved in a ring.[10]

PDMS and multilayer soft-lithography are very well-suited for easy prototyping of a complex lab-on-a-chip and the microfluidic rotary mixer was an appealing microreactor. To develop our microchip for whole transcriptome assays, we thus chose to perform each reaction of the whole protocol in a microfluidic rotary mixer and then to integrate those microreactors. Each chip consisted of two levels of channels: fluids circulate in the top layer (fluidic channels, in black on Figures 16.1 and 16.2), while the bottom channels (control channels, in gray on Figures 16.1 and 16.2) operate the pressure-actuated valves. The rotary mixer (2D-geometry shown on Figure 16.2A) is constituted of a 10 nL annular ring and three microfluidic valves (Figure 16.2A, channels 3, 5, and 7), which constitute a peristaltic pump when actuated with a given sequence and frequency.

Molds for the fluidic channels were made of Shipley SPR 220–7 photoresist spin-coated at 800 rpm for 60 s. Once developed, the resist was heated at 165 °C for 15 min to create channels with a rounded cross-section (as the photoresist reflows). Molds for the control channels were made of SU8-2010 using the MicroChem protocol for 20 μm high channels. The mold for the fluidic channels was exposed to a vapor of trimethylchlorosilane (TMCS) for 2 min. A thick layer of PDMS 5:1 mixture of monomer (GE RTV 615 component A) and hardener (GE RTV 615 component B) was then poured onto the mold placed in a Petri dish and left at room temperature for 15 min to degas. PDMS (a 20:1

Figure 16.1 Schematic of microfluidic valve. (A) Two-dimensional view of superimposed perpendicular fluidic and control channels. (B) Schematic of a section of a microfluidic valve: a thin PDMS membrane separates the control and fluidic channel. This membrane will be deflected and act as a valve when the control channel is pressurized.

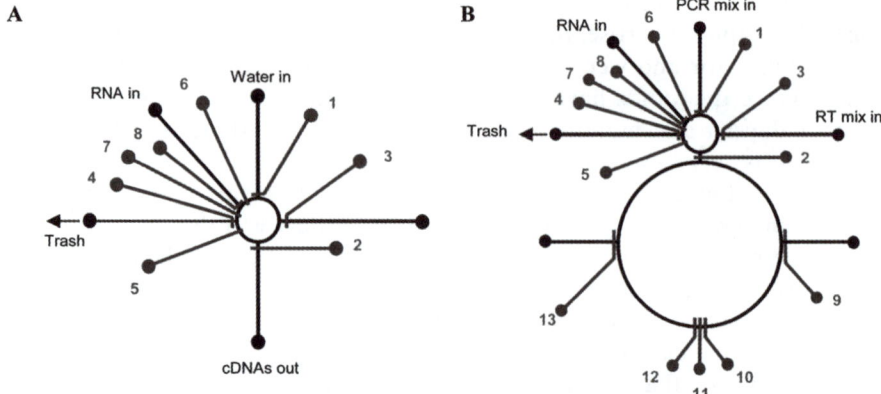

Figure 16.2   Schematics of the microfluics devices. Control channels (10 μm high) are in gray, fluidic channels (15 μm high), in black. (A) For reverse transcription of RNA samples, a 10 nL rotary mixer (2 mm wide ring) was filled up with a mix containing all the reagents except primers and RNA. RT was then performed for 20 min at 37 °C on a flat bed thermocycler while reagents were continuously mixed in the loop by peristaltic pumping using valves 3, 5, and 7. Once the reaction was achieved, cDNAs were collected out by flushing the ring with water. (B) For RT-PCR, an additional 50 nL rotary mixer (larger 10 mm ring) was added to the previous device (A). The RT reaction was performed as described, then cDNAs and PCR reagents were transferred into the 50 nL ring. Reagents were peristaltically pumped using valves 10, 11, and 12 and PCR amplification was performed by cycling the whole chip on a flat bed thermocycler. Amplified cDNAs were then collected out by flushing the ring with water.

mixture) was spin-coated at 2500 rpm for 60 s onto the mold for making the associated control channels. The two layers were cured for 45 min at 78Foot_MrkFoot_MrkSYMBOL 108 °C. The layer for the fluidic channel was then aligned to that for the control channel. The two-layer device was cured overnight. Holes for fluidic and control channels were then punched. The device was eventually sealed onto a pre-cleaned glass slide after a 1 minute plasma treatment and left overnight at 78 °C. To avoid any cross-contamination, microchips were single-used.

16.3   Integrating Reverse Transcription on a Chip

Transcriptome profiling of single cells requires first the capture and lysis of the cell, then the reverse transcription of mRNAs into cDNAs, followed by an amplification step of all cDNAs. In most of the protocols,[15,16,20] these reactions are conducted in standard microtubes, in volumes as large as 5 to 10 μL. The initial genetic material of the cell, contained in a few picoliters, is thus diluted a million times to be analyzed. However, more sensitive approaches have also been reported, such as the protocol developed by Klein et *al.*[2] which allowed for both mRNA and genomic DNA profiling, with a much greater sensitivity. This gain in sensitivity may be due to the fact that all the reactions were performed on paramagnetic beads, thus in a much higher local RNA concentration. To test whether increasing the RNA/cDNA concentration during RT-TSPCR would indeed improve the efficiency of reactions, we decided to perform those reactions in nanoliter volumes, *i.e.* in much smaller volumes than previously reported.

16.3.1 Gene Expression Profiling of Single-Cell Scale Amounts of RNA

We first focused on the integration of the RT reaction into a nanoliter rotary mixer as this step is critical for further gene expression analysis. Indeed, mRNAs are converted to cDNAs by RT and the genetic information lost at this point cannot be recovered: the PCR step can only amplify the cDNAs produced. Also, it is well known that the reverse transcriptase has a low efficiency (around 10–20%), even in the presence of relatively high amounts of mRNA. Last, it has been reported that this efficiency is directly proportional to the RNA concentration and thus even lower when transcripts are in low abundance[30–32] (which is the case when dealing with single cells).

Taking all these data into account, we first reverse transcribed 10 pg of purified mouse forebrain RNA (corresponding to the RNA content of a single cell) either in a microcentrifuge tube (10 μL) or in a rotary mixer (10 nL) (Figure 16.2A), all other concentration of reagents being the same.[9] Both cDNA samples were then amplified in tube by TS-PCR, in a 50 μL final volume. The overall efficiency of the reverse-transcription (RT) was assessed by analyzing the quantity and size of the amplified products, after electrophoretic separation on an agarose gel (Figure 16.3).

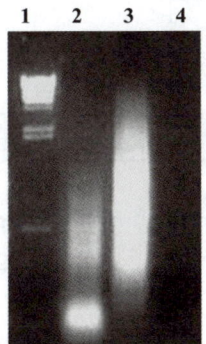

1. Lambda Hind III DNA digest
2. RT performed in tube (10 μl)
3. RT performed in the device (10 nl)
4. Negative control

Figure 16.3 Electrophoretic assessment of cDNAs obtained from 10 pg of RNA, after reverse transcription performed in a tube or in the rotary mixer followed by TS-PCR amplification.

We found that performing reverse transcription in a lab-on-a-chip helped to achieve full-length transcript conversion into cDNAs. Indeed, the size distribution of amplified cDNAs, spanning from about 100 bp to 6 kb, as expected for mammalian full-length cDNAs, was only observed for samples reverse transcribed in the rotary mixer. The overall yield of amplification was also increased as can be inferred from the smear intensities.

We next analyzed the whole transcriptome of the RT-TSPCR products generated from 10 pg of RNA, with the RT being conducted in tube or in the microfluidic device, using pangenomic DNA microarrays[33] containing 25 000 mouse genes. Gene expression profiles obtained from 10 pg of RNA were compared to those determined without amplification, using 15 µg of the same RNA sample. Analyzing the microarrays intensities, we showed that the signals obtained with 10 pg of RNA processed in the rotary mixer were similar to those obtained using 15 µg of RNA (Figure 16.4), whereas they were five-fold lower when RT was performed in tube. Consequently, the number of transcripts detected was increased from 719 to 12 632 (more than 17-fold) with our microfluidic approach compared to the conventional method. Transcriptome analysis proved tremendously difficult with 10 pg RNA reverse-transcribed in tube, as less than 4% of the genes expressed in the control experiments (15 µg of RNA without amplification) could be detected (Figure 16.5). However, our

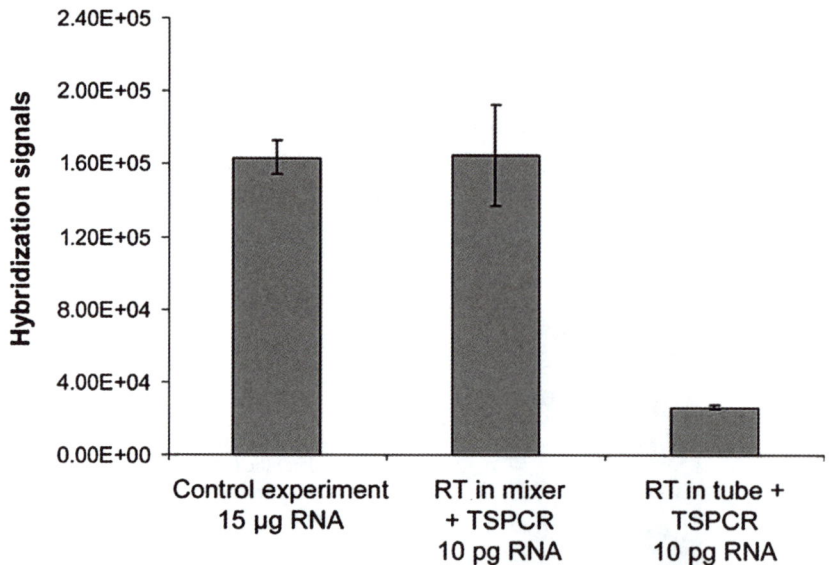

Figure 16.4   Sensitivity of the RT-TSPCR reaction when performed on 10 pg of total RNA. Hybridization signals (signal/background ratios) were obtained from the microarray experiments performed either from 15 µg of RNA (control experiment) or from 10 pg of RNA, after reverse transcription in the rotary mixer (microfluidic protocol) or in a tube (conventional protocol) and TS-PCR amplification.

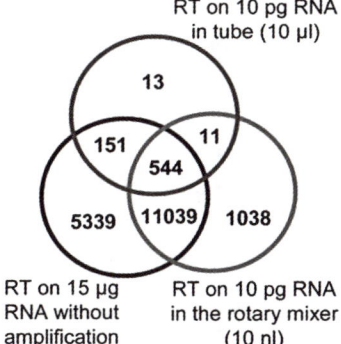

RT on 10 pg RNA
in tube (10 µl)

13

151

11

544

5339

11039

1038

RT on 15 µg
RNA without
amplification

RT on 10 pg RNA
in the rotary mixer
(10 nl)

Figure 16.5 Venn diagram illustrating the gene expression profiles obtained on microarrays. Comparison of the expression profiles obtained on 15 µg amounts of RNA (black circle) or at the single-cell level with RT being performed either in a tube (dark gray circle) or in a microfluidic device (gray circle).

microfluidic approach turned out to be extremely sensitive as it allowed for successfully detection of 74% of the control-identified transcripts.

These results validated our microfluidic system as they showed that RNA amounts equivalent to the ones of single cells could be amplified more efficiently and reproducibly by TS-PCR when the RT reaction was conducted in nanolitre rather than in microlitre volumes.

16.3.2 Gene Expression Profiling of Single Cells

We then applied our new microfluidic approach to the characterization of individual mouse neuronal progenitors from the caudal ganglionic eminence (CGE), a proliferative zone of the ventral telencephalon, at embryonic day 14 (E14). For these single cell experiments, the device designed for RT (Figure 16.2A) was modified to integrate a trapping cross, made of two channels and four microfluidic valves, and a second peristaltic pump, used to gently pump the cells into the microfluidic device (Figure 16.6).

Embryonic neuronal cells were dissociated from CGE explants and injected into the new RT rotary mixer. Four individual cells were successfully captured into the trapping cross (Figure 16.6B) and the RT reaction for each of these cells was performed on-chip as described above for RNA samples. The cDNAs were then amplified in a tube by TS-PCR and used for pan-genomic microarray experiments. As previously observed on total RNA, reverse transcription was more efficient and more accurate when performed in the 10 nL rotary mixer than in a tube (Figure 16.7). Gene expression profiles of the four individual cells were successfully obtained and consisted of around 5000 expressed genes (5035, 5173, 4712, and 4480, respectively). Their transcriptome analysis allowed for the discovery of new cell lineage specific markers as only 2445 genes were

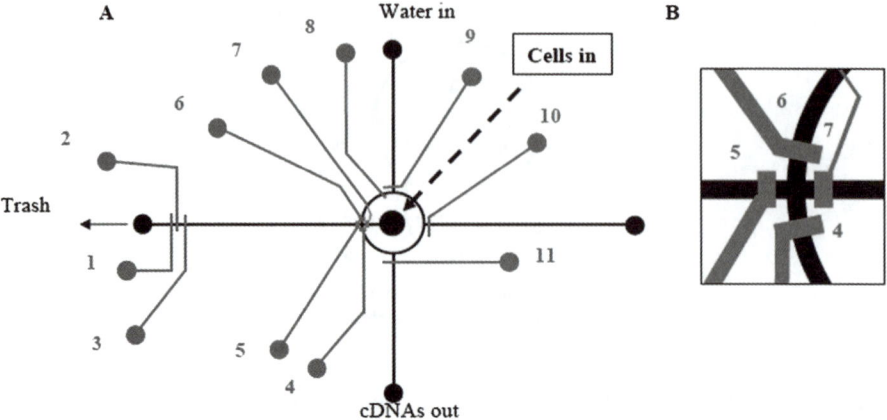

Figure 16.6 Schematic of the device used for single cells experiments. (A) Neuronal cells were injected in the "cells in" entry port and peristaltically pumped using valves 1, 2, and 3 while all valves except 4 and 6 were closed. Individual single cells were trapped between closed valves 4, 5, 6, and 7. Valves 8 and 10 were then opened and the RT reagents were injected into the rest of the 10 nL rotary mixer. The RT reaction was next performed for 25 min at 37 °C on a flat bed thermocycler while reagents were continuously mixed by peristaltic pumping using valves 6, 7, and 8. The cDNAs were collected out as described in Figure 2. (B) Close-up view of the cell trapping region of the device.

common between all the cells. This result indicates a high degree of cellular heterogeneity in the caudal ganglionic eminence and emphasizes the necessity to work at the single cell level.

In conclusion, using our microfluidic approach for RT coupled to TS-PCR amplification in tube, we were able to successfully obtain the transcriptome of all the cells analyzed. This represents a significant improvement in the single cell analysis field since conventional protocols in tube reported only 18% success[20] (16 cells over 90 analyzed).

16.4 Amplifying the Transcriptome on a Chip

In parallel, we investigated the integration of both RT and PCR steps on a single chip. A rotary mixer for PCR was made five times larger than the one for RT to allow further integration of RT and PCR (Figure 16.2B). Indeed, RT mixture is known to inhibit the PCR efficiency,[34,35] but a dilution to the fifth of RT reagents prior to PCR was proven experimentally sufficient to prevent most of this inhibition.

We focused on integrating the PCR amplification step into the large rotary mixer, studying the amplification of short (250 bp) and long (6 kb) DNA fragments from single cell amounts of DNA to ensure that all sizes of cDNA could be amplified in the rotary mixer. Our first experiments indicated that a third layer of PDMS channels should be added to our two-layer PDMS device to prevent evaporation during the denaturing phase of PCR. A hydration layer was thus designed and integrated (Figure 16.8).

Starting from 0.16 pg of lambda phage DNA, *i.e.* a quantity comparable to a single cell cDNA content, we first investigated the amplification of a 250 bp and a 6 kb long DNA fragments in the three-layer rotary mixer.[9] Using quantitative

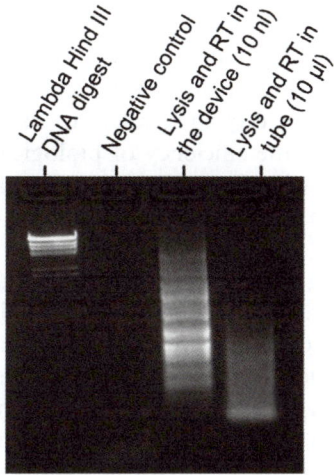

Figure 16.7 Electrophoretic assessment of cDNAs synthesized by RT-TSPCR from an individual neuronal cell. The cDNAs were obtained after lysis and RT performed in the rotary mixer or in a tube followed by TS-PCR amplification, starting from an individual neuronal cell.

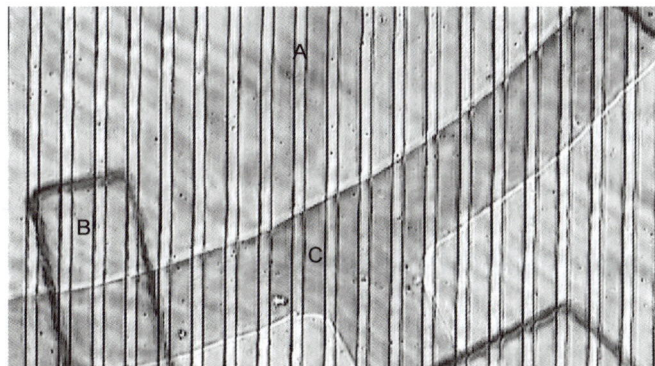

Figure 16.8 Picture of our three-layer PDMS device for PCR amplification. The two top layers form the rotary mixer (B and C). The bottom layer (A) is a hydration channel (30 μm wide and 15 μm high to prevent evaporation in the mixer during PCR).

PCR (qPCR), we demonstrated the successful amplification of these two fragments in the rotary mixer (Figure 16.9A).

Once the PCR had been validated, we performed the reverse transcription and amplification of two highly expressed genes (GAPDH and HPRT) from 10 pg of RNA. Analysis by qPCR showed specific amplification of both genes, indicating that our RT-PCR approach allowed for single-cell gene analysis (Figure 16.9B). We then assessed the efficiency of whole transcriptome amplification (RT-TSPCR) in the device, starting from 10 pg of RNA. Surprisingly, we did not observe any amplification when analyzing the sample after electrophoretic separation on agarose gel or after qPCR.

Our main hypothesis to explain this result is a "molecular crowding" effect. In the cell, macromolecules (proteins and RNA) occupy an important part of the cytoplasm volume. Molecular crowding refers to the effect of high concentrations of macromolecules on the efficiency of biological reactions. This effect is quite complex because crowding increases thermodynamic activities and at the same time reduces diffusion of the macromolecules, leading to an increase of the viscosity of the medium.[36–38] The net result of these two opposite effects depends on the size of the crowding and the type of reaction, and is not linear with the concentration of molecules. Indeed, regarding the DNA polymerase activity, it has been reported that molecular crowding might enhance the Taq processivity at low concentration and then decrease it at higher rates.[37,39,40] Therefore, in our RT-TSPCR experiment on-chip, we could consider that the absence of amplification would be related to this effect (*i.e.* too high concentration of molecules).

To investigate this hypothesis, we performed various TS-PCR amplifications in tube starting with the same quantity of cDNAs but using a range of dilutions between RT and TS-PCR reaction volumes. The influence of molecular crowding on TS-PCR efficiency was then assessed by comparing the profiles of

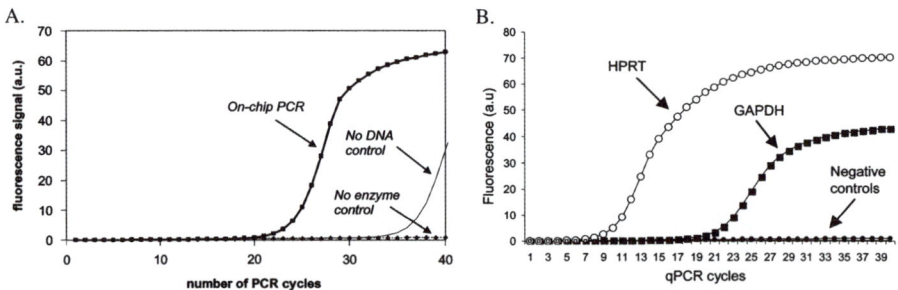

Figure 16.9   Quantitative PCR assessment of the efficiency of PCR amplification in the microfluidic device. (A) PCR amplification of a 6 kb long DNA fragment on-chip. Negative controls are respectively shown by black circles (no enzyme) and the solid black line (no DNA). (B) On-chip RT-PCR of single genes starting from 10 pg of RNA. RT was performed in the rotary device and GAPDH (white circles) or HPRT (black squares) fragments were then amplified on-chip with 40 cycles of PCR. Negative controls (no PCR amplification) are respectively shown by black circles and the solid black line.

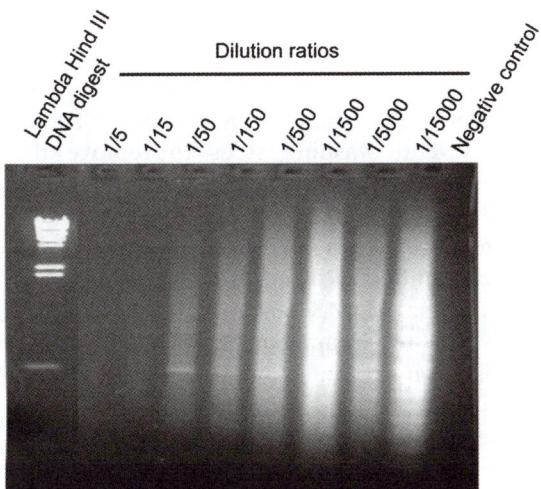

Figure 16.10 Electrophoretic assessment of molecular crowding effect. TS-PCR amplification was conducted on the same quantity of cDNAs but with different dilution ratios between RT and PCR volumes.

PCR products obtained on agarose gel (Figure 16.10, where the dilution ratio = RT volume/final volume of the PCR reaction). When the RT mixture was diluted only five-fold, which corresponds to the condition tested on-chip, amplification of the cDNAs was too poor to be detected. The TS-PCR seemed to be really efficient when the dilution factor of RT products was at least 500. Thus, the reaction volume for TS-PCR should be at least 5 µL when performing RT in a 10 nL rotary mixer.

These results indicate that the PCR reactor has to be optimized for single cell whole transcriptome analysis to avoid inhibition by molecular crowding. In particular, the geometry has to be modified to increase the reaction volume to a few microliters while ensuring good mixing of the cDNAs and PCR reagents. Another great improvement would be to integrate temperature control on-chip for RT and TS-PCR reactions so as to fine tune reaction times and cycling conditions.

16.5 Detecting the Transcriptome on a Chip

16.5.1 Microfluidics and Conventional Microarrays

To obtain an integrated system for single cell transcriptome analysis, one should integrate gene expression profiling on chip. In this regard, DNA microarrays have proved to be an extremely powerful tool as commercial arrays allow the analysis of the whole transcriptome with a density close to 20 000 transcripts cm^{-2} (http://nimblegen.com, http://affymetrix.com).

Using microarray technology to detect gene expression in a microfluidic device would require the integration of the following steps after TS-PCR: (1) labeling of the cDNAs with a fluorescent dye followed by a purification step to remove the non-incorporated dyes, (2) hybridization with the gene-specific probes, (3) several washing steps to remove the non-hybridized cDNAs, and (4) detection of the specific hybridization signals by fluorescence readout. The main difficulty would lie in the purification step as this proved quite complex to implement in a lab-on-a-chip. Unfortunately, this step is crucial and cannot be omitted as it ensures the removal of fluorescent dyes that could adsorb on the PDMS surface and generate high background noise. One way, however, to circumvent this problem would be to detect the probe–target hybridization in real time or to use a label-free detection approach.

Regarding the detection of cDNAs, it should be noted that moving labeled cDNAs along a serpentine channel above the spots of microarrays not only allows but also increases the hybridization kinetics. One way to improve the detection step could thus be to reduce the microarray surface for faster hybridization kinetics. This could be done, for example, by using the micro-fabricated fountain pens developed by Reese and colleagues[41] that allow for printing arrays with a density as high as $25\,000$ spots cm^{-2}.

16.5.2 Microarray Development Using DNA Immobilization onto Microchannels

Many groups have been working on integrating microarrays into microfluidics devices[42] as these provide the ability to work with smaller volumes while reducing reaction time and reagents consumption. Various procedures for immobilization of oligonucleotide probes onto microchannels have been reported. Soper and co-workers described protocols to covalently attach amino-modified oligonucleotides in fluidic channels using UV patterning and micro-printing, which could achieve medium density microarrays (thousands to tens of thousands of probes).[43,44] In 2001, Lee and colleagues described a new method for fabricating DNA arrays.[45] They first created a "one-dimensional (1D) DNA line array" with parallel microchannels PDMS device attached to a modified gold surface. 5'-Thiol-modified DNA probes injected in individual channels reacted with the gold surface. After removing the PDMS mold, a second layer of channels was then created perpendicularly to the 1D DNA line array to deliver the targets. This led to a "2D hybridization array" in which the target–probe hybridization could be detected by surface plasmon resonance imaging. As a proof of concept, they used this method to measure 20 fmol of *in vitro* transcribed RNA from *Arabidopsis thaliana* in volume of 1 μL. However, the main limitation of this system was the small number of DNA probes (< 100 lines) that could be part of the 1D DNA line array. More recently, Canadian researchers improved this system using electrowetting-on-dielectric (EWOD) digital microfluidics, a technique that allows manipulation of reagents in individual droplets[46] (see Chapter 17 for more details on EWOD). DNA probes were dispensed on the gold surface using droplets and covalently attached under an electric field. Targets were then sequentially injected into the channels, hybridization was achieved in 8 min and specific signals were measured by surface plasmon resonance imaging. This technology looks sensitive and interesting as it offers a label-free and real-time detection system. However, it should be improved for whole transcriptome analysis to allow the immobilization of 25 000 specific probes and measurement of hybridization signals with their complementary targets. Another promising system has been developed by Suter *et al.* using a liquid core optical ring resonator (LCORR) sensor.[47] This platform[48] integrates microfluidics and photonics for detection of low DNA concentration into nanoliter volumes. Amino-modified DNA probes (25–mer, 50–mer or 100–mer) can be attached on the activated surface of a circular channel, the label-free targets are then injected through the capillary and can hybridize to their complementary stands. The DNA interaction at the surface of the ring resonator can be measured and quantified by photonic sensing technology. The authors demonstrated that the limit of detection is lower than 10 pM. This system takes advantage of the high detection sensitivity of the ring resonator and the increase of hybridization kinetics due to small reaction volume and large surface-to-volume ratio into the circular fluidic channel. It has to be optimized for DNA microarray development but it is estimated that thousands of ring resonators can be packed onto a $1 \, cm^2$ chip.

16.5.3   Towards Transcriptome Analysis in the Liquid Phase

Another possible design for an integrated platform to analyse gene expression would be to detect DNA–RNA hybridization in solution as this would favor the interaction probe–target and therefore increase the reaction kinetics. Different groups have been working on interesting approaches for transcriptome analysis using microfluidic bead arrays for rapid DNA detection and quantitative gene profiling.[49–51] These methods rely on the immobilization of specific DNA probes on microbeads through biotin–streptavidin linkage. These functional beads are injected into microfluidic channels and trapped into microchambers. When labeled targets flow through the channels, they can hybridize to their complementary strands and the quantification is done by real-time measurement of fluorescence using a CCD camera. This technology takes advantage of high density probes on the bead surface that enables kinetics and sensitivity improvement.

Finally, the technology developed by Han and co-workers in 2001[52] could be extremely interesting for whole transcriptome analysis on chip. This approach is based on the optical properties of quantum dots (QDs). QDs are nanocrystal fluorescent semiconductors highly luminescent, stable against photobleaching and their emission wavelengths can be tuned by changing their size. Han *et al.*[52] packed multicolor QDs into microbeads to create a nanobarcode: using six different QDs with 10 intensity levels for each should create a barcode with a million possibilities. Using only three QDs, the authors demonstrated the hybridization of labelled target DNA with specific probes linked onto the beads. As illustrated by the results obtained by Eastman and colleagues in 2006, this approach could be used for accurate gene expression profiling.[53] Four different QDs with 12 levels of intensity for each were used to generate barcoded microbeads: each bead is coated with a combination of the four QDs allowing identification by a specific spectral nanobarcode and is conjugated with a specific DNA probe. Biotinylated targets are synthesized from RNA samples and hybridized to the probes on the beads. The authors used another QD linked to streptavidin (that can interact with the biotin of the target) as a reporter for gene expression quantification. This system provides simultaneously gene identification by reading the nanobarcode and quantification by measuring the fluorescence level of the streptavidin-conjugated QD. Using this method, they analysed the variations in the expression of around 100 genes in TGFα-treated cells versus untreated. The results obtained are in agreement ($R = 0.86$) with those obtained for the the same experiments done on Affymetrix microarrays.

This approach proved highly sensitive and offers the advantages of increased hybridization kinetics and reduction of sample and reagents consumption. All this makes it a potential method of choice for whole trancriptome analysis in integrated microfluidic platform.

16.6 Some Practical Conclusions

We have demonstrated that is now possible to generate gene expression profiles of single cells using microfluidic devices in which we perform the first step, *i.e.* the reverse transcription of RNA into cDNA. If we perform the reverse transcription in microliter volumes using standard techniques instead of nanoliters with microfludic devices gene expression profiles are not complete and reliable. We have discussed the possibility of integrating the other reactions in order to integrate all steps on the same lab-on-chip. This would also allow include a better quantification of gene expression. At this point using our PDMS device for reverse transcription and TS-PCR for amplification we get a more qualitative than quantitative view of gene expression levels.

The next generation of microfluidic device for single cell whole transcriptome analysis might be based on digital microfluidics that could bring high throughput to our technology. Indeed we recently showed that reverse transcription can be performed in droplets with the same efficiency than in rotary mixer. This would be the way towards analysis of a high number of cells in on experiment (at least 96) starting from a few thousands of cells. This would pave the way to many applications in medical applications and diagnosis.

References

1. E. S. Kawasaki, *Ann. N. Y. Acad. Sci.*, 2004, **1020**, 92.
2. C. A. Klein, S. Seidl, K. Petat-Dutter, S. Offner, J. B. Geigl, O. Schmidt-Kittler, N. Wendler, B. Passlick, R. M. Huber, G. Schlimok, P. A. Baeuerle and G. Riethmuller, *Nat. Biotechnol.*, 2002, **20**, 387.

3. H. Andersson and A. van den Berg, *Lab Chip*, 2006, **6**, 467.
4. M. A. Northrup, M. T. Ching, R. M. White and R. T. Watson, *Proceedings of the 7th International Conference on Solid State Sensors and Actuators, Yokohama, Japan*, 1993, p. 924.
5. P. J. Obeid, T. K. Christopoulos, H. J. Crabtree and C. J. Backhouse, *Anal. Chem.*, 2003, **75**, 288.
6. N. R. Beer, B. J. Hindson, E. K. Wheeler, S. B. Hall, K. A. Rose, I. M. Kennedy and B. W. Colston, *Anal. Chem.*, 2007, **79**, 8471.
7. J. S. Marcus, W. F. Anderson and S. R. Quake, *Anal. Chem.*, 2006, **78**, 3084.
8. L. Warren, D. Bryder, I. L. Weissman and S. R. Quake, *Proc. Natl Acad. Sci. U. S. A.*, 2006, **103**, 17807.
9. N. Bontoux, L. Dauphinot, T. Vitalis, V. Studer, Y. Chen, J. Rossier and M. C. Potier, *Lab Chip*, 2008, **8**, 443.
10. H. P. Chou, M. A. Unger and S. R. Quake, *Biomed. Microdev.*, 2001, **3**, 323.
11. J. W. Hong, V. Studer, G. Hang, W. F. Anderson and S. R. Quake, *Nat. Biotechnol.*, 2004, **22**, 435.
12. R. N. Van Gelder, M. E. von Zastrow, A. Yool, W. C. Dement, J. D. Barchas and J. H. Eberwine, *Proc. Natl Acad. Sci. U. S. A.*, 1990, **87**, 1663.
13. L. Petalidis, S. Bhattacharyya, G. A. Morris, V. P. Collins, T. C. Freeman and P. A. Lyons, *Nucleic Acids Res*, 2003, **31**, e142.
14. I. Tietjen, J. Rihel and C. G. Dulac, *Int. J. Dev. Biol.*, 2005, **49**, 201.
15. S. Gustincich, M. Contini, M. Gariboldi, M. Puopolo, K. Kadota, H. Bono, J. LeMieux, P. Walsh, P. Carninci, Y. Hayashizaki, Y. Okazaki and E. Raviola, *Proc. Natl Acad. Sci. U. S. A.*, 2004, **101**, 5069.
16. K. Kurimoto, Y. Yabuta, Y. Ohinata and M. Saitou, *Nat. Protoc.*, 2007, **2**, 739.
17. K. B. Jensen and F. M. Watt, *Proc. Natl Acad. Sci. U. S. A.*, 2006, **103**, 11958.
18. K. Kurimoto, Y. Yabuta, Y. Ohinata, Y. Ono, K. D. Uno, R. G. Yamada, H. R. Ueda and M. Saitou, *Nucleic Acids Res.*, 2006, **34**, e42.
19. B. Seshi, S. Kumar and D. King, *Blood Cells Mol. Dis.*, 2003, **31**, 268.
20. I. Tietjen, J. M. Rihel, Y. Cao, G. Koentges, L. Zakhary and C. Dulac, *Neuron*, 2003, **38**, 161.
21. A. Kawaguchi, T. Ikawa, T. Kasukawa, H. R. Ueda, K. Kurimoto, M. Saitou and F. Matsuzaki, *Development*, 2008, **135**, 3113.
22. K. Kurimoto, Y. Yabuta, Y. Ohinata, M. Shigeta, K. Yamanaka and M. Saitou, *Genes Dev.*, 2008, **22**, 1617.
23. K. Kurimoto, M. Yamaji, Y. Seki and M. Saitou, *Cell Cycle*, 2008, **7**, 3514.
24. M. Yamaji, Y. Seki, K. Kurimoto, Y. Yabuta, M. Yuasa, M. Shigeta, K. Yamanaka, Y. Ohinata and M. Saitou, *Nat. Genet.*, 2008, **40**, 1016.
25. N. N. Iscove, M. Barbara, M. Gu, M. Gibson, C. Modi and N. Winegarden, *Nat. Biotechnol.*, 2002, **20**, 940.
26. T. Subkhankulova and F. J. Livesey, *Genome Biol.*, 2006, **7**, R18.
27. H. A. Stone, A. D. Stroock and A. Ajdari, *Annu. Rev. Fluid Mech.*, 2004, **36**, 381.

28. M. A. Unger, H. P. Chou, T. Thorsen, A. Scherer and S. R. Quake, *Science*, 2000, **288**, 113.
29. V. Studer, A. Pepin, Y. Chen and A. Ajdari, *Analyst*, 2004, **129**, 944.
30. J. Curry, C. McHale and M. T. Smith, *Biotechniques*, 2002, **32**, 770.
31. J. P. Levesque-Sergerie, M. Duquette, C. Thibault, L. Delbecchi and N. Bissonnette, *BMC. Mol. Biol.*, 2007, **8**, 93.
32. A. Stahlberg, M. Kubista and M. Pfaffl, *Clin. Chem.*, 2004, **50**, 1678.
33. K. Le Brigand, R. Russell, C. Moreilhon, J. M. Rouillard, B. Jost, F. Amiot, V. Magnone, C. Bole-Feysot, P. Rostagno, V. Virolle, V. Defamie, P. Dessen, G. Williams, P. Lyons, G. Rios, B. Mari, E. Gulari, P. Kastner, X. Gidrol, T. C. Freeman and P. Barbry, *Nucleic Acids Res.*, 2006, **34**, e87.
34. D. P. Chandler, C. A. Wagnon and H. Bolton Jr., *Appl. Environ. Microbiol.*, 1998, **64**, 669.
35. L. N. Sellner, R. J. Coelen and J. S. Mackenzie, *Nucleic Acids Res.*, 1992, **20**, 1487.
36. N. A. Chebotareva, B. I. Kurganov and N. B. Livanova, *Biochemistry (Moscow)*, 2004, **69**, 1239.
37. R. J. Ellis, *Trends Biochem. Sci.*, 2001, **26**, 597.
38. D. Miyoshi and N. Sugimoto, *Biochimie*, 2008, **90**, 1040.
39. R. R. Lareu, K. S. Harve and M. Raghunath, *Biochem. Biophys. Res. Commun.*, 2007, **363**, 171.
40. Y. Sasaki, D. Miyoshi and N. Sugimoto, *Biotechnol. J.*, 2006, **1**, 440.
41. M. O. Reese, R. M. van Dam, A. Scherer and S. R. Quake, *Genome Res.*, 2003, **13**, 2348.
42. C. Situma, M. Hashimoto and S. A. Soper, *Biomol. Eng.*, 2006, **23**, 213.
43. M. Hashimoto, F. Barany and S. A. Soper, *Biosens. Bioelectron.*, 2006, **21**, 1915.
44. C. Situma, Y. Wang, M. Hupert, F. Barany, R. L. McCarley and S. A. Soper, *Anal. Biochem.*, 2005, **340**, 123.
45. H. Jin-Lee, T. T. Goodrich and R. M. Corn, *Anal. Chem.*, 2001, **73**, 5525.
46. L. Malic, T. Veres and M. Tabrizian, *Biosens. Bioelectron.*, 2009, **24**, 2218.
47. J. D. Suter, I. M. White, H. Zhu, H. Shi, C. W. Caldwell and X. Fan, *Biosens. Bioelectron.*, 2008, **23**, 1003.
48. I. M. White, H. Oveys and X. Fan, *Opt. Lett.*, 2006, **31**, 1319.
49. J. Wen, X. Yang, K. Wang, W. Tan, L. Zhou, X. Zuo, H. Zhang and Y. Chen, *Biosens. Bioelectron.*, 2007, **22**, 2759.
50. M. F. Ali, R. Kirby, A. P. Goodey, M. D. Rodriguez, A. D. Ellington, D. P. Neikirk and J. T. McDevitt, *Anal. Chem.*, 2003, **75**, 4723.
51. J. Kim, J. Heo and R. M. Crooks, *Langmuir*, 2006, **22**, 10130.
52. M. Han, X. Gao, J. Z. Su and S. Nie, *Nat. Biotechnol.*, 2001, **19**, 631.
53. P. S. Eastman, W. Ruan, M. Doctolero, R. Nuttall, G. de Feo, J. S. Park, J. S. Chu, P. Cooke, J. W. Gray, S. Li and F. F. Chen, *Nano Lett.*, 2006, **6**, 1059.

CHAPTER 17
Tiny Droplets for High-throughput Cell-based Assays

J.-C. BARET AND V. TALY

Institut de Science et d'Ingénierie Supramoléculaires (ISIS), Université de Strasbourg, CNRS UMR 7006, Strasbourg, France

Abstract

In order to perform a high number of assays on biological objects or chemical compounds which are sometimes impossible to obtain in large quantities, miniaturizing experiments has been the main issue. Standard technologies, such as microtiter plates have nearly reached the smallest sizes of samples they can handle. In order to miniaturize the assays, a further decrease in the size of the reactors in which reactions are performed is required. The use of small droplets provides new ways to miniaturize assays; moreover droplets are ideal for compartmentalizing biological objects or compounds. Droplets of equal sizes are produced in series and manipulated on demand providing quantitative and miniaturized versions of the microtiter plate assays and enables the parallelization of the assay.

In this chapter we will review droplet-based microfluidics: droplet production, flow, fusion, sorting, and detection. We will discuss how to design reactions in droplets and particularly how to maintain living cells in droplets and run biological reactions.

RSC Nanoscience & Nanotechnology No. 15
Unravelling Single Cell Genomics: Micro and Nanotools
Edited by Nathalie Bontoux, Luce Dauphinot and Marie-Claude Potier
© Royal Society of Chemistry 2010
Published by the Royal Society of Chemistry, www.rsc.org

17.1   Introduction

In chemistry and biology, assays have always been performed by manipulation of finite volumes of reagents in well-controlled environments. Indeed from the test tubes of the nineteenth century to the microtiter plates of the pharmaceutical industries, reagents are brought together by dripping or pipetting one into another before being mixed, and the evolution of sciences and technologies has pushed the assays to controlled containers and always smaller sizes. This quest of miniaturization is mainly governed by the need to perform an increasing number of assays on biological objects which are sometimes impossible to obtain in large quantities (*e.g.* stem cells) or on chemical compounds produced by the pharmaceutical industries in small quantities for economic reasons. Nowadays, the technology used to assay, *e.g.* by the pharmaceutical industry, enables about one test per second to be performed on volumes of the order of 1 µL.

Although this volume is relatively small – 1 µL of water represents a droplet of about 1 mm diameter – it is huge when biological material is involved. As an example, considering cell cultures, with a typical density of about 10^7 cells per mL, an assay performed on 1 µL represents the averaged response of 10^4 cells. Under those conditions, the presence of an individual with specific characteristics will be hidden in the statistical response of the population. Of course, by simple dilutions, one cell per microliter is achievable but the cell density becomes so low that the concentrations of molecules released by the single cell are very small. Alternatively the amount of reagents to be used for millions of tests of 1 µL each becomes the bottleneck to perform the assays with reasonable costs. One microliter is also huge compared to the sizes at which biological reactions occur in nature. Indeed a bacterium cell such as *Escherichia coli* has a typical size of about 1–2 µm, which represents a volume of a few femtoliters. Therefore several orders of magnitude in volumes are available to miniaturize biological assays.

Standard technologies, such as microtiter plates have nearly reached the smallest sizes of samples they can handle. In order to miniaturize the assays, a further decrease of the size of the reactors in which reactions are performed is required. The vessels must keep their content independent from each other, and this content must be accessible to the operator on demand. Simple tasks have to be performed, such as mixing of reagents or dilutions, aliquoting or detection and selection of specific variants encapsulated in the vessel.

The use of small droplets provides new ways to miniaturize assays. Indeed, water droplets – liquid structures with perfectly defined boundary – are ideal for compartmentalizing compounds. This well-defined interface is the result of the cohesion of the fluid structure linked to the interactions of the fluid molecules.[1] Production and manipulation of droplets will require the ability to overcome on-demand the cohesion of the fluid structure. Droplets are produced from a bulk solution by the use of various forces (see, for example, the review by Basaran[2]): gravity in the simple case of dripping, another interfacial force in the case of spotting or microcontact printing, inertia in the case of inkjet printing,[3,4] electrostatic force[5] in the case of electrospraying, shear forces in emulsion production.[6] Practically, droplet manipulation is also influenced by surface energies linked to the fluid cohesion. In the absence of other energies, the minimization of energy gives a single solution for a two-phase system: one single large droplet. An ensemble of small liquid structures is not an equilibrium solution and has to be stabilized. This is achieved by separating the small structures spatially (*e.g.* for arrays spotted with a large distance between the droplets) or by the use of a physical barrier such as walls (*e.g.* the wells of a microtiter plate). A physicochemical way to stabilize the system is to add surfactant molecules which prevents droplet coalescence. This approach enables foams (liquid–gas interfaces) and emulsion (liquid–liquid interfaces) stabilization.[6] An approach based on emulsions also permits to reduce evaporation. Evaporation of small liquid structures – sometimes as fast as the typical timescale of a biochemical assay[7] – modifies the concentration of the species and impacts the assay. By immersing the droplet in a saturated environment,[8] evaporation can be reduced. Alternatively, the immersion of the droplets in an immiscible phase, coupled to the use of surfactant molecules – the basis of emulsions[6,8] – enables the production of independent micro-reactors which leads to the first attempts to miniaturize assays in micrometer-sized droplets.[9] Over the past few years the combination of reactions in droplets and controlled microfluidic droplets manipulation has provided new ways of performing quantitative assays at a high-throughput. Droplets of equal sizes are produced in series and manipulated on demand providing quantitative and miniaturized versions of the microtiter plate assays and enables the parallelization of the assay. A series of operations can be performed, just as they are on microtiter plate: splitting of liquid from a reservoir in different compartments, mixing with another reagent, detection and selection of "hits", namely organisms, small molecules or macromolecules with specific desired properties.

17.2 Droplet-based Microfluidics

Droplet-based microfluidic systems enable the control of droplets in or on microfluidic chips through several systems. Among them, the most versatile and flexible solutions are digital microfluidics, a name originally given to the manipulation of droplet using electrowetting on dielectrics (EWOD) and droplet-based microfluidics in microchannels, which is the two-phase flow variation of the standard and well established microfluidics technology.[10,11] These two systems function at different level of operation.

17.2.1 EWOD and "Digital Microfluidics": Tools for High-content Screening

Digital microfluidics is the main example of an array-based manipulation of droplets. Droplets are actuated on a chip that can be configured for different tasks almost in real time[12] (for a review, see Abdelgawad and Wheeler[13]). The technique proves to be very efficient for the manipulation of a relatively small number of droplets in parallel on a surface. The actuation mechanism is based on the local modulation of the wetting properties of a surface by the electrowetting effect.[14–16] Indeed droplets move in gradients of wettability towards the most wettable surface.[17] Here the microfluidic chip used is highly engineered with patterned electrodes, on top of which a dielectric insulating layer is deposited. An additional hydrophobic surface coating is usually added and counter electrodes are positioned to contact the droplet or a second plate closes the system (Figure 17.1A). Finally, the system is sealed to immerse the droplet in oil and reduce contact angle hysteresis. By patterning the surface with arrays of electrodes, the wetting properties of the surface are modulated in space and time in order to move droplets from one electrode patch to the other.[18] This technique enables a series of operations on droplets such as production of small droplet from a reservoir, splitting of droplets in two parts, merging of droplets and droplets actuation[18,19] (Figure 17.1B and C). Extensions or variants of this technique involve other means to manipulate droplets on planar surfaces, for example using dielectrophoresis,[20] magnetic actuation[21] or surface acoustic wave actuation.[22] In all of these cases, the system is interesting for the possibility to reconfigure the operation,[23] to run complex multi-step assays or to repeat multiple operations on a few droplets – ideal for high-content screening – but fails to handle billions of droplets in parallel or at high-throughput.

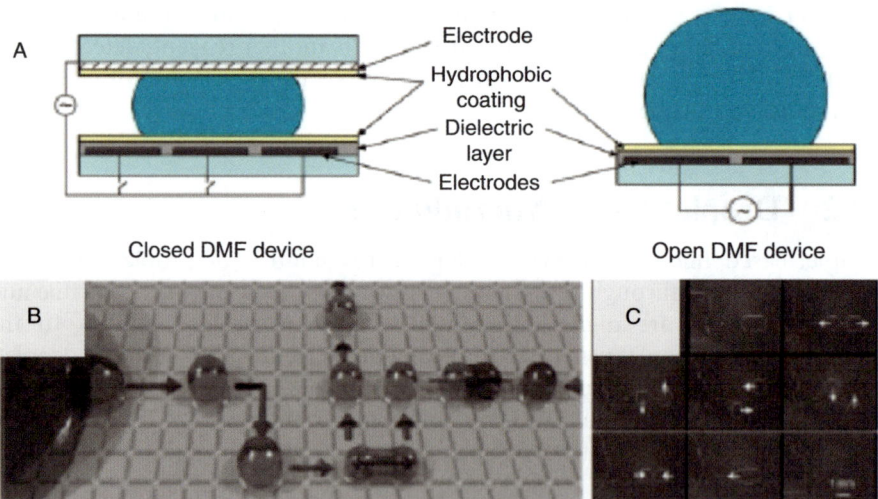

Figure 17.1   Digital microfluidic systems (DMF) for the manipulation of droplets. Droplets are actuated on open or closed surfaces patterned with electrodes (A). Using electrowetting on dielectric, droplets are produced, actuated, split, and fused on a surface patterned with electrodes (B and C) in an array-based system. Reprinted, with permission, from Abdelawad and Wheeler.[13]

17.2.2 Droplet-based Microfluidics: Tools for High-throughput Screening

Droplet-based microfluidics in microchannels is, on the other hand, the ideal tool for high-throughput screening. Over the past 15 years, a series of modules have been developed for droplet manipulation in microfluidic channels, at high-throughput ($\sim$1000 droplets per second, see Figure 17.2). The variety of materials in which these devices can be produced is extremely vast but the most widespread technique used is soft lithography which allows rapid prototyping[24] (see also Chapter 11). Channels are easily produced in poly(dimethylsiloxane) (PDMS) using replica molding of a mould of a photoresist (for example, SU8, microchem) deposited on a silicon wafers. Other materials such as poly(methyl methacrylate) (PMMA), glass or SU8 structures are also usable. Soft-lithography also enables the creation of complex three-dimensional structures and recent developments enabled the production of electrodes as microfluidic channels filled by a low melting temperature solder which improves the basic capabilities of the rapid prototyping method.[25] In these channels, droplet actuation is mainly predetermined by the flow of the droplets along those channels and includes droplet production and reinjection, droplet flow for incubation, droplet splitting, droplet pair fusion, droplet detection and droplet sorting. The droplets will flow in series along the channels and the same type of operation will be performed on each droplet. The integration of droplet-based microfluidic modules makes it a perfect tool to screen large libraries of variants containing small numbers of extraordinary variants,[26,27] typically interesting for high-throughput screening or directed evolution experiments. In the following we will present the progress on droplet manipulation in microchannels, usable for example for high-throughput applications. These modules can be split in two categories, passive modules and active modules. In passive modules, the flow of droplets is predetermined and controlled solely by the hydrodynamics. Three effects are involved: the mechanical pumping of the liquid, viscous dissipation and surface tension. Mechanical pumping is easily achieved using syringe pumps (controlled flow rate) or pressure driven system (controlled pressure). A summary of additional pumping mechanisms can be found, for example, in references 10 and 22. The interplay of channel geometry and the liquid physical characteristics (viscosity and surface tension of the fluid/fluid interface) controls the response of the fluid to the hydrodynamic forcing. In active modules, external forces such as electric or thermocapillary forces, are used to actuate single droplets or groups of droplets among the population of droplets. The combination of passive and active elements provides a high level of control for the operator to process droplets at high-throughput ($>$1000 droplets per second).

17.3 Generating and Manipulating Droplets

17.3.1 Droplet Production

Droplets are produced on-chip by co-flowing two immiscible fluids: the instability of the interface between the two fluids will lead to droplet

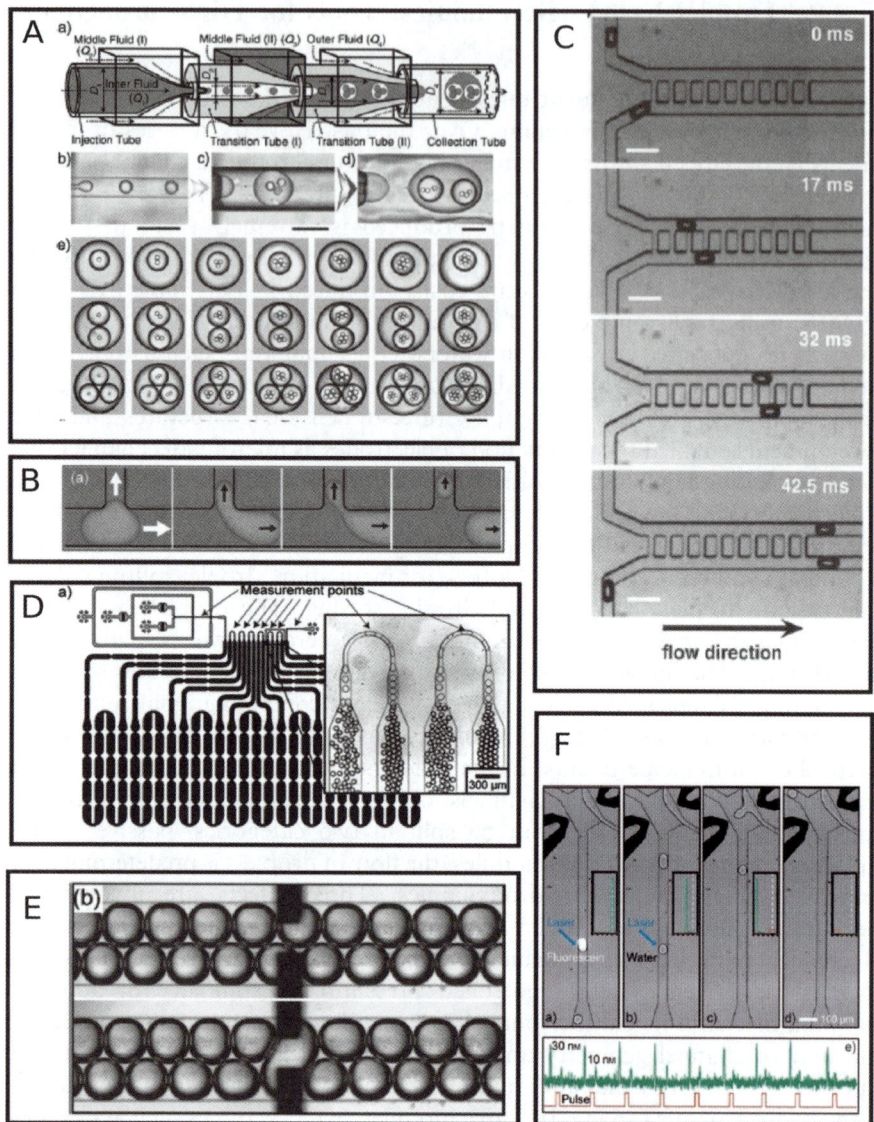

Figure 17.2   Manipulation of droplets in microfluidic channels. Droplets are produced as emulsions by coflow which also enables complex multiple
emulsions to be controlled (A).[42] The modules also enable droplet
splitting (B),[63] droplet synchronization (C),[51] droplet incubation (D),[67]
droplet fusion (E),[78] and droplet detection and sorting (F).[26]

production.[28] Three geometries are mainly used, T-junction,[29] steps,[30,31] and Y-
junction (also known as "flow-focusing junction").[32] Already in simple co-
flowing cylindrical geometries, several regimes of droplet production are
observed depending on the flow conditions and the dynamic response of the

interfaces to perturbations.[28,33,34] The so-called dripping regime leads to the most monodisperse droplet production with standard sizes of 10 µm to several hundreds of micrometers, at rates up to 10 kHz for the smallest droplets. In rectangular channels, the influence of geometry gives another level of complexity[35,36] by the increased number of parameters (*e.g.* aspect ratio of the channels) that can be varied. Besides these geometrical parameters, droplet production is controlled by the interfacial properties of the liquid–liquid and liquid–solid interfaces:[37] single droplets[32] are easily produced but the design of microfluidic junction and the control of wetting properties also enables pairs of droplets,[38,39] multiple emulsions[40–43] or foams[44] to be generated (Figure 17.2A). Finally, the use of external electric fields gives an additional level of control on droplet production, either using the electrowetting effect[45] or electrospraying-based technique.[46] This use of electric fields also enables the creation of charged droplets that can be further actuated in electric fields[47] and of droplets of sizes down to 1 µm[46] hardly accessible by other means. Additionally, the use of surfactants for droplet stabilization enables collection of droplets, for example for incubation of biochemical reactions.[27] Microfluidic modules enable the production of stable and remarkably monodisperse emulsions usable for specific applications but also for fundamental studies in emulsion sciences thanks to the perfect control on the emulsification process and the droplet characteristics (*e.g.* content, size, dispersity).[48,49]

17.3.2 Droplet Division

The flow of droplets at channel junctions enables droplets to be split apart. Such systems enable aliquots from a single mother droplet to be produced in a controlled way. The mechanism of breakup is linked to the destabilization of the interface induced by combined effect of the flow and the geometry of the channels[50–52] (Figure 17.2B). An additional level of control for this operation is achieved by the use of laser controlled break-up.[53] In this case, the volume of the droplets produced from the mother droplet can be actively tuned externally by laser power. This type of device will be ideal to aliquot droplets for example to make two types of assays on the same droplet and is also used to produce small droplets from a large mother droplet.

17.3.3 Droplet Flow, Droplet Synchronization, and Droplet Incubation

After production, droplets flow along the channels of the microfluidic chip. While in single-phase flow the hydrodynamic resistance is simply given by Poiseuille's law,[54] droplets modify this simple law:[55–58] they are a source of pressure drop depending on their size and volume fraction in the oil phase which modifies the pressure distribution and therefore the flow of the continuous oil stream. This results in collective behaviour[59] or coupling between channels that influences droplet production by pressure feedback,[38,56,60] direction[61] or synchronization[31,39,62,63] (Figure 17.2C), which in turn provides additional control of droplets flow. On a practical viewpoint, the flow of droplets along channels is used for incubation of chemical or biochemical reactions[64-67] and is perfectly compatible with high-throughput constraints: the flow of droplet reaches ~ 1000 droplets per second (Figure 17.2D). However, when the incubation requires the readout of complex information (*e.g.* kinetic data, growth rate measurement, imaging, spectroscopy) this system is hardly usable. In this case, droplets are trapped and immobilized in the channels for measurement in wells[68] or by controlling the filling of channel.[69] These systems are then hybrid between array-based systems and continuously flowing systems. Interestingly, besides these systems, other hybrid systems have been developed recently, coupling electrowetting-based droplet manipulation and droplet flow in microfluidic channels.[70] In the future, all of these systems should provide versatile and efficient droplet manipulation, both for high-throughput and high-content screening. They are complemented by active elements that enable direct and accurate manipulation of single droplets or groups of droplets.

17.3.3.1 Using Valves

A first way to act on droplet flow is to modify the geometry of the channels during the flow. Although this looks like a surprising possibility, it is possible when the microfluidic channels are produced in an elastomeric material such as PDMS. Indeed by applying pressure in channels, the channel walls can be deformed which has been used for a long time already in valve-based systems.[71,72] This method is now used to control the droplet production[73] or the orientation of groups of droplets.[74] Considering the time scale of actuation by this mechanism, linked to the mechanical response of the elastic material it cannot actuate single droplets at high-throughput (> 1 kHz) but rather groups of droplets. This approach will enable a fine-tuning of the droplets' behavior by the operator and therefore increase the level of control given by the technology.

17.3.3.2 Droplet Fusion, Coalescence, and Electrocoalescence

When droplets are not stabilized by surfactants, droplet coalescence is straightforward. It has therefore been used to initiate chemical reactions at well defined locations.[75] On the contrary, when droplets are stabilized by surfactants, the mixing of the reagents initially encapsulated in the different droplets has to be induced by an external force. This forcing can be achieved using wetting patches.[76] In this case a first droplet spreading on an hydrophilic patch is coalesced with another following droplet or a group of several droplets. Alternatively, it has been shown recently that geometrical constraints inducing separation of the droplet after a collision promotes droplet fusion.[48] Using such systems will therefore enable fusion of droplet pairs (Figure 17.2E). However systems that can be externally controlled are more suited: the operator can induce externally and on-demand the fusion of droplet pairs, for example to initiate the reactions only when the transient states linked to the starting or ending of an experiment are over. The use of electric fields is one of the most efficient ways to perform droplets coalescence. Electro-coalescence – based on the destabilization of the oil film between two water containers[66,77,78] – enables selective fusion of droplet pairs for example to initiate a chemical reaction between reagents present in the two droplets.[79] Although until now the electro-coalescence has been described only for large number of droplet pairs, it is possible to trigger the electric field on the detection of a specific droplet which could open the door to selective droplet fusion. Other systems have also been developed, for example using destabilization of the interface between two droplets by local heating by a laser focused in the channels.[53] To summarize, fusion of droplet pairs is an essential part for applications requiring several steps; the development of externally controlled modules for droplet fusion is a key to provide flexibility to the droplet-based microfluidic technology.

17.3.4 Droplet Content Detection and Droplet Sorting

When microfluidic modules are produced in transparent materials such as glass and PDMS, they are easily interfaced with optical systems which enable real-time measurement of droplet content as they flow in the channel. Measurement of droplet fluorescence or fluorescence of objects such as beads or cells encapsulated in the droplets can then be performed[26,64,79–81] in a similar manner as in flow cytometers (see, for example, Ashcroft and Lopez[82]). More complex operation can also be performed such as Raman confocal micro-spectroscropy[83] on droplets. Finally, a single droplet can be sorted from the overall population,[26,27,47,84] as a function of the signal detection. Here again, several systems have been described. The principle is to actively control the flow of droplets at a junction by applying on demand a force larger than hydro-dynamic forces. The most developed technique is the use of electric forces. Indeed electrodes are easily patterned on surfaces[84] or as microfluidic channels.[25] Droplets can be charged at production and a d.c. electric field applied across the electrodes will result in net force acting on the charged droplet.[47] However, charged droplets are complex to handle because of repulsion during the flow and leakage of the charges in time which will restrict the use of this technique. Using an a.c. field, it is possible to induce a net dielectrophoretic force on neutral droplets[26,84] (Figure 17.2F). In this case, droplets are actuated at rates up to 2 kHz at a sorting junction when the electric field is triggered on the detection of fluorescence intensities above a given threshold[27] and selection of single event becomes possible on populations of millions of droplets. Other systems have been developed for droplet sorting based on all-optic systems.[53,85] Here local heating of the droplet interface will induce a force that can stop droplets in channels or force them at specific locations. In the future, these systems might provide new ways to obtain reconfigurable channel geometries which might improve the flexibility of the technology, maintaining the throughput of the assays.

In summary, droplet-based technology, based on the integration of modules of droplet manipulation, provides a flexible and versatile way to control droplets at a high-throughput and on small volumes. A large number of operations can be performed on droplets and the numerous applications of the technology drive efficiently the technological developments. The combination of technologies is the source of flexible droplet manipulation that provides tools both for high-content screening and high-throughput screening experiments. Finally, the use of these modules of droplet manipulation to perform biological and chemical assays can revolutionize assays performed in laboratories. In the following we will discuss the recent progress in the use of the modules described above, focusing on the applications of high-throughput manipulation of droplets.

17.4 *In Vitro* Compartmentalization of Biological Reactions

In vitro compartmentalization (IVC) of biological reactions in microdroplets was developed initially for protein directed evolution purposes.[9] The basic idea is to perform billions of experiments by partitioning each experiment into a separate microscopic compartment: each compartment is a water microdroplet, containing all the ingredients for an experiment, which is separated from other microdroplets by an immiscible surrounding fluid phase (oil). This technique has been used to select a range of proteins[9,86–93] and RNAs[94,95] for catalysis, and has also be used to select peptides and proteins for ligand binding,[96–100] and for regulatory activity.[101] Recently direct analysis and sorting of water-in-oil-in-water double emulsions has been described for the directed evolution of *in vitro* expressed β-galactosidases[92] or *in vivo* expressed thiolactonase (PON1).[93]

The combination of IVC and microfluidics has led to the development of droplet-based microfluidic systems that represent a powerful new paradigm in high-throughput screening (HTS) and directed evolution. First, the reduction of reagents volumes, thanks to the small size of the microreactors, greatly reduces expense of screening libraries containing millions of compounds. Second, combined to the potential high-throughput, the high level of precision for reaction and incubation times makes this technology ideal for a rapid, reproducible and quantitative readout of a particular process.[102]

As mentioned earlier in this chapter, conventional methods for studying the effect of environmental stress or drugs on cells behavior, implies the measurement of large number of cells in order to provide information over the population as a whole. However, in microdroplets even single-cells can be analyzed while being at biologically relevant concentrations (due to the small volume of compartment) allowing quantitative biological studies on a single-cell basis for large populations.[103,104] Droplet-based microfluidics represents a high-throughput phenotyping procedure, allowing, for example, the selection of cell-displayed proteins libraries, as well as libraries cloned in heterologous host for cytoplasmic expression, rate enhancement or efficient turnover. The restriction of product diffusion by compartmentalization affords a sensitive and general mode of detection. Moreover, a wide range of experimental conditions can be applied to the cells since they do not have to stay intact for selection (DNA could be recovered and characterized after selection).

An ideal platform[103] for single-cell analysis should allow for: (1) encapsulation of a predefined number of cells per compartment (with the option of encapsulating single cells being highly desirable); (2) incubation of the compartmentalized samples allowing efficient gas-exchange, nutrition, *etc.*; (3) efficient read-out of the results of the experiments and/or recovery of the cells from the compartments in a way that does not abolish cell viability; and (4) facile integration of functional components in biologically relevant platforms that should allow to manipulate the droplets (addition of reagents, fusion, division of droplets, sorting, *etc.*).

17.4.1 Cell Compartmentalization in Aqueous Droplets

Microfluidic systems have been used to controllably compartmentalize both prokaryotic and eukaryotic cells[81,102–107] and even embryos of multicellular organisms[108] within aqueous droplets. Using microfluidic droplet generation modules (see above), the process of loading cells in drops is purely random and, consequently, the number of cells per drops follows a Poisson distribution solely controlled by the cell density.[103–109] Two main approaches have been described to overcome the inherent limitations linked to the variability of the number of cell per droplets due to the stochastic cell loading. The first consists of passively sorting droplets containing single-cells from smaller empty droplets.[110] More recently, self-organization of cells under flow has led to an improved encapsulation.[104] In that system, the cells enter the drop generator with the frequency of drop formation which greatly increases the probability that only one cell is present whenever a drop is generated and thus minimize the number of droplets containing more than one cell or no cells at all.

17.4.2 Incubation and Cell Viability in Droplets

Cell-based assays generally require the read-out of individual samples after an incubation step (to screen the phenotype of individual cell within an heterogeneous population) which implies the need for efficient and biocompatible storage conditions. Perfluorocarbon oils as the continuous phase are perfectly suited for high-throughput cell-based assays. Such oils are compatible with PDMS devices, immiscible with water, and transparent (allowing optical read-out). Moreover most organic molecules are not soluble in fluorinated oil, which limits phase partitioning of the compounds. Individual cells are in their own sterile microenvironment and remain healthy and viable given the remarkable solubility of respiratory gases in the perfluorocarbon carrier fluid.[111] Indeed such fluids can dissolve more than 20 times the amount of O_2 and three times the amount of CO_2 than water and have shown to facilitate the delivery of respiratory gas to both prokaryotic and eukaryotic cells in culture.[65]

Plugs (droplets confined by the microfluidic channel without wetting the walls[112]) provide a simple method for manipulating samples with no dispersion or losses to interfaces.[113] They are suitable for creating large compartments which have been reported as preferential for long-term assays since the cells proliferate during the assay.[103] Systems have been described where such droplets are generated within a microfluidic chip and afterwards flushed into a Teflon capillary tube for cultivation[103,105] or into an open reservoir.[27,103] A droplet-based microfluidic platform has been described which allows the creation of miniaturized reaction vessels in which both adherent (HEK293T) and non-adherent cells (JURKART) can survive for several days.[103] In this study, incubation of the microcompartments in gas-permeable polytetrafluoroethylene (PTFE) tubing allows for cell survival for several days when glass capillaries and vinyl tubing resulted in cell death within 24 h. The authors also described a full life cycle of an encapsulated multicellular organism (the nematode *Caenorhabditis elegans*).[103]

When using surfactant-stabilized droplets, besides the continuous phase, the biocompatibility of the surfactant molecules is a key to the success. In contrast to mineral or organic oils, a few amphiphilic molecules are commercially available for an aqueous–perfluorinated oil interface. By modifying the hydrophilic head (the part in contact with the biological samples) of a commercially available PFPE-based surfactants (poly(perfluoropropylene glycol)-carboxylates sold as Krytox by Dupont) new surfactants have been synthesized[65,103] (Figure 17.3). According to these studies, while ionic surfactants seem to mediate cell lysis, surfactants bearing polyethylene glycol (PEG)[65] or di-morpholino phosphate groups[103] hydrophilic head groups have been shown to exhibit high biocompatibility (did not affect membrane integrity, allowed cell proliferation, and recovery of living cells after growth). These surfactants were used for *in vitro* translation of plasmid DNA encoding *Escherichia coli* β-galactosidase (Lac Z) as well as growth of encapsulated yeast cells,[65] human cells (both adherent and non-adherent)[103] or mammalian hybridoma cells[109] in culture media. Cells remain mobile in the droplet, and do not adhere at the interface of the droplet. Moreover, simple procedures allowing the recovery of cells from droplets that have been stabilized with surfactants have been described without impact on cell viability.[103]

In conclusion, using appropriate surfactant, oil and incubation procedures, cell growth in droplets is mainly limited by gas exchange, lack of nutrition or accumulation of toxic metabolites, as in classical cell culture.

17.4.3 Cell-based Assays and Cell Manipulation

Besides cell survival, cell-based protein expression in microfluidic generated droplets have been demonstrated. The expression of yellow fluorescent protein

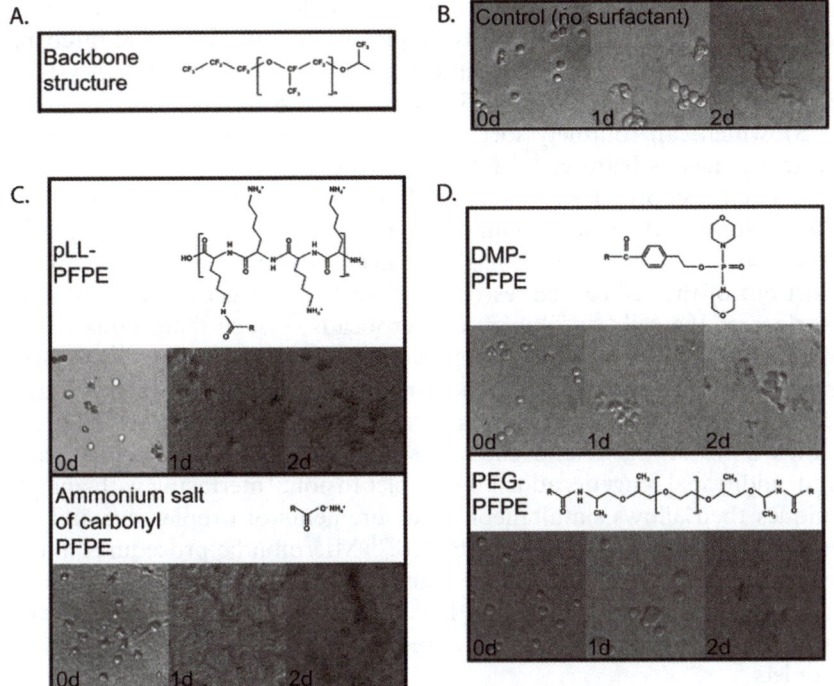

Figure 17.3 Examples of surfactants obtained by modification of the hydrophilic head of commercially available PFPE-based surfactants (poly(per-fluoropropylene glycol)-carboxylates, A) and their effect on long-term survival of eukaryotic cells (modified from Clausell-Tormos *et al.*[103]). For each surfactant, the chemical structure and the results of the biocompatibility assay (microscopical bright-field images) are shown. For the assay, HEK293T cells were incubated for 48 h on a layer of perfluorinated FC40 oil in the presence or absence of the indicated surfactant (0.5% w/w). In the absence of any surfactant, the cells retained an intact morphology and even proliferated (control, B), whereas the ammonium salt of carboxy-PFPE and poly-L-lysine-PFPE (PLL-PFPE) mediated cell lysis (C). However, polyethylene glycol-PFPE (PEG-PFPE) and dimorpholinophosphate-PFPE (DMP-PFPE) showed good biocompatibility, did not affect the integrity of the cellular membrane, and even allowed for cell proliferation (D).

(YFP) in individual *E. coli* cells has been analyzed in microfluidic-generated droplets with simultaneous measurement of droplet size and cell occupancy.[81] Such a system should allow high-throughput protein expression to be performed and the related quantification of the expressed proteins in a highly uniform and reproducible manner.

A challenging area of droplet-based microfluidic is the high-throughput phenotyping by enzymatic markers. To date, the majority of screens for enzymatic activities rely on HTS using chromogenic or fluorogenic substrates.

However, in screening experiments – of either colonies on agar plates or individual clones in microtiter plate wells – typically 10^3–10^4 clones and rarely more than 10^5 clones can be screened, even using sophisticated automated systems.[93] One way of greatly accelerating HTS is to use fluorescence-activated cell sorting (FACS), which can routinely sort $>10^7$ clones per hour, and has a series of other advantageous features.[114] FACS has already proven a highly successful technique to select proteins (notably antibodies) with high binding affinities.[115–122] In addition, FACS has significant potential to select for catalysis;[93,123] however, so far, this approach has only been possible when the diffusion of product out of the cell can be restricted,[124] or the product can be captured on the surface of the cell,[125,126] or onto microbeads.[90] Such limitations could be overcome by double emulsions (water-in-oil-in-water) screening.[92,93] However, droplet-based microfluidic systems generate far less polydisperse droplets, which facilitates quantitative analysis of concentration changes in the droplets and thus more stringent and efficient screening. Moreover, droplets can be steered, additional reagents added by droplet fusion. Interfacing with analytical techniques then allows simultaneous measurements of droplet size and fluorescence with higher precision than FACS.[81] Microfluidic procedures allowing fluorescence analysis of individual compartments (containing a single cell) subsequent to an incubation period have been described recently[103] as well as cell sorting either based on the cell fluorescence[26] or on their enzymatic activity in droplets.[27]

Another interesting feature of the droplet-based procedure will be the tracking of individual cells over time to study phenotypic variations among cell population. A droplet parking device called "dropspots", consisting of a simple microfluidic system that uses an array of well-defined chambers to immobilize thousands of femtoliter- to picoliter-scale aqueous droplets suspended in an inert carrier oil, has been recently described[69] (Figure 17.4). The droplets can be stored, individually monitored, and then recovered and ultimately even sorted. The growth of single yeast cells has been monitored within droplets of water in perfluorocarbon oil parked in the "dropspots". In the same study, the authors monitored enzyme levels in a population of single cells using a fluorogenic assay. Such a device can be used to study dynamic behavior of libraries of individuals allowing, for example, the monitoring of the heterogeneity in individual gene expression.

A platform combining optical trapping and microfluidic-based droplet generation has been developed for single target cell or subcellular structures (mitochondria) analysis in picoliter or femtoliter aqueous droplets.[107] In this study, rapid laser photolysis has been performed on the cells upon encapsulation: the cells are frozen in the state in which they are at the time of photolysis and the lysate is encapsulated within the small volume of the droplet. A fluorogenic assay allowing the detection of enzymatic activity has then been performed on intracellular β-galactosidases within single lysed cells. The key advantage of such lysis over bulk lysis is the confinement of the single-cell content using droplets. This is an important aspect of high-throughput approaches where the copy number present in the cell is low.

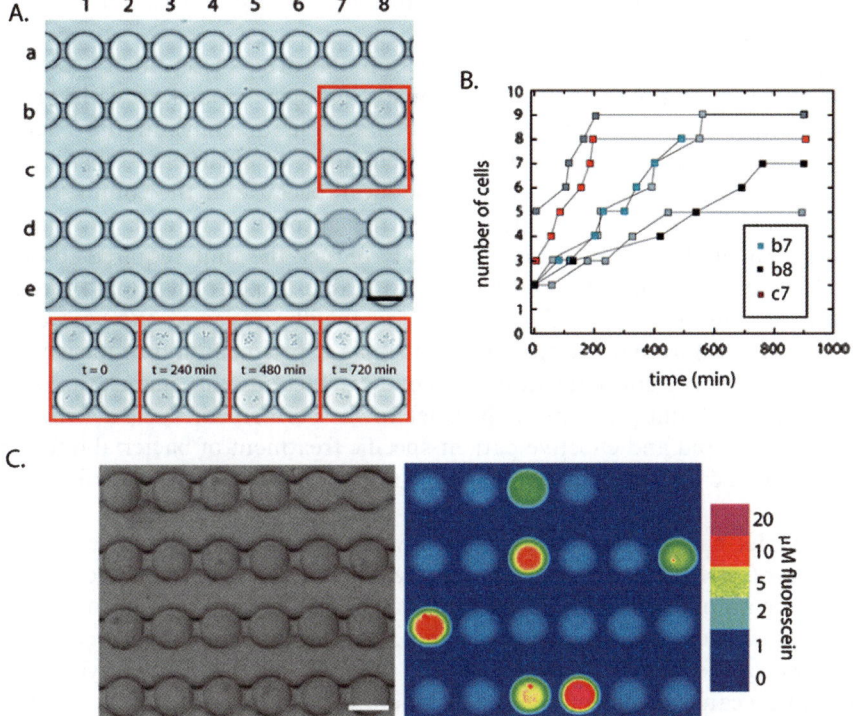

Figure 17.4 Droplet parking device for cell growth studies or cell-based assays (modified from Schmitz *et al.*[69]). (A and B) Monitoring of growth rates of yeast cells in an array of chambers of a "dropspots" device. (A) Bright field image of cells parking in the device at the beginning of the experiment (top) or at specific time over 12 h for a sub-set of droplets (bottom). (B) Tracking of the number of cells in individual droplets over a 15 h incubation period for six individual representative droplets. Colour plots represent droplets identified in the image (panel A, bottom). (C) Monitoring of β-galactosidase activity of cells in droplets in an array of chambers of a dropspots device. Left: Bright field image of the parked cells; Right: Colour map gradient of a fluorescence image at time = 45 min. Scales, 40 μm.

17.5 Towards Integrated Platforms for Cell-based Assays

While controlled manipulation of droplets has been extensively demonstrated, and individual operations (biological material encapsulation, droplet fusion or splitting, droplet sorting, *etc.*) often been described (see above), much work still remains in integrating these different manipulations into a single platform able to address specific biological questions. The studies reported in this section consist of the development of platforms integrating different modules, which constitute proof-of-principles of the pertinence of droplet-based microfluidics for cell-based analysis.

A plug-based platform for rapid detection and drug susceptibility screening of bacteria in samples, including complex biological matrices without pre-incubation, has been described recently.[127] When for conventional bacterial cultures the clinicians have to perform incubation of a sample to increase the concentration of bacteria to a detectable level, because of the confinement of single bacterial cells in nanoliters plugs, the authors were able to eliminate such pre-incubation step and consequently reduce the time required to detect bacteria. More specifically, the authors were able to perform the antibiogram (chart of antibiotic sensitivity) of a methicillin-resistant *Staphylococcus aureus* (MRSA) to many antibiotics in a single experiment and to measure the minimal inhibitory concentration (Mic) of a drug (cefutoxin) against this strain. By permitting multiple tests in parallel to be performed, such a procedure could allow rapid and effective patient-specific treatment of bacterial infections with a significantly decreased detection time from 1–4 days to a few hours (3–7.5 h).

A droplet-based microfluidic method was developed for the detection and analysis of cell-surface protein biomarkers on individual human cells using enzymatic amplification.[128] Such biomarkers have already proven to be useful diagnostic indicators of disease state and clinical outcome.[128] When commonly used approaches (FACS analysis of cells labeled with fluorescent-dye-coupled antibodies) can only detect highly or moderately expressed biomarkers (several hundreds to thousands proteins per cell), the authors are using enzyme-based amplification techniques leading to the detection of low-abundance bio-markers. In addition, by incorporating a basic droplet optical labeling, they paved the way to perform high-throughput sensitive analysis on several cell samples.

The electroporation of single cells within a microfluidic system has been described.[129] In this system, the cell-containing droplets flow between a pair of microelectrodes with a constant voltage established between them. As the oil phase is non-conductive, each flowing droplet experiences a variation in the field intensity that is equivalent to a pulse, while the two electrodes are connected by the droplet resulting in electroporation of the cells contained in the droplets. Plasmids allowing enhanced green fluorescent protein (eGFP) expression have been successfully delivered into Chinese hamster ovary (CHO) cells. Such a technique could lead to droplet-based high-throughput functional genomics studies.

Recently, a novel droplet-based microfluidic system capable of sorting bacterial cells, based upon their enzymatic activity, was developed.[27] This system has been called fluorescence-activated droplet sorting (FADS) (Figure 17.5). The sorting module that has been developed exploits an asymmetric sorting junction and the dielectrophoretic effect to displace specific droplets, containing the bacterial cells, from a flowing stream into a collection channel. The false positive error rate has been determined to be less than 1 in 10^4 analyzed droplets. To validate the platform, two *E. coli* strains have been

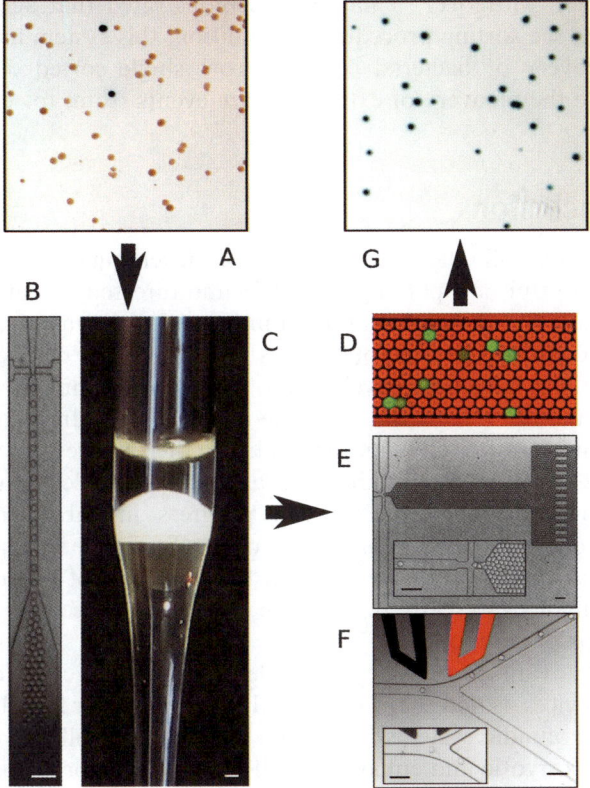

Figure 17.5 Microfluidic cell sorting based on enzymatic activities of the cells. (A) A mixture of beta-galactosidase positive (blue) and negative (white) bacterial cells is encapsulated in droplets with a fluorogenic substrate (B) and incubated in a Pasteur pipette to allow the enzymatic reaction to occur (C). After incubation, the emulsion is reloaded in a microfluidic device. Picture obtained using a fluorescence microscope shows positive droplets (D) in the pool of empty droplets containing either no cells or negative cells. The re-injection of the droplets in a sorter (E) enables the selection of the fluorescent droplets using electric fields (F). The cells in the droplets are recovered and plated back onto agar plates (G).

used: one strain expressing an active β-galactosidase and the other expressing an inactive variant. By encapsulating mixtures of cells with a fluorogenic β-galactosidase substrate (fluorescein-di-β-galactopyranoside, FDG) in a water-in-perfluorocarbon emulsion and sorting the resulting droplets based upon fluorescence, the population have been successfully enriched for active cells with enrichment factors being function of the cell density. The enrichment is here limited by the co-encapsulation of positive and negative cells in the droplet and only positive cells can be recovered for sufficiently low cell dilutions. Throughput was ~400 droplets per second, meaning that 1 000 000 variants can be screened (and selected) in 0.7–7 h, depending on the number of cells per droplet. Moreover, it has been demonstrated that active cells were recovered from the sorting procedure. In addition, this system has allowed the successful recovery of bacterial colonies from single sorted droplets which makes possible the recovery of extremely rare events from, for example, large cell libraries.

17.6 Conclusions

Droplet-based microfluidics has led to the development of systems that represent a powerful new paradigm in HTS and directed evolution where the individual assays are compartmentalized in microdroplet microreactors.

The first advantage of microfluidic is the flexibility of the technology: numerous modules have been developed to make highly uniform droplets, fuse droplet pairs, mix their contents, incubate droplets, split droplets, detect their fluorescence and sort desired "hits" according to their fluorescent signals. All of these modules function at the kilohertz regime on droplet volumes ranging from 1 pL to several nanoliters. Such progress in sub-nanoliter droplet manipulation allows for a level of control of picoliter scale biochemical assays that was hitherto impossible. In addition, the reduction of volume of reagent due to their small size greatly reduces the expense of screening libraries containing millions of compounds.

The second advantage of microfluidic devices, and especially droplet-based microfluidic devices, is that they are perfectly well suited for handling biological materials. Microfluidic systems have been used to controllably compartmentalize both prokaryotic and eukaryotic cells and even embryos of muticellular organisms within aqueous droplets; cell viability and proliferation in droplets as well as protein expression in droplets have been demonstrated using surfactants designed for these purpose and gas permeable systems. Practically, subnanoliter droplets enable statistical studies of single cell rather than population analysis. Thus a droplet-based procedure will enable the tracking of individual cells to study phenotypic variations among cell populations.

In addition, droplets can be steered, additional reagents added by droplet fusion. Interfacing with analytical techniques then allows simultaneous measurements of droplet size and fluorescence with higher precision than FACS. In contrast to FACS, the range of selectable activities is not limited to products remaining in the cell or at the surface of the cells: the activity of molecules secreted by the cells can be assayed and there is no restriction to non-diffusing products. This technology should open the way for quantitative cell-based screening and can use 10^3 to 10^9 smaller assay volumes and around 1000-fold higher throughput than conventional microtiter plate assays. In addition, because of the small volume of the microdroplets, the expense of screening libraries containing millions of compounds will be greatly reduced.

The pertinence of droplet-based microfluidics for high-throughput and high-content quantitative cell screening has definitely been proven. Future studies will consist in the design of integrated platforms able to address specific biological questions in many fields including single-cell analysis, cell populations dynamic probing, drug screening, directed evolution, gene sequencing or functional genomic. Moreover, because of the flexibility and versatility of design and processing of microfluidic devices, they will probably become an essential part of laboratory equipment and procedures.

References

1. T. Young, *Phil. Trans. R. Soc. London*, 1805, **95**, 65–87.
2. O. A. Basaran, *AIChE J.*, 2002, **48**, 1842–1848.
3. P. Calvert, *Chem. Mater.*, 2001, **13**, 3299–3305.
4. E. A. Roth, T. Xu, M. Das, C. Gregory, J. J. Hickman and T. Boland, *Biomaterials*, 2004, **25**(17), 3707–3715.
5. O. Yogi, T. Kawakami, M. Yamauchi, J. Y. Ye and M. Ishikawa, *Anal. Chem.*, 2001, **73**(8), 1896–1902.
6. J. Bibette, F. Leal-Calderon and P. Poulin, *Rep. Prog. Phys.*, 1999, **62**, 696–1033.
7. E. Litborn, A. Emmer and J. Roeraade, *Electrophoresis*, 2000, **21**(1), 91–99.
8. J. Bibette, D. C. Morse, T. A. Witten and D. A. Weitz, *Phys. Rev. Lett.*, 1992, **69**, 2439–2442.
9. D. S. Tawfik and A. D. Griffiths, *Nat. Biotechnol.*, 1998, **16**(7), 652–656.
10. T. M. Squires and S. R. Quake, *Rev. Mod. Phys.*, 2005, **77**(3), 977–1026.
11. G. M. Whitesides, D. Janasek, J. Franzke, A. Manz, D. Psaltis, S. R. Quake, C. Yang, H. Craighead, A. J. deMello, P. El-Ali, P. K. Sorger, K. F Jensen, P. Yager, T. Edwards, E. Fu, K. Helton, K. Nelson, M. R. Tam and B. H. Weighl, *Nature*, 2006, **442**(7101), 367–418.
12. A. R. Wheeler, *Science*, 2008, **322**(5901), 539–540.
13. M. Abdelgawad and A. R. Wheeler, *Adv. Mater.*, 2009, **21**, 920–925.
14. M. G. Lippmann, *Ann. Chim. Phys*, 1875, **5**, 494.
15. B. Berge, *C. R. Acad. Sci. III*, 1993, **317**, 157.
16. F. Mugele and J.-C. Baret, *J. Phys. Condens. Matter*, 2005, **17**, R705–R774.
17. A. A. Darhuber and S. M. Troian, *Annu. Rev. Fluid Mech.*, 2005, **37**, 425–455.
18. S. K. Cho, H. Moon and C. J. Kim. Creating, *J. Microelectromech. Syst.*, 2003, **12**, 70–80.
19. Y. Fouillet and J.-L. Achard, *Comptes rendus physique*, 2004, **5**(5), 577–588.
20. P. R. C. Gascoyne, J. V. Vykoukal, J. A. Schwartz, T. J. Anderson, D. M. Vykoukal, K. W. Current, C. McConaghy, F. F. Becker and C. Andrews, *Lab Chip*, 2004, **4**(4), 299–309.
21. U. Lehmann, C. Vandevyver, V. K. Parashar and M. A. M. Gijs, *Angew. Chem. Int. Ed. Engl.*, 2006, **45**(19), 3062–3067.
22. T. A. Franke and A. Wixforth, *ChemPhysChem*, 2008, **9**(15), 2140–2156.
23. J. A. Schwartz, J. V. Vykoukal and P. R. C. Gascoyne, *Lab Chip*, 2004, **4**(1), 11–17.

24. Y. N. Xia and G. M. Whitesides, *Annu. Rev. Mater. Sci.*, 1998, **28**, 153–184.

25. A. C. Siegel, D. A. Bruzewicz, D. B. Weibel and G. M. Whitesides, *Adv. Mater.*, 2007, **19**, 727–733.

26. L. M. Fidalgo, G. Whyte, D. Bratton, C. F. Kaminski, C. Abell and W. T. S. Huck, *Angew. Chem. Int. Ed.*, 2008, **47**, 2042–2045.

27. J. C. Baret and O. J. Miller OJ, *et al., Lab Chip*, 2009, **9**, 1850–1858.

28. Lord Rayleigh, *Proc. R. Soc. London, Ser A*, 1879, **29**(71).

29. T. Thorsen, R. W. Roberts, F. H. Arnold and S. R. Quake, *Phys. Rev. Lett.*, 2001, **86**(18), 4163–4166.

30. C. Priest, S. Herminghaus and R. Seemann, *Appl. Phys. Lett.*, 2006, **88**(2), 024106.

31. V. Chokkalingam, S. Herminghaus and R. Seemann, *Appl. Phys. Lett.*, 2008, **93**, 254101.

32. S. L. Anna, N. Bontoux and H. A. Stone, *Phys. Lett.*, 2003, **82**, 364–366.

33. A. S. Utada, A. Fernandez-Nieves, H. A. Stone and D. A. Weitz, *Phys. Rev. Lett.*, 2007, **99**(9), 094502.

34. A. S. Utada, A. Fernandez-Nieves, J. M. Gordillo and D. A. Weitz, *Phys. Rev. Lett.*, 2008, **100**(1), 014502.

35. P. Garstecki, H. A. Stone and G. M. Whitesides, *Phys. Rev. Lett.*, 2005, **94**(16), 164501.

36. B. Dollet, W. van Hoeve, J.-P. Raven, P. Marmottant and M. Versluis, *Phys. Rev. Lett.*, 2008, **100**(3), 034504.

37. L. Shui, A. Van den Berg and J. C. T. Eijkel, *Lab Chip*, 2009, **9**(6), 795–801.

38. V. Barbier, H. Willaime, P. Tabeling and F. Jousse, *Rev. E*, 2006, **74**, 046306.

39. L. Frenz, J. Blouwolff, A. D. Griffiths and J.-C. Baret, *Langmuir*, 2008, **24**(20), 12073–12076.

40. E. Lorenceau, A. S. Utada, D. R. Link, G. Cristobal, M. Joanicot and D. A. Weitz, *Langmuir*, 2005, **21**(20), 9183–9186.

41. A. S. Utada, E. Lorenceau, D. R. Link, P. D. Kaplan, H. A. Stone and D. A. Weitz, *Science*, 2005, **308**(5721), 537–541.

42. L.-Y. Chu, A. S. Utada, R. K. Shah, J.-W. Kim and D. A. Weitz, *Angew. Chem. Int. Ed. Engl.*, 2007, **46**, 8970–8974.

43. N. Pannacci, H. Bruus, D. Bartolo, I. Etchart, T. Lockhart, Y. Hennequin, H. Willaime and P. Tabeling, *Phys. Rev. Lett.*, 2008, **101**, 164502.

44. D. Weaire and W. Drenckhan, *Adv. Colloid Interface Sci.*, 2008, **137**(1), 20–26.

45. F. Malloggi, H. Gu, A. G. Banpurkar, S. A. Vanapalli and F. Mugele, *Eur. Phys. J. E. Soft Matter*, 2008, **26**(1–2), 91–96.

46. H. Luo, D. Kim, D. R. Link, D. A. Weitz, M. Marquez and Z. Cheng, *Appl. Phys. Lett.*, 2007, **91**, 133106.

47. D. R. Link, E. Grasland-Mongrain, A. Duri, F. Sarrazin, Z. D. Cheng, G. Cristobal, M. Marquez and D. A. Weitz, *Angew. Chem. Int. Ed.*, 2006, **45**, 2556–2560.

48. N. Bremond, A. R. Thiam and J. Bibette, *Phys. Rev. Lett.*, 2008, **1**, 024501.
49. J.-C. Baret, F. Kleinschmidt, A. El Harrak and A. D. Griffiths, *Langmuir*, 2009, **25**(11), 6088–6093.
50. D. R. Link, S. L. Anna, D. A. Weitz and H. A. Stone, *Phys. Rev. Lett.*, 2004, **92**, 054503.
51. L. Menetrier-Deremble and P. Tabeling, *Phys. Rev. E: Stat. Nonlin. Soft Matter Phys.*, 2006, **74**(3, Pt. 2), 035303.
52. A. M. Leshanski and L. M. Pismen, *Phys. Fluids*, 2009, **21**, 023303.
53. C. N. Baroud, M. R. de Saint Vincent and J.-P. Delville, *Lab Chip*, 2007, **7**, 1029–1033.
54. D. J. Tritton, *Physical Fluid Dynamics*, Oxford Science Publications, 1988 (ISBN: 0 19 854493).
55. M. J. Fuerstman, A. Lai, M. E. Thurlow, S. S. Shevkoplyas, H. A. Stone and G. M. Whitesides, *Lab Chip*, 2007, **7**(11), 1479–1489.
56. M. T. Sullivan and H. A. Stone, *Philos. Trans. A, Math. Phys. Eng. Sci.*, 2008, **366**(1873), 2131–2143.
57. M. Schindler and A. Ajdari, *Phys. Rev. Lett.*, 2008, **100**(4), 044501.
58. T. Beatus, R. Bar-Ziv and T. Tlusty, *Phys. Rev. Lett.*, 2007, **99**, 124502.
59. C. N. Baroud, X. C. Wang and J.-B. Masson, *J. Colloid Interface Sci.*, 2008, **326**, 445–450.
60. N. R. Beer, K. A. Rose and I. M. Kennedy, *Lab Chip*, 2009, **9**, 838–840.
61. F. Jousse, R. Farr, D. R. Link, M. J. Fuerstman and P. Garstecki, *Phys. Rev. E: Stat. Nonlin. Soft Matter Phys.*, 2006, **74**(3 Pt. 2), 036311.
62. B. Zheng, J. D. Tice and R. F. Ismagilov, *Anal. Chem.*, 2004, **76**(17), 4977–4982.
63. M. Prakash and N. Gershenfeld, *Science*, 2007, **315**(5813), 832–835.
64. F. Courtois, L. F. Olguin, G. Whyte, D. Bratton, W. T. S. Huck, C. Abell and F. Hollfelder, *Chem. Bio. Chem.*, 2008, **9**, 439–446.
65. C. Holtze, A. C. Rowat, J. J. Agresti, J. B. Hutchison, F. E. Angile, C. H. J. Schmitz, S. Koster, H. Duan, K. J. Humphry, R. A. Scanga, J. S. Johnson, D. Pisignano and D. A. Weitz, *Lab Chip*, 2008, **8**(10), 1632–1639.
66. K. Ahn, J. Agresti, H. Chong, M. Marquez and D. A. Weitz, *Appl. Phys. Lett.*, 2006, **88**, 264105.
67. L. Frenz, K. Blank, E. Brouze and A. D. Griffiths, *Lab Chip*, 2009, **10**, 1344–1348.
68. A. Huebner, D. Bratton, G. Whyte, M. Yang, A. J. Demello, C. Abell and F. Hollfelder, *Lab Chip*, 2009, **9**(5), 692–698.
69. C. H. J. Schmitz, A. C. Rowat, S. Koster and D. A. Weitz, *Lab Chip*, 2009, **9**(1), 44–49.
70. M. Abdelgawad, M. W. L. Watson and A. R. Wheeler, *Lab Chip*, 2009, **9**, 1046–1051.
71. A. Y. Fu, C. Spence, A. Scherer, F. H. Arnold and S. R. Quake, *Nat. Biotechnol.*, 1999, **17**(11), 1109–1111.
72. A. Y. Fu, H. P. Chou, C. Spence, F. H. Arnold and S. R. Quake, *Anal. Chem.*, 2002, **74**, 2451–2457.

73. A. R. Abate, M. B. Romanowsky, J. J. Agresti and D. A. Weitz, *Appl. Phys. Lett.*, 2009, **94**, 023503.

74. A. R. Abate and D. A. Weitz, *Appl. Phys. Lett.*, 2008, **92**, 243509.

75. L. H. Hung, K. M. Choi, W. Y. Tseng, Y. C. Tan, K. J. Shea and A. P. Lee, *Lab Chip*, 2006, **6**(2), 174–178.

76. L. M. Fidalgo, C. Abell and W. T. S. Huck, *Lab Chip*, 2007, **7**(8), 984–986.

77. M. Chabert, K. D. Dorfman and J.-L. Viovy, *Electrophoresis*, 2005, **26**, 3706–3715.

78. C. Priest, S. Herminghaus and R. Seemann, *Appl. Phys. Lett.*, 2006, **89**(2), 134101.

79. L. Frenz, A. El Harrak, M. Pauly, S. Begin-Colin, A. D. Griffiths and J. C. Baret, *Angew. Chem. Int. Ed. Engl.*, 2008, **47**(36), 6817–6820.

80. N. R. Beer, K. A. Rose and I. M. Kennedy, *Lab Chip*, 2009, **9**, 841–844.

81. A. Huebner, M. Srisa-Art, D. Holt, C. Abell, F. Hollfelder, A. J. deMello and J. B. Edel, *Chem. Commun.*, 2007, **12**, 1218–1220.

82. R. G. Ashcroft and P. A. Lopez, *J. Immunol. Methods*, 2000, **243**, 13–24.

83. G. Cristobal, L. Arbouet, F. Sarrazin, D. Talaga, J.-L. Bruneel, M. Joanicot and L. Servant, *Lab Chip*, 2006, **6**, 1140–1146.

84. K. Ahn, C. Kerbage, T. P. Hunt, R. M. Westervelt, D. R. Link and D. A. Weitz, *Appl. Phys. Lett.*, 2006, **88**, 024104.

85. C. N. Baroud, J.-P. Delville, F. Gallaire and R. Wunenburger, *Phys. Rev. E: Stat. Nonlin. Soft Matter Phys.*, 2007, **75**, 046302.

86. Y.-F. Lee, D. S. Tawfik and A. D. Griffiths, *Nucleic Acids Res.*, 2002, **30**(22), 4937–4944.

87. H. M. Cohen, D. S. Tawfik and A. D. Griffiths, *Protein Eng. Des. Sel.*, 2004, **17**, 3–11.

88. F. J. Ghadessy, J. L. Ong and P. Holliger, *Proc. Natl. Acad. Sci. U.S.A.*, 2001, **98**(8), 4552–4557.

89. F. J. Ghadessy, N. Ramsay, F. Boudsocq, D. Loakes, A. Brown, S. Iwai, A. Vaisman, R. Woodgate and P. Holliger, *Nat. Biotechnol.*, 2004, **22**(6), 755–759.

90. A. D. Griffiths and D. S. Tawfik, *EMBO J.*, 2003, **22**(1), 24–35.

91. N. Doi, S. Kumadaki, Y. Oishi, N. Matsumura and H. Yanagawa, *Nucleic Acids Res.*, 2004, **32**, e95.

92. E. Mastrobattista, V. Taly, E. Chanudet, P. Treacy, B. T. Kelly and A. D. Griffiths, *Chem. Biol.*, 2005, **12**(12), 1291–1300.

93. A. Aharoni, A. D. Griffiths and D. S. Tawfik, *Curr. Opin. Chem. Biol.*, 2005, **9**, 210–216.

94. J. J. Agresti, B. T. Kelly, A. Jaschke and A. D. Griffiths, *Proc. Natl. Acad. Sci. U. S. A.*, 2005, **102**, 16170–16175.

95. M. Levy, K. E. Griswold and A. D. Ellington, *RNA*, 2005, **11**(10), 1555–1562.

96. A. Sepp, D. S. Tawfik and A. D. Griffiths, *FEBS Lett.*, 2002, **532**(3), 455–458.

97. M. Yonezawa, N. Doi, Y. Kawahashi, T. Higashinakagawa and H. Yanagawa, *Nucleic Acids Res.*, 2003, **31**(19), e118.
98. M. Yonezawa, N. Doi, T. Higashinakagawa and H. Yanagawa, *J. Biochem.*, 2004, **135**(3), 285–288.
99. J. Bertschinger and D. Neri, *Protein Eng. Des. Sel.*, 2004, **17**, 699–707.
100. A. Sepp and Y. Choo, *Mol. Biol.*, 2005, **354**(2), 212–219.
101. K. Bernath, S. Magdassi and D. S. Tawfik, *J. Mol. Biol.*, 2005, **345**, 1015–1026.
102. A. Huebner, S. Sharma, M. Srisa-Art, F. Hollfelder, J. B. Edel and A. J. Demello, *Lab Chip*, 2008, **8**(8), 1244–1254.
103. J. Clausell-Tormos, D. Lieber, J.-C. Baret, A. El-Harrak, O. J Miller, L. Frenz, J. Blouwolff, K. J. Humphry, S. Koster, H. Duan, C. Holtze, D. A. Weitz, A. D. Griffiths and C. A. Merten, *Chem. Biol.*, 2008, **15**, 427–437.
104. J. F. Edd, D. Di Carlo, K. J. Humphry, S. Koster, D. Irimia, D. A. Weitz and M. Toner, *Lab Chip*, 2008, **8**(8), 1262–1264.
105. K. Martin, T. Henkel, V. Baier, A. Grodrian, T. Schon, M. Roth, J. M. Kohler and J. Metze, *Lab Chip*, 2003, **3**(3), 202–207.
106. S. Sakai, K. Kawabata, T. Ono, H. Ijima and K. Kawakami, *Biotechnol. Prog.*, 2005, **21**(3), 994–997.
107. M. He, J. S. Edgar, G. D. M. Jeffries, R. M. Lorenz, J. P. Shelby and D. T. Chiu, *Anal. Chem.*, 2005, **77**(6), 1539–1544.
108. A. Funfak, A. Brosing, M. Brand and J. M. Kãhler, *Lab Chip*, 2007, **7**(9), 1132–1138.
109. S. Koster, F. E. Angile, H. Duan, J. J. Agresti, A. Wintner, C. Schmitz, A. C. Rowat, C. A. Merten, D. Pisignano, A. D. Griffiths and D. A. Weitz, *Lab Chip*, 2008, **8**(7), 1110–1115.
110. M. Chabert and J.-L. Viovy, *Proc. Natl. Acad. Sci. U. S. A.*, 2008, **105**, 3191–3196.
111. K. C. Lowe, M. R. Davey and J. B. Power, *Trends Biotechnol.*, 1998, **16**(6), 272–277.
112. J. D. Tice, H. Song, A. D. Lyon and R. F. Ismagilov, *Langmuir*, 2003, **19**, 9127–9133.
113. H. Song, D. L Chen and R. F. Ismagilov, *Angew. Chem. Int. Ed.*, 2006, **45**(44), 7336–7356.
114. G. Georgiou, *Adv. Protein Chem.*, 2000, **55**, 293–315.
115. M. J. Feldhaus, R. W. Siegel, L. K. Opresko, J. R. Coleman, J. M. Weaver Feldhaus, Y. A. Yeung, J. R. Cochran, P. Heinzelman, D. Colby, J. Swers, C. Graff, H. S. Wiley and K. D. Wittrup, *Nat. Biotechnol.*, 2003, **21**(2), 163–170.
116. G. Chen, A. Hayhurst, J. G. Thomas, B. R. Harvey, B. L. Iverson and G. Georgiou, *Nat. Biotechnol.*, 2001, **19**, 537–542.
117. A. Hayhurst and G. Georgiou, *Curr. Opin. Chem. Biol.*, 2001, **5**(6), 683–689.
118. K. D. Wittrup, *Curr. Opin. Biotechnol.*, 2001, **12**(4), 395–399.
119. A. Wentzel, A. Christmann, T. Adams and H. Kolmar, *J. Bacteriol.*, 2001, **183**(24), 7273–7284.

120. B. R. Harvey, G. Georgiou, A. Hayhurst, K. Jun Jeong, B. L. Iverson and G. K. Rogers, *Proc. Natl. Acad. Sci. U. S. A.*, 2004, **101**(25), 9193–9198.

121. E. V. Shusta, P. D. Holler, M. C. Kieke, D. M. Kranz and K. D. Wittrup, *Nat. Biotechnol.*, 2000, **18**(7), 754–759.

122. A. W. Nguyen and P. S. Daugherty, *Nat. Biotechnol.*, 2005, **23**(3), 355–360.

123. S. Becker, H.-U. Schmoldt, T. M. Adams, S. Wilhelm and H. Kolmar, *Curr. Opin. Biotechnol.*, 2004, **15**, 323–329.

124. Y. Kawarasaki, K. E. Griswold, J. D. Stevenson, T. Selzer, S. J. Benkovic, B. L. Iverson and G. Georgiou, *Nucleic Acids Res.*, 2003, **31**(21), e126.

125. M. J. Olsen, D. Stephens, D. Griffiths, P. Daugherty, G. Georgiou and B. L. Iverson, *Nat. Biotechnol.*, 2000, **18**(10), 1071–1074.

126. N. Varadarajan, J. Gam, M. J. Olsen, G. Georgiou and B. L. Iverson, *Proc. Natl. Acad. Sci. U. S. A.*, 2005, **102**(19), 6855–6860.

127. J. Q. Boedicker, L. Li, T. R. Kline and R. F. Ismagilov, *Lab Chip*, 2008, **8**, 1265–1272.

128. H. N. Joensson, M. L. Samuels, E. R. Brouzes, M. Medkova, M. Uhlen, D. R. Link and H. Andersson-Svahn, *Angew. Chem. Int. Ed. Engl.*, 2009, **48**(14), 2518–2521.

129. Y. Zhan, J. Wang, N. Bao and C. Lu, *Anal. Chem.*, 2009, **81**(5), 2027–2031.

CHAPTER 18

New Detection Methods for Single Cells

EMMANUEL FORT

Centre d'Imageries Plasmoniques Appliquées, Institut Langevin, ESPCI ParisTech, CNRS UMR 7587, Université Paris Diderot, 10 rue Vauquelin, 75 231 Paris Cedex 05, France

Abstract

The major factors that have limited the use of DNA microarrays in research and diagnostics are the amount of target needed, the detection specificity, as well as the cost and reliability of detection equipment and assays. While the current gold standard for detection is fluorescence technology, the emphasis on more efficient and sensitive instrumentation has spurred the development of a number of new labeling and detection methodologies. Recent reports have demonstrated that alternative techniques like semiconductor or metal nanoparticle labels or electrochemical detection techniques could eliminate the need for target amplification steps such as PCR. Besides, the improvements in labeling technology enabled multiplexing by increasing the number of detectable colour channels and direct tracking of gene expression inside living cells.

This chapter is devoted to the latest developments in terms of hybridization techniques and read-out technologies that are promising for enhanced detection and quantification of specific gene sequences. These techniques are widely used in two emerging trends: the bio-barcode strategy and live-cell imaging.

RSC Nanoscience & Nanotechnology No. 15
Unravelling Single Cell Genomics: Micro and Nanotools
Edited by Nathalie Bontoux, Luce Dauphinot and Marie-Claude Potier
© Royal Society of Chemistry 2010
Published by the Royal Society of Chemistry, www.rsc.org

18.1   Introduction

Any population of cells will exhibit some degree of variability. Genetic differences are one of the main factors responsible for cellular heterogeneity. Cellular heterogeneity has been observed in a wide variety of cell types ranging from simple bacterial cells[1,2] to more complex mammalian cells.[3] However, variation is also present in genetically identical cell populations, even when the cells have been exposed to the same environment and have the same history.[4-7] Random fluctuations in the process of gene expression are thought to contribute to this phenotypic variation. Thus, analyzing gene expression at the single-cell level provided crucial insight into oscillatory or nonlinear behavior in asynchronous cells and has revealed the cell-to-cell variability that arises owing to the stochastic nature of gene expression.[8]

A wide variety of new tools have allowed investigators to monitor gene expression with single-cell resolution. Owing to differences in characteristics such as cellular throughput, number of gene products analyzed, sensitivity, and temporal resolution, the gene expression analysis method utilized depends on the objective of the study.

Flow cytometry is a technique that monitors gene expression at the protein level and is very high throughput in terms of the number of cells analyzed. It has proven to be useful for monitoring a large number of gene products.[9] However, this technique gives only snapshots of gene expression patterns in individual cells.

More recently, microarrays technology has emerged as a robust methodology for quantitatively analyzing a large number of nucleic acid sequences in parallel.[10] Differential gene expression analysis is used to determine which gene is up-regulated or down-regulated during specific cellular processes or in response to environmental stimuli. Cellular responses triggered during specific disease states, or by exposure to drugs, toxins, or other molecules of interest is the subject of numerous studies. In addition, microarrays have found applications in identifying single-nucleotide polymorphisms (SNPs) or other genetic variations. This ability has led to the development of microarray for diagnostic and pharmacogenomic applications.

The major factors that have limited the use of microarrays in the research and diagnostic applications are the amount of target needed, the detection specificity, as well as the cost and reliability of detection equipment and assays. A critical determinant of these parameters is the labeling and detection methodology. While the current "gold standard" is fluorescence technology, the emphasis on more efficient and sensitive instrumentation has spurred the development of a number of new labeling and detection methodologies. Recent reports have demonstrated that alternative techniques like semiconductor or metal nanoparticle labels or electrochemical detection techniques could eliminate the need for target amplification steps such as PCR. This is of major importance in the need for new labels since amplification steps are time consuming and considerably increase the cost and complexity of microarrays.

Besides, the improvements in labeling technology enabled multiplexing by increasing the number of detectable color channels.

This chapter is devoted to the latest developments in terms of hybridization techniques and read-out technologies that are promising for enhanced detection and quantification of specific gene sequences. These techniques are massively used in two emerging trends: the bio-barcode strategy and live-cell imaging. These two trends will be respectively addressed in the following sections before describing the details of the most promising technologies. Following sections will be devoted to specific techniques that are fast-growing like quantum dots (QDs), gold nanoparticles for resonant light scattering, molecular rulers, molecular beacon or enhanced surface raman spectroscopy.

18.2 Bio-barcode Strategy

18.2.1 Principle

Most biomolecular detection systems (like microarrays for example), regardless of their need for target amplification, require steps complicated by solution and surface binding kinetics. New strategies have been developed to circumvent these limitations. The basic concept is to develop smart nanostructures that possess not only molecular recognition abilities but also built-in codes for rapid target identification. Each nanosensor will usually include a target-specific probe and a built-in barcode. These barcode sensors can be tracked directly in the cell environment by means of optical microscopy for instance or can be immobilized on a substrate for subsequent reading.

Optical reporting methods are usually privileged. Recent advances in several groups have led to a burst of activities in this field based on various techniques like segmented nanorods,[11,12] porous silicon,[13] rare-earth doped glass,[14] fluorescent silica colloids,[15,16] photobleached patterns,[17] oligonucleotide-linked colloidal gold,[18,19] enhanced Raman nanoparticles,[20,21] and semiconductor quantum dots (QDs).[22–26]

We will present briefly one of the many very-recent innovative technique based on DNA origami.[27] Although, this technique is not mature and needs many improvements, it exemplifies the barcode strategy.

18.2.2 An Example: DNA Origami

DNA origami, in which long, single-stranded DNA segments are folded into shapes by short staple segments, is used to create nucleic acid probe tiles that are molecular analogs of macroscopic DNA chips.[27] One hundred trillion probe tiles were fabricated in one step and bear pairs of 20-nucleotide-long single-stranded DNA segments that act as probe sequences. These tiles can hybridize to their targets in solution and, after adsorption onto mica surfaces, can be examined by atomic force microscopy in order to quantify binding events, because the probe segments greatly increase in stiffness upon hybridization. The nucleic acid probe tiles have been used to study position dependent hybridization on the nanoscale and have also been used for label-free detection of RNA. Figure 18.1 shows schematics of the DNA tile and the process for the use of this barcode. Although some technical issues need to be overcome, future prospects are promising.

18.3   Imaging Gene Expression in Living Cells

18.3.1   Motivations

The development of new labeling techniques as well as bio-imaging techniques has also made possible new biosensing real-time spatially resolved strategies of specific relevance for single-cell gene expression analysis. The use of a solid substrate for gene detection implies a step during which the cell is usually lysed, delivering the intracellular components to the surrounding solution. This lysis allows a subsequent separation step that is highly advantageous for resolving the high complexity of analytes that can be found in a single cell. However, (1) subsequent separation steps may be critical due to analyte adsorption on surfaces and the low abundance of many analytes of interest; and (2) it gives a static image of the cell activity condensing it into a single snapshot. Hence, *in vitro* methods such as real-time PCR, northern blotting and DNA microarrays which rely on the use of cell lysates, thereby lose the temporal and spatial information. However, it is now possible to follow changes in the spatial distribution of molecules as a function of time and of perturbations in the cellular environment. The kinetics of these movements, as well as their specific interactions with cellular structures, can be analyzed using the power of optical microscopy and new development in labeling technology.[28]

18.3.2   Improvements in Photonic Microscopy

Crucial improvements have occurred in photonic microscopy over the past decade. For example, in detection devices, the technology of electron-multiplying CCDs (EMCCDs) provides higher levels of sensitivity in photon collection that can detect extremely weak signals while maintaining a low noise. Developments in optical filters technology and the emergence of tunable filters, which can scan across a spectrum of wavelengths with high precision, enable the combinatorial use of many more dyes.[29,30] The introduction of wavelengths into the imaging process adds a fifth dimension to space and time.

The improvements in electronic devices for unloading and storage of the collected data, together with the fast increase in the speed of computers and in their storage capacity, now allow large-format (mega-pixel) images to be captured and stored rapidly. Frame rates within the range of a few milliseconds are now achievable. This technological revolution allows most biological trafficking events in cells to be recorded in real time.

The development of various optical techniques has provided insights into the kinetics and trafficking of subsets of the studied proteins and their accessibility to different cellular compartments. Fluorescence recovery after

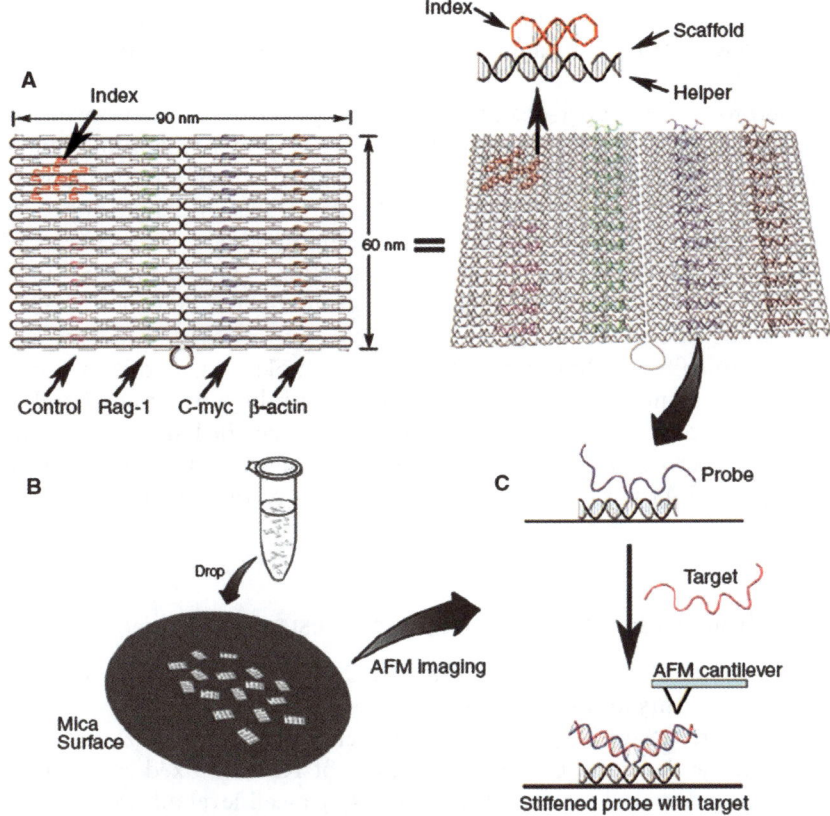

Figure 18.1 (A) *Left*: Schematic layout of the indexed nucleic acid probe tiles bearing three different probes and a control probe. A simple, rectangular-shaped DNA origami tile was used, in which a circular single-stranded M13 viral DNA (black lines), is folded and stapled, with the help of short synthetic DNA strands, to form the desired two-dimensional tile. Helper strands without probe modifications are shown (gray lines), and helper strands modified with probe sequences and control sequences are shown in different colors. There are 12 copies of the specific probes on the tile. Six control probes are arranged in a shorter line. An index spot, composed of six closely packed dumbbell-shaped bulge loops at the top left corner, is designed to give the AFM topological feature with which to orient the AFM image of each individual tile. *Right*: Tilted view of the tile illustrating the 3D view of the probe tile layout. (B) Illustration of the process for the use of probe tiles for target detection. Probe tiles are self-assembled in solution, hybridized with targets, and then dropped onto the mica surface for AFM imaging. (C) Probe design and detection mechanism. A pair of neighboring helper strands is extended out of the surface of the tile, with each 20-base-long extension bearing a half of the target sequence. These single-stranded probes are flexible and do not produce a visible feature under AFM imaging. Upon target hybridization with the pair of half-probes, the double helix of the DNA–RNA duplex forms, and the stiffer V-shaped junction is readily detected with the AFM cantilever. Reproduced, with permission, from Ke *et al.*[27]

photobleaching (FRAP), fluorescence loss in photobleaching (FLIP), inverse-FRAP (i-FRAP) and uncaging by photoactivation are commonly used techniques.[28]

Furthermore, new super-resolution techniques have been developed using mathematical and physical approaches.[31,32] For super-resolution imaging, the key to overcoming the diffraction limit is to spatially and/or temporally modulate the transition between two molecular states of a fluorophore (for example, a dark and a bright state). Some techniques allow a super resolution by narrowing the microscope point spread function. These techniques include stimulated emission depletion (STED),[33] ground-state depletion (GSD),[34] and saturated structured illumination microscopy (SSIM).[35–37] Other super-resolution imaging techniques detect single molecules and rely on the principle that a single emitter can be localized with high accuracy if sufficient numbers of photons are collected.[38] These techniques include photoactivated localization microscopy (PALM),[39] fluorescence photoactivation localization microscopy (FPALM)[40] and stochastic optical reconstruction microscopy (STORM).[41]

18.3.3 Improvements in Fluorophore Design

Another avenue of progress over the past decade has been the quality and diversity of reagents and labelling strategies.

The ability to design reagents for fluorescence *in situ* hybridization (FISH) has allowed the detection of single molecules of RNA in fixed cells. The first attempts to understand transcription at the single-cell level allowed the detection of a gene that actively produces mRNA at the site of transcription and provided a first description of the transcriptional–activation events that take place at a gene locus.[42] The ability to visualize the site of gene expression allowed for the single-cell analysis of several genes that are active at the time of fixation. A mixture of probes could be designed to obtain a "spectral barcode" by conjugating several dyes to different oligonucleotide probes. In this way, each gene was uniquely labelled with a different set of dyes. The simultaneous identification of several transcription sites in a single cultured cell provided a snapshot of the transcriptome at the time of fixation and showed extensive variability in single-cell gene-expression profiles.[43]

Detailed analyses of the patterns of gene expression in simple single-cell systems such as bacteria and yeast have revealed the stochastic nature of gene expression.[4,44,45] These techniques for transcriptional visualization in fixed cells are important in understanding complex genetic regulation such as that associated with imprinting, cellular responses to positional information within organs, or co-regulation pathways. These methods are useful in medical diagnostics by addressing specific gene-expression patterns at the cellular level in tissue biopsies.[46]

Only methods that can follow gene expression in individual living cells can address dynamic nature of transcription. Imaging biological processes *in vivo* has been revolutionized since the introduction of genetically encoded fluorescent reagents. Fluorescent proteins (FPs), in particular the green fluorescent protein (GFP) and its variants, have allowed a variety of proteins to be "fluorescentized", in many cases with minimal functional consequences.[47] It is now feasible to co-express several fusion proteins in one cell, each of which is tagged with a different fluorescent protein, and then to follow their subcellular paths in living cells.

Most super-resolution imaging techniques need fluorophores with specific characteristics. These techniques determine the nanoscale localization of individual fluorescent molecules by sequentially switching them on and off using light of different wavelengths. In contrast to STED, in which the switching is predefined in space through the superimposition of the two laser beams, the switching is usually done stochastically in single-molecule-based super-resolution methods. In each imaging cycle, most molecules remain dark, but a small percentage of molecules are stochastically switched on, imaged, and then localized. Repeating this process for many cycles allows the reconstruction of a super-resolution image. Two classes of probes are used for super-resolution imaging: fluorescent proteins (FPs) and non-genetically encoded probes, such as organic small-molecule fluorophores and quantum dots.

Although simple FPs, such as GFP and yellow FP (YFP), have been used in STED imaging[48,49] most super-resolution imaging techniques exploit the intrinsic ability of certain FPs to change their spectral properties on irradiation with light of a specific wavelength. There are three main classes of FPs used in super-resolution imaging (Figure 18.2): (i) FPs that convert from a dark state to a bright fluorescent state (photoactivatable FPs (PA-FPs)). This class be can split into two depending on the fact that the PA-FPs can photoactivate either reversibly or irreversibly. (ii) FPs that change fluorescence wavelength on irradiation called photoshiftable FPs (PS-FPs). All known PS-FPs shift their wavelength irreversibly.[50,51]

Besides, three main classes of non-genetically encoded probes have been used in super-resolution imaging: inorganic quantum dots (see section 18.3.3), reversible photoactivatable fluorophores (also known as photoswitchers) and irreversible photoactivatable fluorophores (also known as photocaged fluorophores). The small-molecule analogues to reversible PA-FPs (such as dronpa and rsCherry) are photochromic probes, including rhodamines and diarylethenes, as well as photoswitchable cyanines. New probe based on rhodamine 590 (PC-RH590), whose extra rigidity improves the cross-section compared to the original PC-RHB for two-photon absorption have been recently developed.[52]

Although non-genetically encoded probes have generally increased fluorescence properties (brightness and photostability), they require a means to target them to the biomolecule of interest inside the cells. Traditionally, targeting strategies use antibody conjugation which has many disadvantages. For instance, antibodies are not membrane permeable and consequently cannot be used for intracellular labelling of living cells. Besides, antibody staining has a usually low labelling efficiency and the large size of antibodies adds uncertainty ($\sim$10–20 nm) to the spatial relationship between the label and its target.[53]

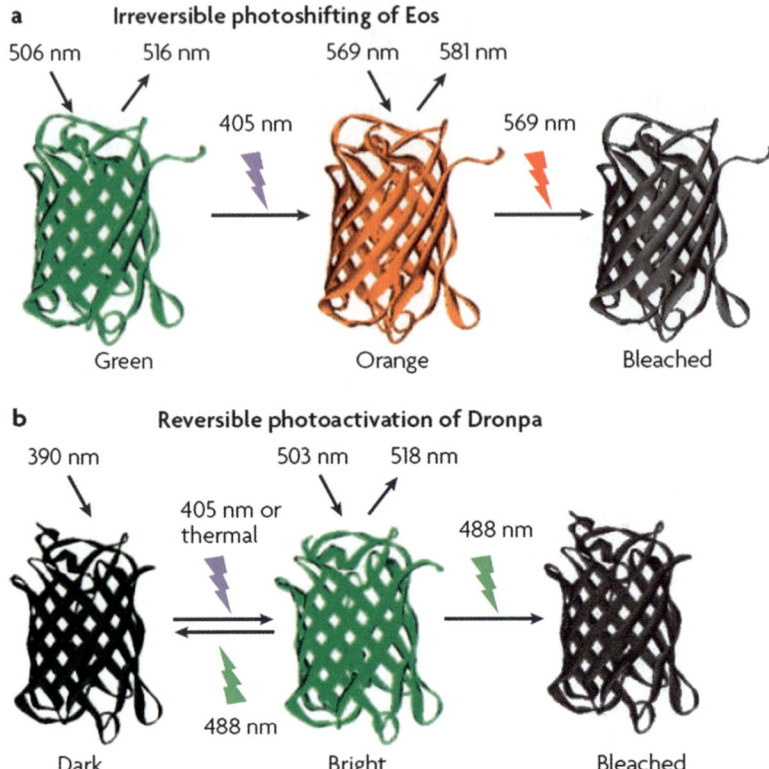

a   Irreversible photoshifting of Eos

506 nm   516 nm   569 nm   581 nm

405 nm   569 nm

Green   Orange   Bleached

b   Reversible photoactivation of Dronpa

390 nm   503 nm   518 nm

405 nm or
thermal   488 nm

488 nm

Dark   Bright   Bleached

Figure 18.2   Examples of fluorescent proteins (FPs) that are used for super-resolution imaging. (A) exposure to ultraviolet (UV) or blue light causes an irreversible spectral shift in the PS-FP Eos from a green state to an orange state. (B) The reversible PA-FP Dronpa fluorescense green in its bright state. Prolonged or intense irradiation with green light leads to a non-fluorescent form with absorption maximum at 390 nm, which can then be reversibly photoactivated back to the green-emitting form by irradiation with 405 nm light. Dronpa can undergo 100 cycles of activation–quenching with only a 25% loss of its original fluorescence. Reproduced, with permission, from Fernández-Suárez and Ting.[53]

Several methods have been developed for targeting of organic fluorophores to specific proteins in live cells. Theses methods can be gathered in two approaches for site-specific targeting. The first approach (Figure 18.3A) consists in fusing the target protein to a peptide or protein recognition sequence, which then recruits the small molecule. In general, protein recognition domains offer greater targeting specificity but larger bulk than peptide recognition domains. The second strategy is enzyme-mediated protein labelling (Figure 18.3B). A recognition peptide is fused to the protein of interest and a natural or engineered enzyme ligates the small-molecule probe to the recognition peptide. This approach can give highly specific and rapid labelling, with the benefit of a small directing peptide sequence.

A **Protein/peptide-directed labelling**

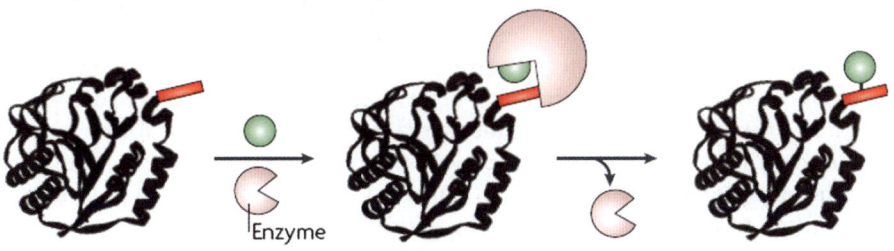

B **Enzyme-mediated protein labelling**

Figure 18.3 Strategies for site-specific targeting of small-molecule probes to cellular proteins. (A) fusion of the target protein to a peptide or protein recognition sequence, which then recruits the small molecule. (B) Enzyme-mediated protein labeling. Reproduced, with permission, from Fernández-Suárez and Ting.[53]

All these recent developments in term of targeting strategies and fluorescent probe designs have transformed microscopes into powerful biometric readers with nanometric resolution that allow cellular compartments to be used for the study of gene expression directly into a single living cell.

18.4 Quantum Dots-based Techniques

Although fluorescence-dependent signal generation has disadvantages with regard to its use in multiplexed assays, fluorescence is still, mainly because of its high sensitivity, a major transduction modality in biosensors. Semiconductor quantum dots (QDs) have recently provided an attractive field of research for biosensor development because they have high photostability and narrow emission peaks, enabling greater multiplexing potential. QDs are ideal fluorophores for optical labeling and bar coding because their fluorescence emission wavelengths can be tuned continuously by changing the particle size, and a single wavelength can be used for simultaneous excitation of different-sized QDs. High-quality CdSe, CdTe, and alloyed QDs are stable against photobleaching and have narrow and symmetrical emission peaks.

QD development has also triggered new gene detection approaches like bead-based assays or single QD-based DNA sensors.

18.4.1 Quantum Dots Bead-based Assays

The unique optical properties of QDs have been used in bead-based assays with optically encoded microbeads for massively parallel and high-throughput analysis of biological molecules. This encoded bead technology is based on the optical properties of semiconductor QDs and the ability to incorporate multicolor QDs into small polymer beads at precisely controlled ratios.

Bead-based assays are increasingly being used for analytical detection and screening of biological molecules, cells, and pathogens.[54,55] Unlike solid-phase biochips and microarrays, microbeads (micrometer-sized spherical particles) are easy to manipulate and provide much faster reaction kinetics due to their suspension in homogeneous solution and the associated advantages in diffusion rates. In addition, microbeads have large surface areas and can be used to enrich target molecules and to increase the dynamic range of target detection.[56] A major application of these encoded beads is multiplexed chemical analysis, in which several target molecules are measured in a single sample. Recent research has led to the development of encoded microbeads using organic dyes[57] or metallic nanoparticles.[58] However, QD-based microbeads offer many advantages in terms of photostability and multiplexing cababilities.[22,26,59] Besides, contrary to standard organic dyes, embedded QDs that are spatially separated from each other do not undergo fluorescence resonance energy transfer.

In principle, multiple QD colors and intensities can be used to encode thousands or even millions of genes, proteins, and small-molecule compounds.[22] However, the brightness and uniformity of bar coding signals is critical and often limits bead identification at high speed and with accuracy. The use of encoded mesoporous silica beads provide sufficient high uniformity in size and internal structure as well as coding signal brightness to be identified by using a standard flow cytometer at a readout speed as high as 1000 beads per second.[60]

Progress has also been made in developing multifunctional beads that would allow simultaneous target encoding, enrichment, and separation. For instance, the development of dual-function beads using both semiconductor QD and iron oxide nanocrystals permits magnetic bead separation.[59] These encoded beads can be integrated with microfluidic devices for multiplexed and ultra-sensitive gene expression detection.

18.4.2 Single Quantum Dots-based DNA Nanosensors

Numerous studies have shown that biocompatible surface modified single QD are viable substrates for fluorescence-based DNA detection.[61] The specificity of the hybridization target probe is the basis of many DNA sensing approaches using different linkage strategies of the probe oligonucleotides on the surface of QDs (*e.g.* amide, streptavidin). The quenching/recovery of fluorescence when QDs and fluorophore quenchers are brought into close proximity or spaced apart illustrates the variety of changes in the fluorescent signals used to discriminate the presence of the target DNA.

An example made use of two probes: a reporter probe labeled with an organic fluorophore and a capture probe labeled with biotin (Figure 18.4). In the presence of the target DNA, these two target-specific oligonucleotide probes formed a sandwich hybrid that could be captured by streptavidin-capped QDs.[62] The formation of the nanoassembly enabled FRET to occur from the QD to the fluorophores in close proximity, and the system revealed some major advantages. Several sandwiched hybrids were captured by a single QD, so QDs function as nanoscaffolds, resulting in amplified signals. In addition, the presence of targets was detected by the co-localization of two fluorescent signals, and because unhybridized probes do not participate in FRET, background fluorescence was negligible.

A similar principle of QD-based single-molecule coincident detection has been proposed but using two QDs with different wavelength emissions instead. In this case, two biotinylated oligonucleotide probes hybridized with complementary target DNA to create a sandwich hybrid as described above. The DNA hybrids were captured at the surface of the QDs through streptavidin–biotin binding, forming complexes with the two-color cross-linked QDs.[63]

18.4.3 Quantum Dots for Super-resolution Microscopy

Irvine and co-workers recently reported the ability to switch on and off a certain kind of QD, thus rendering this type of QD suitable for super-resolution imaging.[64] They showed that the fluorescence of manganese (Mn)-doped ZnSe QDs can be reversibly depleted with $\sim 90\%$ efficiency using a continuous-wave modulation laser of $\sim 2\,MW\,cm^{-2}$. The main novelty of this report is that modulation is achieved directly by light and relies only on internal electronic transitions, without the need for an external photochromic activator or quencher. Thus, this type of QD can be used in super-resolution imaging in the same way as small-molecule organic photoswitchers.[50]

18.5 Gold Nanoparticle-based Detection Methods

In the 1850s, Michael Faraday discovered that light incident on metallic nanoparticles induces their conduction electrons to oscillate collectively with a resonant frequency.[65] This unique spectral property is called localized surface plasmon resonance (LSPR). At this frequency, the nanoparticles absorb and scatter light so intensely that a single nanoparticle can be easily observed by eye using dark-field optical scattering microscopy.[66–70] This property has been used in stained glasses where a few ppm quantities of silver or gold nanoparticles induce intense yellow and red colors respectively. A single 80 nm silver nanosphere, for instance, scatters 445 nm blue light with a scattering cross-section of $3 \times 10^{-2}\,\mu m^2$, a million-fold greater than the fluorescence cross-section of a fluorescein molecule, and a thousand-fold greater than the cross-section of a similarly sized nanosphere filled with fluorescein to the self-quenching limit.[68] This phenomenon enables noble-metal nanoparticles to serve as extremely

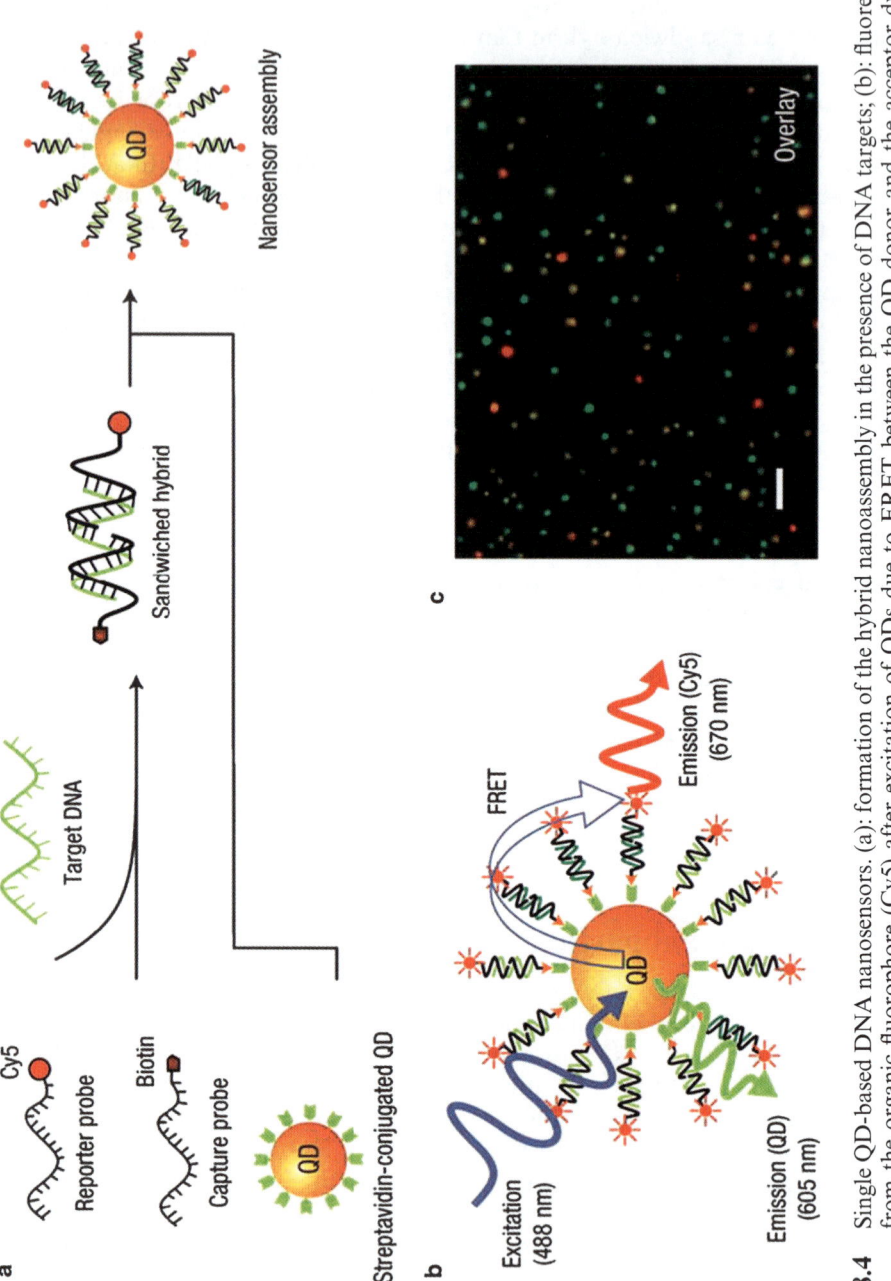

Figure 18.4 Single QD-based DNA nanosensors. (a): formation of the hybrid nanoassembly in the presence of DNA targets; (b): fluorescence from the organic fluorophore (Cy5) after excitation of QDs due to FRET between the QD donor and the acceptor dye; (c): detection of single-QD FRET signals in the presence of targets by the co-localization of QDs (green) and Cy5 (red) originating a fluorescence image with blended colors in yellow and orange (scale bar 10 μm). Reproduced, with permission, from Zhang et al.[62]

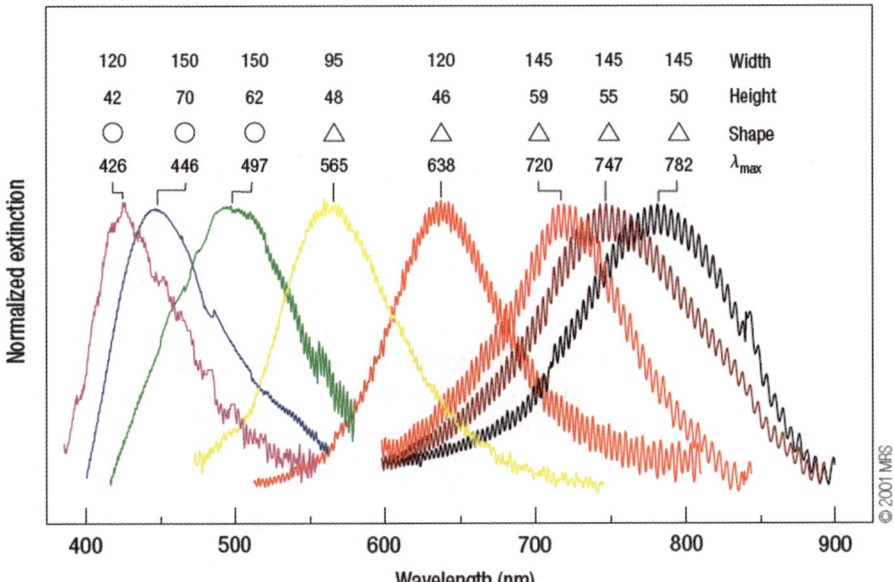

Figure 18.5 Effect of size and shape on LSPR extinction spectrum for silver nanoprisms and nanodiscs formed by nanosphere lithography. The high-frequency signal on the spectra is an interference pattern from the reflection at the front and back surfaces of the mica. Reproduced, with permission, from Haes *et al.*[77]

intense labels for immunoassays,[68,70,71] biochemical sensors,[72,73] and surface-enhanced spectroscopies.[74–76]

The shape of the nanoparticle extinction and scattering spectra, and in particular the peak wavelength λ_{max}, depends on nanoparticle composition, size, shape, orientation, and local dielectric environment. The LSPR can be tuned during fabrication by controlling these parameters with a variety of chemical syntheses and lithographic techniques. The spectrum of plasmonic nanoparticles can be tuned from the near UV (380 nm for silver) to IR by varying the material, their size and shape. For instance, varying the relative thickness of the gold shell in particles composed of a silica core and a gold shell allows one to tune the LSPR from the visible to a few micrometers wavelengths. Figure 18.5 shows the extinction spectra for a series of silver nanoprisms and nanohemispheres made by nanosphere lithography with λ_{max} varying between 426 and 782 nm.[77]

Plasmonic nanoparticles can act as transducers that convert small changes in the local refractive index into spectral shifts in the intense nanoparticle extinction and scattering spectra. Most organic molecules have a higher refractive index than buffer solution; thus, when they bind to nanoparticles, the local refractive index increases, causing the extinction and scattering spectrum to redshift. Molecular binding can be monitored in real time with high sensitivity by using simple and inexpensive transmission spectrometry, which measures extinction, the sum of absorption and scattering. It is noteworthy that real-time LSPR-shift assays are based on a similar principle to the commercial surface plasmon resonance (SPR) instrument based on propagating surface plasmons on thin gold film. This technique is thus often designated as nano-SPR. However, LSPR possesses less interference with the bulk refractive index and greater spatial resolution, both lateral and normal, when compared with standard SPR.[78] Nano-SPR can be applied to a chip format, in which nanoparticles with different optical signatures and coated with different antibody specificities[11] are immobilized in an array format for high throughput screening of biomolecular binding interactions in a real-time, label-free manner. For instance, LPR-shift assays have been used for ultrasensitive quantification of proteins. They could measure concentrations of amyloid-derived diffusible ligands (ADDLs), a neurotoxin that is thought to be important in the pathology of Alzheimer's disease, at concentrations down to 100 fM.[73]

Besides, LSPR can also be dispersed in cells to obtained real-time and local information on live cell activity. A single plasmonic nanoparticle will thus give precious information with nanometric spatial resolution and locally minute concentration of biomolecules.

18.5.1   Resonant Light Scattering Detection

Detection measuring the resonant light scattering (RLS) of the metallic nanoparticle labels provides dramatically enhanced sensitivity over optical absorbance (typically >3 orders of magnitude) in various bioassay applications. The potential for RLS particles as highly sensitive tracers in cell biology has been previously documented.[70] For instance, Mirkin and co-workers reported the use of light-scattering particles for enhanced discrimination of single-nucleotide polymorphisms in a model oligonucleotide system.[79] The application of RLS for high-sensitivity detection of DNA hybridization on cDNA microarrays has been demonstrated.[80] RLS labeling technology is significantly more sensitive than Cy3 labeling for DNA microarray applications. This higher sensitivity allows the identification of very low expressors not detectable by conventional (Cy3/Cy5) fluorescence and the use of less material per assay while maintaining a strong signal-to-background ratio. Among them, a novel assay based on gold nanoparticle-promoted reduction of silver was reported to detect target DNA down to a concentration of 50 fM.[81]

18.5.2   Molecular Beacons with Gold Nanoparticles

Molecular beacons are single-stranded oligonucleotide hybridization probes that form a stem-and-loop structure. The loop contains a probe sequence that is complementary to a target sequence, and the stem is formed by annealing the complementary arm sequences that are located on either side of the probe sequence. A fluorophore is covalently linked to the end of one arm and a quencher is covalently linked to the end of the other arm. Gold nanoparticles are example of good quencher.[82] Molecular beacons do not fluoresce when they are free in solution. However, when they hybridize to a nucleic acid strand containing a target sequence they undergo a conformational change that enables them to fluoresce brightly. This property enables molecular beacons to be used as detector probes in diagnostic assays where it is not necessary to isolate the probe–target hybrids. Molecular beacons are added to the assay mixture before carrying out gene amplification and fluorescence is measured in real time.[83]

Molecular beacons are powerful probes which allow the detection of gene expression *in vivo* with a high sensitivity and specificity. It is used in applications like: (1) monitoring in real-time the changes in mRNA expression level due to disease states, toxic assaults, stimuli, drug leads, *etc*; (2) studying mRNA processing, localization, and transport, and to quantify the knock-down effect of siRNA; and (3) performing disease detection and diagnosis.

18.5.3 Molecular Plasmonic Rulers

LSPR sensitivity can be used to detect changes in the extension of single molecules by sandwiching molecules between two nanoparticles. Such plasmonic molecular ruler structures are extremely sensitive to distance-dependent plasmonic coupling between the nanoparticle pairs,[84,85] enabling high-resolution monitoring of molecular conformation.[86,87] The magnitude and direction of the spectral shift depend on the orientation of the nanoparticle pair with respect to the polarization axis of the incident light. With unpolarized light, when a nanoparticle label binds to a fixed nanoparticle sensor to form a nanoparticle pair, the scattering intensity increases markedly (by about fourfold compared with a single particle) and a large spectral redshift is observed, about 75 nm for 40-nm gold nanoparticles. This shift depends on the distance between nanoparticles, decreasing approximately exponentially with distance. For 40-nm gold particles, the resonance shifts by about 10 nm per nanometer separation for small separations. This distance dependence is roughly proportional to particle radius, so smaller particles will be more sensitive to small changes in plasmonic spacing, although with a reduced dynamic range.

Alivisatos, Liphardt and co-workers have developed a molecular ruler to monitor the separation between single pairs of nanoparticles.[86,87] A molecular ruler was used to detect the hybridization of DNA oligonucleotides complementary to the single-stranded DNA (ssDNA). Gold nanospheres, 40 nm in diameter, were functionalized with streptavidin and immobilized on glass slides. The immobilized nanoparticles were then exposed to nanoparticles functionalized with biotinylated ssDNA, allowing the anchored nanoparticles to capture the ssDNA-functionalized nanoparticles and form pairs (Figure 18.6). This binding event caused an immediate redshift and an increase in scattering intensity of the immobilized nanoparticles as a result of plasmon resonance coupling between the nanoparticles. The scattering spectra of single nanoparticles and nanoparticle pairs were monitored by dark-field microscopy. Because the ssDNA chain is relatively flexible, introducing the complementary DNA to the ssDNA linkers causes hybridization to a more extended form,

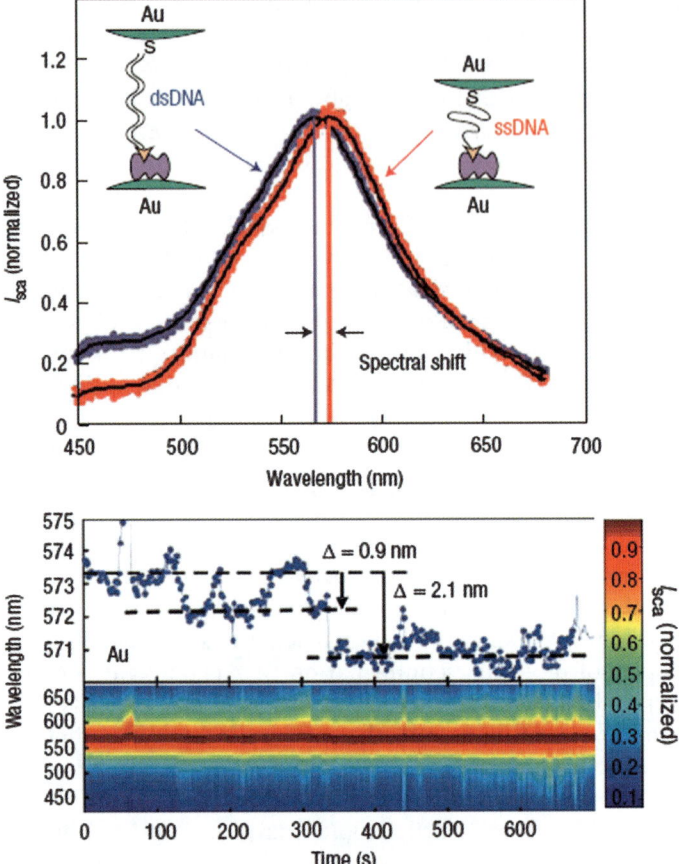

Figure 18.6   Molecular plasmonic ruler. *Top*: Example of a spectral shift between a gold nanoparticle pair connected with single stranded DNA (red) and double-stranded DNA (blue). *Bottom*: Spectral position as a function of time after the addition of complementary DNA. The scattered intensity (I_{sca}) is shown color-coded on the bottom; the peak position obtained by fitting each spectrum is traced on the top. Discrete states are observed, indicated by horizontal dashed lines Reproduced, with permission, Sonnichsen *et al.*[87]

pushing the nanoparticles apart and resulting in a blueshift in the scattering spectrum of the nanoparticle pair. By continuously monitoring the spectrum of nanoparticle pairs, one can observe the dynamics of DNA hybridization.

18.5.4   Surface-enhanced Raman Scattering Detection

Raman spectroscopy provides information concerning the vibrational quantum states of a molecule. The spectrum is usually composed of sharp and narrow peaks that constitute a signature of a given molecule. This spectroscopy has two main advantages over fluorescence: (1) it is label-free, although specific Raman dyes may be used as labels; and (2) it can be multiplexed easily. Parallel detection of multiple analytes is possible because Raman bands are sufficiently narrow to reduce the likelihood of spectral overlap.[88]

Raman cross-sections of molecules are, however, many orders of magnitude lower than their fluorescence cross-section which means that Raman spectroscopy cannot compete with fluorescence sensitivity unless associated with nanostructures exhibiting plasmonic properties. The plasmonic modes associated with huge local electromagnetic field enhancements induce a dramatic amplification of the Raman signal. This technique is called surface-enhanced Raman scattering (SERS). It is noteworthy that fluorescence detection can also be enhanced by plasmonic modes.[89,90]

SERS-active platforms are typically composed of self-assembled monolayers, nanostructured rough metallic substrates or colloidal Au or Ag nanoparticle clusters. Raman enhancement factors as high as 10^{14} have been reported in single-molecular-level detections.[91–93] Colloidal gold and silver nanoparticles are also utilized in SERS cellular imaging to enhance signal intensity and increase image contrast.[94] However, conventional nanoparticles have inherent limits for *in vivo* biomolecular SERS imaging for three reasons. Firstly, strong enhancement dispersion occurs. SERS enhancement depends strongly on the local geometry of the nanostructured material which creates local field enhancements called "hot spots". These hot spots are not controlled in randomly formed nanoparticle clusters. Secondly, the spatial imaging resolution degrades with increasing size of nanoparticles clusters. Thirdly, the random distribution of nanoparticles within the biological cell voids the spatial specificity. Smart plasmonic nanoparticles with various designed shapes, like nanoshells,[95] nanotips,[96] and nanorings,[97] and nanocrescents,[98] have thus been used to circumvent these limitations. Besides, these excellent stand-alone SERS probes can be controlled in orientation and position by means of magnetic schemes.[99]

Alternative SERS strategies have been developed using multiplexed detection of oligonucleotide targets with gold nanoparticle probes labeled with oligonucleotides and Raman-active dyes as a barcode.[18] The gold nanoparticles facilitate the formation of a silver coating that acts as a surface enhanced Raman scattering promoter for the dye-labeled particles that have been captured by target molecules and an underlying chip in microarray format (Figure 18.7). Although oligonucleotides can be directly detected by SERS on aggregated particles,[100,101] the structural similarities of oligonucleotides with different sequences result in spectra that were too difficult to distinguish. To show multiplexing and ratioing capabilities, six dissimilar DNA targets with six Raman-labeled nanoparticle probes were distinguished, as well as two RNA targets with single nucleotide polymorphisms. The unoptimized detection limit of this method was 20 fM.

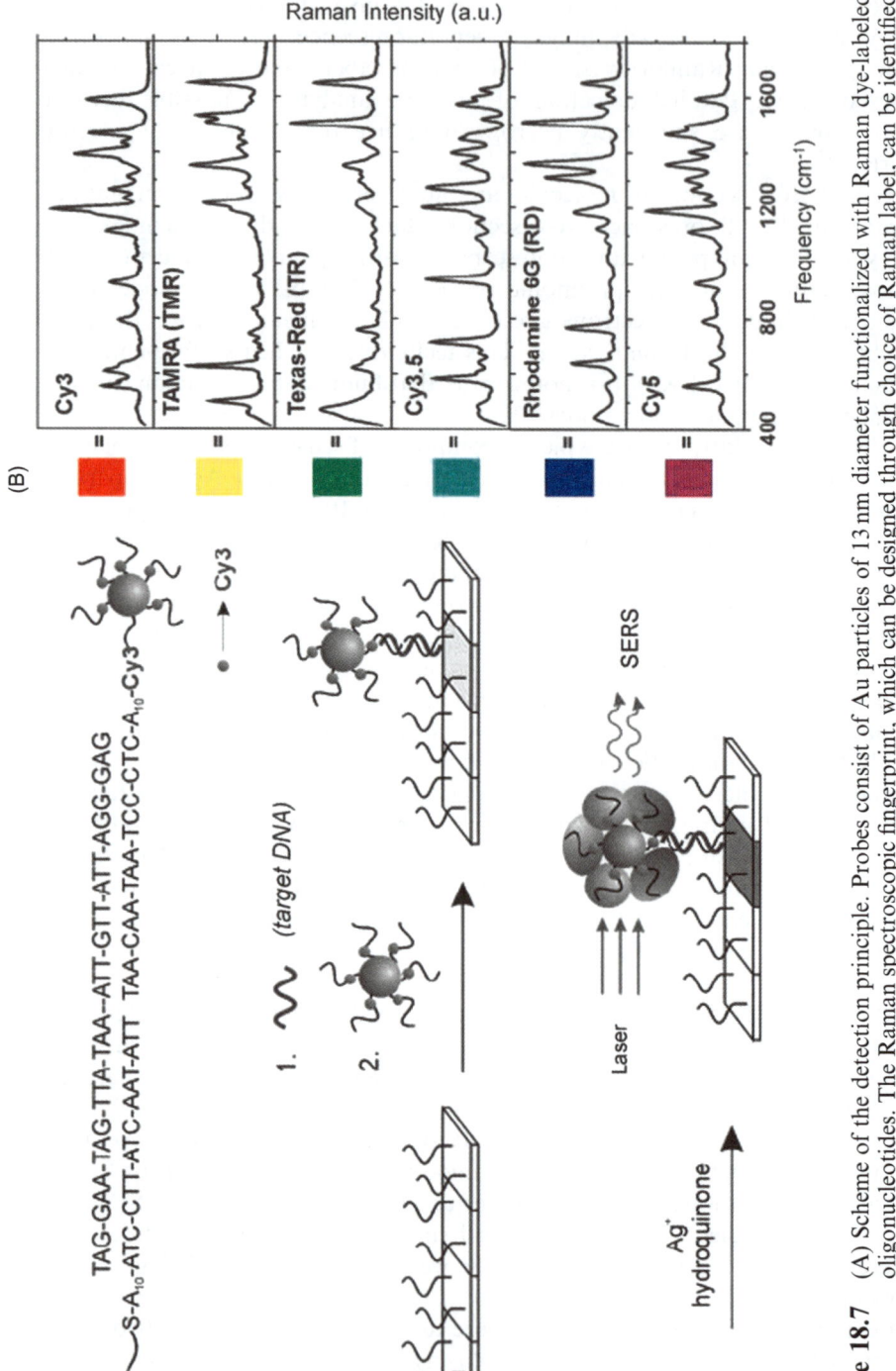

Figure 18.7 (A) Scheme of the detection principle. Probes consist of Au particles of 13 nm diameter functionalized with Raman dye-labeled oligonucleotides. The Raman spectroscopic fingerprint, which can be designed through choice of Raman label, can be identified after Ag enhancing by scanning Raman spectroscopy. (B) The Raman spectra of six dye-labeled nanoparticle probes after Ag enhancing on a chip (after background subtraction). Each dye correlates with a different color. Reproduced, with permission, from Cao *et al.*[18]

18.6 Electrochemical Sensors

An interesting alternative to optical based sensors is the development of electrochemical DNA sensors with high potential for miniaturization and integration. This field has become a subject of intense research, with the hope to make sophisticated and challenging molecular diagnostics available for low-cost routine clinical practice. Electronic detection methods based on electrodes,[102,103] CMOS field effect transistor (FET),[104–106] and the more recent carbon nanotube (CNT) DNA sensors[107–109] have been reported, which have shown great promise in higher sensitivity and large-scale arrayability.

Most DNA biosensors are developed based on the immobilization of single-stranded DNA onto the electrode surface labeled with an electrochemical indicator to recognize its complementary target sequence.[110] The unique properties of CNTs can be united with the specific molecular-recognition features of DNA by coupling CNTs to peptide nucleic acid and hybridizing these macromolecular wires with complementary DNA.[111] Both covalent and non-covalent linkage of DNA with CNTs have been reported where the former provide the best stability, accessibility, and selectivity during competitive hybridization.[112] The covalent attachment was found to provide accessibility for the DNA molecules to hybridization and to strongly favor hybridization with molecules having complementary sequences compared with non complementary sequences. The integration of CNTs with other materials has been also used for the immobilization of DNA like ZrO_2 or ZnO nanoparticles.[113,114] Recently, Ma *et al.* fabricated an electrochemical DNA biosensor based on LB self-assembly of CNTs and gold nanoparticles (GNPs) *via* covalent-bonding interaction.[115] The biosensor showed an improved sensitivity with an excellent reproducibility due to the high catalytic activities of GNPs and the ability of CNTs to promote electron-transfer reactions. A wide linear response range from 0.5 to 0.01 nM with a detection limit of 7.5 pM for target DNA was achieved. Recently, Niu *et al.* used a manganese complex of rutin as a redox intercalator with carboxylic acid group-functionalized CNTs and fabricated DNA biosensor for DNA hybridization detection.[116] The modified electrode dramatically increased the amount of DNA attachment and the sensitivity of the complementary ssDNA detection mostly due to the large surface area and good charge–transport characteristics of CNTs. A DNA biosensor based on the enhancement of the guanine signal at a CNTs-modified pencil graphite electrode (PGE) was also proposed.[117] PGE behaved as a microelectrode array coupled with its higher porosity and showed improved performance.

Another new DNA biosensor based on electrochemical impedance was developed using a composite material of PPy and carboxylic group-terminated CNTs. A probe with an amino group termination was linked onto the PPy/MWCNTs-COOH/GCE by using EDAC and it was found that the hybridization reaction with its complementary decreased the electron transfer resistance.[118] Ethidium bromide was used as an intercalator and the current change generated from it was monitored. Only the complementary DNA, compared to the five-point mismatched and non-complementary sequences, gave an obvious current flow and a detection limit of 0.85 pM was obtained. A lower detection limit of 120 fM for the target DNA was achieved using an electrochemical DNA biosensor based on palladium nanoparticles combined with CNTs.[119] The improved sensitivity was attributed to the ability of CNTs promoting electron-transfer process and the high catalytic activities of palladium nano-particles for electrochemical reaction of methylene blue which was used as an indicator.

18.7   Concluding Remarks

All the presented ultrasensitive methods are still under development and their performances in terms of multiplexing and sensitivity are constantly improving. Some, like DNA origami, are in their infancy and need further technical developments to be validated and meet the demands of single cell gene analysis.

Depending on the particular application, the analyses of real samples with high throughput and low cost are critical to the validation and routine appli-cations. In particular, the integration of these techniques in microfluidic devices is crucial.

The selected gene expression analysis method is highly dependent on the characteristics of a particular application such as cellular throughput, number of gene products analyzed, sensitivity, and temporal resolution. However, it is clear that the preliminary amplification step will be needed less and less due to the rapid sensitivity improvements. Besides, imaging directly into the living cell will avoid the lysis of the cell for the separation step. Furthermore, imaging techniques enabling gene expression to be followed in individual living cells can address the dynamic nature of transcription which opens an avenue of exciting possibilities.

References

1. E. M. Ozbudak, M. Thattai, I. Kurtser, A. D. Grossman and A. van Oudenaarden, *Nat. Genet.*, 2002, **31**, 69.
2. P. S. Swain, M. B. Elowitz and E. D. Siggia, *Proc. Natl Acad. Sci. U. S. A.*, 2002, **99**, 12795.
3. S. Ramsey, A. Ozinsky, A. Clark, K. D. Smith, P. de Atauri, V. Thorsson, D. Orrell and H. Bolouri, *Philos. Trans. R. Soc. London B, Biol. Sci.*, 2006, **361**, 495.
4. M. B. Elowitz, A. J. Levine, E. D. Siggia and P. S. Swain, *Science*, 2002, **297**, 1183.
5. C. V. Rao, D. M. Wolf and A. P. Arkin, *Nature*, 2002, **420**, 231.
6. W. J. Blake, M. Kaern, C. R. Cantor and J. J. Collins, *Nature*, 2003, **422**, 633.
7. J. M. Raser and E. K. O'Shea, *Science*, 2005, **309**, 2010.
8. D. Longo and J. Hasty, *Nat. Mol. Syst. Biol.*, 2006, msb4100110.
9. J. R. Newman, S. Ghaemmaghami, J. Ihmels, D. K. Breslow, M. Noble, J. L. Derisi and J. S. Weissman, *Nature*, 2006, **441**, 840.

10. R. L. Stears, T. Martinsky and M. Schena, *Nat. Med.*, 2003, **9**, 140.
11. S. R. Nicewarner-Pena, R. G. Freeman, B. D. Reiss, L. He, D. J. Pena, I. D. Walton, R. Cromer, C. D. Keating and M. J. Natan, *Science*, 2001, **294**, 137.
12. I. D. Walton, S. M. Norton, A. Balasingham, L. He, D. F. Oviso Jr., D. Gupta, P. A. Raju, M. J. Natan and R. G. Freeman, *Anal. Chem.*, 2002, **74**, 2240.
13. F. Cunin, T. A. Schmedake, J. R. Link, Y. Y. Li, J. Koh, S. N. Bhatia and M. J. Sailor, *Nat. Mater.*, 2002, **1**, 39.
14. M. J. Dejneka, A. Streltsov, S. Pal, A. G. Frutos, C. L. Powell, K. Yost, P. K. Yuen, U. Muller and J. Lahiri, *Proc. Natl Acad. Sci. U. S. A.*, 2003, **100**, 389.
15. L. Grondahl, B. J. Battersby, D. Bryant and M. Trau, *Langmuir*, 2000, **16**, 9709.
16. B. J. Battersby, D. Bryant, W. Meutermans, D. Matthews, M. L. Smythe and M. Trau, *J. Am. Chem. Soc.*, 2000, **122**, 2138.
17. K. Braeckmans, S. de Smedt, C. Roelant, M. Leblans, R. Pauwels and J. Demeester, *Nat. Mater.*, 2003, **2**, 169.
18. Y. C. Cao, R. Jin and C. A. Mirkin, *Science*, 2002, **297**, 1536.
19. Y. C. Cao, R. Jin, J. M. Nam, C. S. Thaxton and C. A. Mirkin, *J. Am. Chem. Soc.*, 2003, **125**, 14676.
20. S. P. Mulvaney, M. D. Musick, C. D. Keating and M. J. Natan, *Langmuir*, 2003, **19**, 4784.
21. W. E. Doering and S. Nie, *Anal. Chem.*, 2003, **75**, 6171.
22. M. Y. Han, X. H. Gao, J. Z. Su and S. M. Nie, *Nat. Biotechnol.*, 2001, **19**, 631.
23. X. H. Gao, W. C. W. Chan and S. M. Nie, *J. Biomed. Opt.*, 2002, **7**, 532.
24. X. H. Gao and S. M. Nie, *J. Phys. Chem. B*, 2003, **107**, 11575.
25. N. Gaponik, I. L. Radtchenko, G. B. Sukhorukov, H. Weller and A. L. Rogach, *Avd. Mater.*, 2002, **14**, 879.
26. H. Xu, M. Y. Sha, E. Y. Wong, J. Uphoff, Y. Xu, J. A. Treadway, A. Truong, E. O'Brien, S. Asquith, M. Stubbins, N. K. Spurr, E. H. Lai and W. Mahoney, *Nucleic Acids Res.*, 2003, **31**, 43.
27. Y. Ke, S. Lindsay, Y. Chang, Y. Liu and H. Yan, *Science*, 2008, **319**, 180.
28. Y. Shav-Tal, R. H. Singer and X. Darzacq, *Nat. Rev. Mol. Cell Biol.*, 2004, **5**, 855.
29. T. Zimmermann, J. Rietdorf and R. Pepperkok, *FEBS Lett.*, 2003, **546**, 87.
30. G. Patterson, R. N. Day and D. Piston, *J. Cell Sci.*, 2001, **114**, 837.
31. W. A. Carrington, R. M. Lynch, E. D. Moore, G. Isenberg, K. E. Fogarty and F. S. Fay, *Science*, 1995, **268**, 1483.
32. S. W. Hell, *Nat. Biotechnol.*, 2003, **21**, 1347.
33. S. W. Hell and J. Wichmann, *Opt. Lett.*, 1994, **19**, 780.
34. S. W. Hell and M. Kroug, *Appl. Phys. B*, 1995, **60**, 495.
35. M. G. Gustafsson, *Proc. Natl Acad. Sci. U. S. A.*, 2005, **102**, 13081.

36. R. Heintzmann, T. M. Jovin and C. Cremer, *J. Opt. Soc. Am.*, 2002, **19**, 1599.
37. L. Shao, B. Isaac, S. Uzawa, D. A. Agard, J. W. Sedat and M. G. Gustafsson, *Biophys. J.*, 2008, **94**, 4971.
38. R. E. Thompson, D. R. Larson and W. W. Webb, *Biophys. J.*, 2002, **82**, 2775.
39. E. Betzig, G. H. Patterson, R. Sougrat, O. W. Lindwasser, S. Olenych, J. S. Bonifacino, M. W. Davidson, J. Lippincott-Schwartz and H. F. Hess, *Science*, 2006, **313**, 1642.
40. S. T. Hess, T. P. Girirajan and M. D. Mason, *Biophys. J.*, 2006, **91**, 4258.
41. M. J. Rust, M. Bates and X. Zhuang, *Nat. Methods*, 2006, **3**, 793.
42. A. M. Femino, F. S. Fay, K. Fogarty and R. H. Singer, *Science*, 1998, **280**, 585.
43. J. M. Levsky, S. Shenoy, R. Pezo and R. H. Singer, *Science*, 2002, **297**, 836.
44. M. B. Elowitz and S. Leibler, *Nature*, 2000, **403**, 335.
45. J. M. Raser and E. K. O'shea, *Science*, 2004, **304**, 1811.
46. E. Le Moal, E. Fort, S. Lévêque-Fort, A. Janin, H. Murata, F. P. Cordelieres, M.-P. Fontaine-Aupart and C. Ricolleau, *J. Biomed. Opt.*, 2007, **12**, 024030.
47. J. C. Simpson, V. E. Neubrand, S. Wiemann and R. Pepperkok, *Histochem. Cell Biol.*, 2001, **115**, 23.
48. K. I. Willig, R. R. Kellner, R. Medda, B. Hein, S. Jakobs and S. W. Hell, *Nat. Methods*, 2006, **3**, 721.
49. B. Hein, K. I. Willig and S. W. Hell, *Proc. Natl Acad. Sci. U. S. A.*, 2008, **105**, 14271.
50. J. Lippincott-Schwartz and G. H. Patterson, *Methods Cell Biol.*, 2008, **85**, 45.
51. K. A. Lukyanov, D. M. Chudakov, S. Lukyanov and V. V. Verkhusha, *Nat. Rev. Mol. Cell Biol.*, 2005, **6**, 885.
52. J. Fölling, V. Belov, D. Riedel, A. Schönle, A. Egner, C. Eggeling, M. Bossi and S. W. Hell, *ChemPhysChem*, 2008, **9**, 321.
53. M. Fernández-Suárez and A. Y. Ting, *Nat. Rev.*, 2008, **9**, 929.
54. D. R. Walt, *Science*, 2000, **287**, 451.
55. R. S. Rao, S. R. Visuri, M. T. McBride, J. S. Albala, D. L. Matthews and M. A. Coleman, *J. Proteome Res.*, 2004, **3**, 736.
56. P. S. Eastman, W. Ruan, M. Doctolero, R. Nuttall, G. de Feo, J. S. Park, J. S. Chu, P. Cooke, J. W. Gray, S. Li and F. F. Chen, *Nano Lett.*, 2006, **6**, 1059.
57. L. Wang and W. H. Tan, *Nano Lett.*, 2006, **6**, 84.
58. T. R. Sathe, A. Agrawal and S. Nie, *Anal. Chem.*, 2006, **78**, 5627.
59. R. C. Jin, Y. C. Cao, C. S. Thaxton and C. A. Mirkin, *Small*, 2006, **2**, 375.
60. X. H. Gao and S. Nie, *Anal. Chem.*, 2004, **76**, 2406.
61. M. F. Frasco and N. Chaniotakis, *Anal. Bioanal. Chem.*, 2010, **396**, 229.
62. C.-Y. Zhang, H.-C. Yeh, M. T. Kuroki and T.-H. Wang, *Nat. Mater.*, 2005, **4**, 826.

63. C.-Y. Zhang and L. W. Johnson, *Analyst*, 2006, **131**, 484.
64. S. E. Irvine, T. Staudt, E. Rittweger, J. Engelhardt and S. W. Hell, *Angew. Chem. Int. Ed. Engl.*, 2008, **47**, 2685.
65. M. Faraday, *Philos. Trans. R. Soc. London*, 1857, **147**, 145.
66. S. Eustis and M. A. El-Sayed, *Chem. Soc. Rev.*, 2006, **35**, 209.
67. K. A. Willets and R. P. Van Duyne, *Annu. Rev. Phys. Chem.*, 2007, **58**, 267.
68. S. Schultz, D. R. Smith, J. J. Mock and D. A. Schultz, *Proc. Natl Acad. Sci. U. S. A.*, 2000, **97**, 996.
69. C. F. Bohren and D. R. Huffman, *Absorption and Scattering of Light by Small Particles,* Wiley, New York, 1983.
70. J. Yguerabide and E. Yguerabide, *Anal. Biochem.*, 1998, **262**, 157.
71. J. M. Nam, C. S. Thaxton and C. A. Mirkin, *Science*, 2003, **301**, 1884.
72. C. R. Yonzon, D. A. Stuart, X. Zhang, A. D. McFarland, C. L. Haynes and R. P. Van Duyne, *Talanta*, 2005, **67**, 438.
73. A. J. Haes, L. Chang, W. L. Klein and R. P. Van Duyne, *J. Am. Chem. Soc.*, 2005, **127**, 2264.
74. D. L. Jeanmaire and R. P. Van Duyne, *J. Electroanal. Chem. Interface Electrochem.*, 1997, **84**, 1.
75. C. L. Haynes, C. R. Yonzon, X. Zhang and R. P. Van Duyne, *J. Raman Spectrosc.*, 2005, **36**, 471.
76. J. A. Dieringer, A. D. McFarland, N. C. Shah, D. A. Stuart, A. V. Whitney, C. R. Yonzon, M. A. Young, X. Zhang and R. P. Van Duyne, *Faraday Discuss.*, 2006, **132**, 9.
77. J. Haes, C. L. Haynes and R. P. Van Duyne, *Mater. Res. Soc. Symp. Proc.*, 2001, **636**, D4.8.
78. C. R. Yonzon, E. Jeoung, S. Zou, G. C. Schatz, M. Mrksich and R. P. Van Duyne, *J. Am. Chem. Soc.*, 2004, **126**, 12669.
79. T. A. Taton, G. Lu and C. A. Mirkin, *J. Am. Chem. Soc.*, 2001, **123**, 5164.
80. P. Bao, A. G. Frutos, C. Greef, J. Lahiri, U. Muller, T. C. Peterson, L. Warden and X. Xie, *Anal. Chem.*, 2002, **74**, 1792.
81. T. A. Taton, C. A. Mirkin and R. L. Letsinger, *Science*, 2000, **289**, 1757.
82. B. Dubertret, M. Calame and A. J. Libchaber, *Nat. Biotech.*, 2001, **19**, 365.
83. D. P. Bratu, *Methods Mol. Biol.*, 2006, **319**, 1.
84. R. Elghanian, J. J. Storhoff, R. C. Mucic, R. L. Letsinger and C. A. Mirkin, *Science*, 1997, **277**, 1078.
85. J. N. Anker, W. P. Hall, O. Lyandres, N. C. Shah, J. Zhao and R. P. Van Duyne, *Nat. Mater.*, 2008, **7**, 442.
86. B. M. Reinhard, S. Sheikholeslami, A. Mastroianni, A. P. Alivisatos and J. Liphardt, *Proc. Natl Acad. Sci. U. S. A.*, 2007, **104**, 2667.
87. C. Sonnichsen, B. M. Reinhard, J. Liphardt and A. P. Alivisatos, *Nat. Biotechnol.*, 2005, **23**, 741.
88. J. Ni, R. J. Lipert, G. Brent Dawson and M. D. Porter, *Anal. Chem.*, 1999, **71**, 4903.

89. E. Fort and S. Grésillon, *J. Phys. D*, 2008, **41**, 013001.
90. E. Le Moal, S. Lévêque-Fort, M.-C. Pottier and E. Fort, *Nanotechnology*, 2009, **20**, 225502.
91. S. M. Nie and S. R. Emery, *Science*, 1997, **275**, 1102.
92. K. Kneipp, Y. Wang, H. Kneipp, L. T. Perelman, I. Itzkan, R. Dasari and M. S. Feld, *Phys. Rev. Lett.*, 1997, **78**, 1667.
93. M. Culha, D. Stokes, L. R. Allain and T. Vo-Dinh, *Anal. Chem.*, 2003, **75**, 6196.
94. K. Kneipp, A. S. Haka, H. Kneipp, K. Badizadegan, N. Yoshizawa, C. Boone, K. E. Shafer-Peltier, J. T. Motz, R. R. Dasari and M. S. Feld, *Appl. Spectrosc.*, 2002, **56**, 150.
95. S. J. Oldenburg, S. L. Westcott, R. D. Averitt and N. J. Halas, *J. Chem. Phys.*, 1999, **111**, 4729.
96. A. Hartschuh, E. J. Sanchez, X. S. Xie and L. Novotny, *Phys. Rev. Lett.*, 2003, **90**, 095503.
97. J. Aizpurua, P. Hanarp, D. S. Sutherland, M. Kall, G. W. Bryant and F. J. G. de Abajo, *Phys. Rev. Lett.*, 2003, **90**, 057401.
98. Y. Lu, G. L. Liu, J. Kim, Y. X. Mejia and L. P. Lee, *Nano Lett.*, 2005, **5**, 119.
99. G. L. Liu, Y. Lu, J. Kim, J. C. Doll and L. P. Lee, *Adv. Mater.*, 2005, **17**, 2683.
100. K. Kneipp, H. Kneipp, I. Itzkan, R. R. Dasari and M. S. Feld, *Chem. Rev.*, 1999, **99**, 2957.
101. L. A. Gearheart, H. J. Ploehn and C. J. Murphy, *J. Phys. Chem. B.*, 2001, **105**, 12609.
102. J. Wang, *Trends Anal. Chem.*, 2002, **21**, 226.
103. P. Singhal and W. G. Kuhr, *Anal. Chem*, 1997, **69**, 4828.
104. J. Wang, *Electroanalysis*, 2005, **17**, 7.
105. M. J. Schoning and A. Phoghossian, *Analyst*, 2002, **127**, 1137.
106. F. Pouthas, C. Gentil, D. Cote and U. Bockelmann, *Appl. Phys. Lett.*, 2004, **84**, 1594.
107. A. Star, E. Tu, J. Niemann, J. P. Gabriel, C. S. Joiner and C. Valcke, *Proc. Natl Acad. Sci. U. S. A.*, 2006, **104**, 921.
108. J. Hahm and C. Lieber, *Nano Lett.*, 2004, **4**, 51.
109. X. Tang, S. Bansaruntip, N. Nakayama, E. Yenilmez, Y. Chang and Q. Wang, *Nano Lett.*, 2006, **6**, 1632.
110. A. J. S. Ahammad, J. -J. Lee and Md. Aminur Rahman, *Sensors*, 2009, **9**, 2289.
111. K. A. Williams, P. T. M. Veenhuizen, B. G. Torre, R. Eritja and C. Dekker, *Nature*, 2002, **420**, 761.
112. S. E. Baker, W. Cai, T. L. Lasseter, K. P. Weidkamp and R. J. Hamers, *Nano Lett.*, 2002, **2**, 1413.
113. Y. Yang, Z. Wang, M. Yang, J. Li, F. Zheng, G. Shen and R. Yu, *Anal. Chim. Acta*, 2007, **584**, 268.
114. W. Zhang, T. Yang, D. M. Huang and K. Jiao, *Chin. Chem. Lett.*, 2008, **19**, 589.

115. H. Ma, L. Zhang, Y. Pan, K. Zhang and Y. A. Zhang, *Electroanalysis*, 2008, **20**, 1220.
116. S. Niu, M. Zhao, L. Hu and S. Zhang, *Sens. Actuators, B*, 2008, **135**, 200.
117. A. Erdem, P. Papakonstantinou and H. Murphy, *Anal. Chem.*, 2006, **78**, 6656.
118. H. Qi, X. Li, Chen and C. Pei Zhang, *Talanta*, 2007, **72**, 1030.
119. Z. Chang, H. Fan, K. Zhao, M. Chen, P. He and Y. Fang, *Electro-analysis*, 2008, **20**, 131.

Subject Index

Page references to *figures* and *tables* are shown in *italics*.